Fascinating Life Sciences

This interdisciplinary series brings together the most essential and captivating topics in the life sciences. They range from the plant sciences to zoology, from the microbiome to macrobiome, and from basic biology to biotechnology. The series not only highlights fascinating research; it also discusses major challenges associated with the life sciences and related disciplines and outlines future research directions. Individual volumes provide in-depth information, are richly illustrated with photographs, illustrations, and maps, and feature suggestions for further reading or glossaries where appropriate.

Interested researchers in all areas of the life sciences, as well as biology enthusiasts, will find the series' interdisciplinary focus and highly readable volumes especially appealing.

Charles E. Rupprecht
Editor

History of Rabies in the Americas: From the Pre-Columbian to the Present, Volume I

Insights to Specific Cross-Cutting Aspects of the Disease in the Americas

Springer

Editor
Charles E. Rupprecht
LYSSA LLC
Lawrenceville, GA, USA

ISSN 2509-6745 ISSN 2509-6753 (electronic)
Fascinating Life Sciences
ISBN 978-3-031-25051-4 ISBN 978-3-031-25052-1 (eBook)
https://doi.org/10.1007/978-3-031-25052-1

This Springer imprint is published by the registered company Springer Nature Switzerland AG
The registered company address is: Gewerbestrasse 11, 6330 Cham, Switzerland

Foreword

Rabies is very well known since ancient times. The Codex Eshnunna (ca. ~2000 BCE) is a frequently quoted document suggesting evidence that the disease was widespread in antiquity, and that it was associated with dogs. Since then, rabies in dogs, and sometimes wildlife, in Europe and the Middle East was reported sporadically. However, this book is not about rabies in the earliest Orient and Europe – it is about the disease in the Americas. Here, it was long recognized that rabies was spread by carnivores, including dogs. The disease transmitted by hematophagous bats ("vampires") was, and still is, of great concern. Vampires not only infect people but also predominantly livestock, having an important economic impact, as well as a role in conservation biology, considering effects upon other wildlife. Throughout the twentieth century, it became evident that rabies in bats was not limited to vampires but is also very widespread in insectivorous and other bats throughout the Americas. After some delay, rabies was also found in other wild mammalian species.

Today, rabies epidemiology as described in wildlife disease handbooks is often too simplistic: the virus transforms the infected animal into a berserk biting machine that attacks every moving object. The reality is much more complicated. Susceptibility, patterns of virus excretion, length of incubation, and duration of clinical disease must be correlated with the population biology of the principal host to maintain the epizootic. Nevertheless, some highly susceptible taxa (e.g., rodent and lagomorph spp.) are highly susceptible to experimental infection but are hardly ever found rabid among naturally diseased wildlife. The behavior of potential vectors and victims is also significant. Avoidance of diseased (conspecific and allospecific) vectors must be of influence, but the opposite occurs as well, as European roe deer were observed to expose themselves by harassing paralyzed foxes, similar to New World livestock investigating rabid skunks.

Though it was obvious for quite some time that different species of Carnivora and Chiroptera were hosting different epidemiological variants of the virus, the discovery of a wide variety with newly developed techniques confirmed this notion. The twentieth century brought progress in the understanding of disease agents in general and in methods for studying these viruses. Inclusion bodies (Negri bodies) in nerve cells were long considered pathognomonic for rabies. However, other infections also cause the formation of inclusion bodies in neurons. Hence, the introduction of immunofluorescence, which combined morphology and a specific immunological reaction, was most important and facilitated the discovery of the disease in numerous mammalian species. The detection of rabies virus antibodies in human and animal sera with neutralization tests is significant, particularly to verify vaccination success. Numerous other tests for immunity were developed in recent years. Many of them have issues with sensitivity and specificity that are not completely solved. While properly collected sera in the laboratory usually have a satisfactory ratio between sensitivity and specificity, this is quite a different matter with blood taken under less-than-ideal field conditions. I recommend consuming with a grain of salt reports on the high frequency of rabies virus antibodies in sera of wildlife, free-ranging dogs, and fluids that could contain impurities that may interfere with the test.

During the twentieth century, we learned how to identify virus variants in the laboratory. Molecular biology initiated a new era of rabies research. Gene sequencing permitted not only more precise differentiation between different virus variants, but also allowed the reconstruction of their phylogenetic relationships. This combined with phylogeographic analysis has greatly enhanced our understanding of the evolution of these viruses and aspects of their transmission dynamics. The development of monoclonal antibodies is also an achievement of the past century. Monoclonal antibodies, often applied in indirect immunofluorescence to acetone-fixed brain smears, are sometimes called the workhorses for rapid viral variant identification, especially within Latin America.

The culling of rabies vector populations is a procedure practiced since antiquity to the present. In the Americas, it was mostly applied to dog populations, sometimes also to wildlife. For example, vampire bat colonies are destroyed in the vicinity of outbreaks of bovine paralytic rabies. With growing opposition to poisoning rabies vectors, the use of vaccination became more attractive. Vaccines for domestic animals were incorporated into disease prevention and control schemes. The late Dr. George Baer's discovery that foxes can be immunized orally with the attenuated live vaccine ERA was a major contribution. Baits laced with ERA vaccine distributed in fox rabies areas in Europe and North America did indeed help to control the disease. However, the ERA rabies virus has some residual pathogenicity. It was therefore important to develop other vaccines avoiding this problem. The solution was recombinant pox- and adenoviruses expressing the rabies virus glycoprotein gene.

The advancement in our understanding of rabies viruses and the biology of their hosts is due to the efforts of multiple, individual scientists. The progress is also the outcome of superior cooperation encouraged by international organizations. A crucial stimulus that was initiated by Dr. Baer is the annual "Rabies in the Americas Conferences." Research in all fields is still progressing. This book is composed of accounts describing the status quo, past efforts, and current projects. It will also stimulate and encourage new projects. Its editor, Dr. Charles Rupprecht, deserves topmost praise for undertaking this effort.

Alexander I. Wandeler
Centre of Expertise for Rabies
Canadian Food Inspection Agency
Ottawa Laboratory-Fallowfield (retired)
Ottawa, ON, Canada

Preface

> ...those who do not remember the past are condemned to repeat it... – Jorge Agustín Nicolás Ruiz de Santayana y Borrás, 1905

With pandemic fatigue all around, is now the moment for an exposition about rabies in the "New World"? Well, that depends – our book, while focused on the basic roots of hydrophobia, has an embedded, recurring theme of "things change, stuff happens." As such, such a tome was a long time coming, personally and professionally. As a New Jersey kid, was hard enough to wait for the next holiday and summer vacations. Clearly, given human limitations, no one could fathom the implications of immense passages of geological periods, yet still hoped those grand, monstrous "terrible lizards" might still be found alive in some far-off, poorly explored oasis (together with just about every other child, I suppose, then and now, considering my own now captivated grandchildren). Where did they come from and why did they vanish? Fascination was obvious – on the topic of "beginnings" and any related discussions on "transitions" in the natural sciences, such as "what came first, the chicken or the egg" (of course, we all know t'was the egg!). Initiated as a "forever" student from kindergarten to date, I was a huge fan of anything "prehistoric," initiated basic conversations with my elders on the "why things are" (even briefly flirted with becoming a paleontologist in first grade). During the 1960s, a cherished train trip with my father to the American Museum of Natural History in New York City's Central Park was a dream come true. Unfortunately, this gradual realization that all dinos were extinct (with the exception of avian ancestors) landed hard, so I focused upon the extant instead, content with a wonderful collection of outdoor pet turtles in elementary school. About this time I have my earliest recollections about the inklings of "the rabies" – first when chased by a free-ranging neighborhood dog and secondly when my older brother found a bat roosting under the cedar shingles of our modest Cape Cod (quickly dispatched of course because of our mom's exaggerated concerns over what we would "catch from those things").

Thereafter, a professional path toward the rabies field was a tad circuitous and accidental. In high school health class, I annoyed our gym coach by trying to understand the origins of STDs. As an undergrad at the state university, attempting to distance myself from the highly competitive "pre-med throats," I was torn between invertebrate and vertebrate zoology, so compromised on discipline, and majored in ecology, after some gentle prodding by my low-key faculty advisor. Naïve to the focused biopolitics of academia, at graduate school in Wisconsin, I philandered with limnology, ornithology, fisheries biology, herpetology, etc., but finally settled on zoology, with a bent toward mammalogy. This decision was all the more cemented after meeting Merlin Tuttle – being re-introduced to that exciting and somewhat mysterious world of bats (a focus of my Master of Science degree). Along came my first of many later rabies vaccinations, to preclude the occupational risk from my subjects (which Merlin of course poo-pooed and suggested if ever bitten to "just squeeze out the blood"). After a stint in Panama, working on Neotropical bats with the Smithsonian Tropical Research Institute (where I received my first bite from a vampire bat), and meeting a researcher investigating owl monkeys (*Aotus* spp.) as models of falciparum malaria, a light bulb went off. Although "One Health" was not yet born, I mused on embarking into "applied disease ecology and epidemiology" (AKA veterinary medicine) – focused upon non-domestic species.

At the University of Pennsylvania, I was quite fortunate as one of the self-professed "odd-balls" to be mentored by faculty that shared a passion for the non-traditional (such as the late parasitologist Gerry Schad, for whom I nearly missed a Pathology exam, because I was duly collecting discarded mesocarnivore intestines for him from the offal at fur dealers in the hopes of finding a hint of echinococcus), challenged by ever attempting towards anything new on rabies by epidemiologist Larry Glickman ("everything has been done!"), and gelling with contributions down the street from the vet school by key staff at the Wistar Institute of Anatomy and Biology (where the first human tissue culture anti-rabies vaccine and monoclonal antibodies were created under the Old World team of the late Tad Wiktor and Hilary Koprowski).

A career as a budding "rabiesologist" was born of these wellsprings, initially focused as an evolving "gradual student" on the detection, characterization, and appreciation of the dynamics of wildlife rabies virus variants (raccoon rabies was now exploding throughout the mid-Atlantic states having just emerged in Pennsylvania), and cemented during a post-doc opportunity on the development of a wildlife rabies vaccine, while in the preternatural shadow of that "father of oral vaccination," the late CDC-ite, Jorge Baer. One would have never predicted that I too would be recruited from north of the Mason-Dixon line to eventually serve in that same institution in Atlanta. Like true genetic relations and parenting, you may never truly cherish your familial and academic mentors until they are gone, and you have become one yourself, like an antediluvian "old crocodile." Introspective appreciation of such spatio-temporal personal/professional voyages tends to mire one in the throes of history, by definition. Queries such as "how did we arrive here" again came naturally – reinspired by twenty-first century documentation of Native American ancestry, hybridized with nineteenth century Old World genes, morphed with a fascination for the true past, an affinity with indigenous peoples, and a quest for specters of the truth in a chosen field as a born-again provocateur of same – AKA la rage, la rabia, a raiva, die Tollwut, etc.

Given its age as one of the oldest documented infectious diseases and global prominence of burden in agriculture, public health, veterinary medicine, and conservation biology, much has already been written about rabies history. Subjectively, some of us feel that if the field of rabies was led clearly by Old World scientists throughout the nineteenth century. New World contemporaries benefited greatly from such immense background and expertise and offered their own unique accounting only during the later twentieth century. In kind, by comparison, the partial genesis of this present work owes much to the former "Historical Perspective of Rabies in Europe and the Mediterranean Basin" (AA King, AR Fooks, M Aubert, AI Wandeler, eds. World Organization for Animal Health (OIE): Paris, France. 2004, 384 pp.). Hence, what seemed to be lacking was a New World perspective, as a global companion piece. As I hope you may agree, such biomedical and natural history remains relevant today – and is continually being re-written.

Was rabies present in the pre-colonial Americas? Did the Meso-American mythological tales of the bat god, Camazotz serve an allegory to death and destruction, potentially representing rabies virus transmission from vampire bats? (Figs. 1 and 2). Rather curiously, facts such as when, where, and how rabies originated here are simply unknown.

Truth be told, all surveillance is biased. History is skewed substantially by the victorious, in a revisionist version of *spolia opima* simply on a grand scale along the civilization, and colonization, exploitation themes. For the first several hundred years after the fifteenth century colonization of the Americas, most of the appreciation for the new norm, such as various maladies, was penned by privileged white male European colonists, including the clergy. Indigenous languages were suppressed, and with them baseline oral recollections. The same fate befell written vestiges and codices from prior civilizations in Mesoamerica and the Andes. Monuments to the old were destroyed. Temples were razed, and new churches were built over ancient foundations. Public non-converts, keepers of knowledge, and traditional healers were killed brutally. While major human communicable diseases of the time, such as measles, smallpox, influenza, and tuberculosis, may have been recorded by the New Order, other pathogens, particularly zoonoses, may have been very much under appreciated.

As a fundamental disease of nature, such a base appreciation of the introduction and perpetuation of rabies is somewhat akin to our conjecture over the detection and characteristics of "the sound of very softly falling trees in poorly visited forests." Despite our hubris, speculation, and much-vaulted technical prowess, we are left wanting, as COVID-19 so aptly demonstrated. Still, while recognizing such limitations, we must celebrate our "facts" limited as they are, strive to improve upon our re-interpretations of the past, and lastly should never ignore those platitudes to our ancestors (as well as our deserved late contemporaries responsible for these collective fruits of our labors), to relish the here and now at this place and time – such is the background for the anticipated – because rabies is the true king of zoonoses.

I would be remiss to fail on my Acknowledgments – it is not so simple to get a project completed during a pandemic! Nevertheless, I was cheered that Springer agreed to take on this concept, as well as to the anonymous reviewers that offered

Fig. 1 Depiction of the Mayan hero twins, Hunahpú and Xbalanqué, attacked by the giant "bat god" Camazotz (also known as Zotzilaha Chimalman), in the Popol Vuh text. (*In the House of Bats*, by William Sewell, in: L. Spence. The Myths of Mexico and Peru. Thomas J. Crowell Company, New York, 1913)

early guidance for my draft outline. Gratitude extends to my editor, Alison Ball and other staff, for their immense patience, waiting anew for the promised. I am incredibly thankful to my many colleagues from throughout the globe (living or working in this Hemisphere) for devoting their countless hours, creative energies, and experienced insights as co-authoring contributors, particularly those whose mother tongue was not "American English." Attesting to the intricacies of this topic, these represented a diverse cadre of professionals, including anthropologists,

Fig. 2 Terracotta figurine, Zapotec, sixth–ninth century, Oaxaca, Mexico, thought to depict a bat god. (https://commons.wikimedia.org/wiki/File:Bat_god-ETHAM_L000043-IMG_2409-black.jpg, by Rama)

diagnosticians, economists, epidemiologists, historians, physicians, research scientists, veterinarians, virologists, wildlife biologists, etc.

For the audience at large, I sincerely hope you may find something of relevance in these interwoven themes of "host, agent, and environment" between the volumes of this contribution. Lastly, but certainly not the least, my understated but heartfelt devotedness to Nancy and "the kids," for all of their collective love, patience and support, tolerance for those countless hours during 2020–2022 that "der Opa" was shuttered in his dark office cave, with his seemingly omnipresent "PLEASE-DO-NOT-DISTURB" sign up, readin', ritin', reviewin', and editing these many drafts, that would eventually became this final product.

Pax vobiscum.

Lawrenceville, GA, USA
10 October 2022

Charles E. Rupprecht

Contents

Chapter 1
A World in Flux: The Columbian Exchange

Davide Domenici

Abstract In a path-breaking volume published in 1972, Alfred Crosby coined the expression the "Columbian Exchange" to refer to the global circulation of people, animals, plants, and microbes initiated by the European "discovery" of America, an historical phenomenon which shaped the modern world in a radical and unprecedented way. In this chapter, I review the impact the Columbian Exchange had in food habits and health, two fields where the entanglement between biological and cultural facts is of outmost relevance, as shown by the intimate relationship linking epidemics, colonial violence, and demographic collapse. The Columbian Exchange, inducing the global circulation of several animal species, also had a profound impact on human-animal relationships. Given the relevance that European dogs had in the spread of rabies in the Americas, in the last part of the chapter I summarize the results of scholarly studies which explored the changing relationships between animals and humans in the post-colonial Americas.

Keywords Biogeography · Columbian exchange · Dogs · Globalization · Livestock · Non-indigenous species

Introduction

Decades after the Spanish conquest of Mexico, the dread caused by the arrival of peoples, animals and materials never seen before was still vividly inscribed within Indigenous social memory. According to the Nahua interlocutors of the Franciscan friar Bernardino de Sahagún:

> …[the Spaniards] set off on their way to Mexico, coming gathered and bunched, raising dust. Their iron lances and halberds seemed to sparkle, and their iron swords were curved

D. Domenici (✉)
Department of History and Cultures, University of Bologna, Bologna, Italy
e-mail: davide.domenici@unibo.it

C. E. Rupprecht (ed.), *History of Rabies in the Americas: From the Pre-Columbian to the Present, Volume I*, Fascinating Life Sciences,
https://doi.org/10.1007/978-3-031-25052-1_1

> like a stream of water. Their cuirasses and iron helmets seemed to make a clattering sound. Some of them came wearing iron all over, turned into iron beings, gleaming, so that they aroused great fear and were generally seen with fear and dread. Their dogs came in front, coming ahead of them, keeping to the front, panting, with their spittle hanging down…[1]

Mesoamericans had never practiced iron metallurgy in pre-Colonial times, and their dogs were very different from the large hounds that the Spanish conquistadors employed in hunting and war. The strong sensorial emphasis of the Indigenous account powerfully conveys the emotional effect of that momentous and frightful encounter.

The dreadful appearance of iron weapons and war dogs in the lands of Central Mexico was but a glimpse of a much wider phenomenon, that is, the unprecedented global circulation of people, animals, plants and microbes initiated by the European "discovery" of America, commonly known as the Columbian Exchange. This label was coined by the historian Alfred Crosby with the publication of the famous volume, *The Columbian Exchange: Biological and Cultural Consequences of 1492* [2]. Other historians had studied related issues before, but the centrality that Crosby accorded to the biological dimension in his historical essay had no precedents, so that his work is today perceived as a forerunner of what is now called environmental history. Similarly, the vast geographical scope of Crosby's study, mostly centered on the Atlantic world but also considering – albeit in a still limited fashion – Africa and Asia, makes the *Columbian Exchange* a pioneering example of the now burgeoning field of global history.

Since Crosby's path-breaking volume, many other studies – including a second book by Crosby himself [3] – have greatly enriched our perception of the relevance that the movements of plants, animals and germs had in shaping what we call the modern world. As often stressed both in academic and popular writings, it is hard to believe that before the European invasion of the Americas, say, Italians had never heard of tomatoes or that no cattle had ever put hooves on the grasslands of the Argentinian pampas. As shown by Crosby, the transformations induced by the Columbian Exchange went far beyond the realm of food habits, since whole landscapes were changed dramatically by the introduction – either voluntary or involuntary – of new plants and animals, whose wider environmental impact has been so vast that it is even difficult to perceive. Sheep flocks, for example, caused the drying-up of vast areas of highland Central Mexico [4], while many areas with temperate climates such as ample regions of North America, especially prone to be targeted by settler colonialism, were literally invaded by European plants and animals to the degree that – in Crosby's words – they were transformed into veritable Neo-Europes [3]. One of the saddest dimensions of the Columbian exchanges involved the spread of pathogens: diseases as smallpox or measles crossed the Atlantic on European ships and swept the American lands, contributing to the massive demographic decline of Indigenous groups, while going in the opposite direction, syphilis made sexual intercourse in Europe much less light-hearted than before [5]. Other infectious diseases, such as rabies, whose precise presence in the New World before

[1] *Codex Florentinus*, Book XII, Chap. 1. English translation of original Nahuatl text by James Lockhart, in [1].

colonization is uncertain, were strongly exacerbated by the introduction of European rabid dogs and by abundant livestock, acting as it were a veritable feasting-place for vampire bats [6].

It is not by chance that the all the above-mentioned examples have to do with those spheres of human life – food, death and sex – which more than any other blur the divide between biological and cultural phenomena. A nuanced understanding of the entanglements between the biological and cultural dimensions of the Columbian Exchange is, indeed, one of the most difficult results to attain. As the reader would have noticed, the title of Crosby's book mentioned both biological and cultural consequences. Despite this equal pairing of the two terms in the title, it is clear that in his interpretation of the Columbian Exchange, Crosby accorded primacy to biological phenomena, perceiving cultural facts as mostly epiphenomenal reactions to unstoppable biological flows, whose furious current could in no way be stopped or altered by social actors. Additionally, he was very explicit in declaring his interests: at the very beginning of *The Columbian Exchange* Crosby wrote that "…Man is a biological entity before he is a Roman Catholic or a capitalist or anything else…" [2], a position later reiterated when, while discussing the history of the Canary Islands, Crosby plainly stated to be "…more interested, say, in the rabbit propagation than in manifestations of Our Lady of Candelaria…" [3]. As already stated, the centrality accorded to biological and environmental phenomena was *the* strength of Crosby's approach, especially in a time when historians used to pay limited attention to them. For this reason, it would be unfair and anachronistic to accuse Crosby of biological determinism, a criticism that, on the other hand, could be made probably with more reason to some of Crosby's epigons, such as Jared Diamond, author of the Pulitzer Prize-winning *Guns, Germs and Steel* [7]. Now that almost half a century has passed since the publication of Crosby's book, we are in the position to examine the problem with new eyes and to observe how these two dimensions were entangled inextricably (Fig. 1.1).

Fig. 1.1 Allegorical depiction by Johannes Stradanus of 'first contact', East meets West, with Amerigo Vespucci awakening the 'sleeping' America, espousing dual biological and cultural aspects. (https://en.wikipedia.org/wiki/Amerigo_Vespucci#/media/File:Stradanus_America.jpg)

Tasting the Columbian Exchange

Food studies is probably the field in which the transformative effects of the Columbian Exchange have been more thoroughly researched by a recent burgeoning academic literature.[2] Besides a host of works devoted to the history of specific foodstuffs, important advancements have been made, for example, in the exploration of the effects of the Columbian Exchange in non-European regions of Africa and Asia, thus strongly enriching the global scope already present in Crosby's volume (e.g [18, 19].). If some issues already observed by Crosby, as the role that American starches, such as maize and potatoes, had in fueling the European demographic growth and the Industrial Revolution, are still object of debates, other topics were explored in greater detail providing us with a much more complex understanding of the relevance of the political, economic, and cultural frameworks in which such biological phenomena took place. In his veritable masterpiece *Sweetness and Power* [20], for example, Sidney Mintz gave a thick political and economic reading of the history of sugar, further enriched by later contributions by scholars such as Stuart Schwarz [21]. More recently, a similarly dense political reading of the history of potatoes has been provided by Rebecca Earle [22]. The strong political value that foods had on both sides of the Atlantic in pre-fifteenth century times charged food-related colonial encounters with dense political meanings, also influencing the modalities through which gastronomic information was exchanged and recorded in early colonial times [23].

The role of cultural preferences or idiosyncrasies, a topic which Crosby tackled in a rather marginal way, has been explored more thoroughly in recent times, changing our view of the attitude that European consumers had toward new foods: traditionally described in terms of suspicion – if not of stubborn and ultimately ridiculous resistance, as in the infamous case of the aversion that the English herbalist John Gerard manifested toward maize – European reactions are now understood as much more complex and diverse, not being limited to plain refusal or sudden transformation to better meet European tastes and practices. As hilariously put by Rebecca Earle, we must recognize that "…Early modern Europeans were not all the culinary troglodytes they have been depicted as being…" [10]. Studying the consumption of chocolate in early modern Spain, for example, Marcy Norton showed that Europeans were actually prone not only to ingest new foods but also to indulge in novel gustative experiences, even adopting indigenous modes of preparation [24, 25]. An outstanding demonstration of the relevance of wider cultural notions is Rebecca Earle's volume *The Body of the Conquistador*,

[2] Useful, recent syntheses on the Columbian Exchange include [8–12]. Among the many possible relevant references, also including those on specific foodstuffs, see [13–17]; see further references in the main text.

where the author shows that cultural-specific concerns with health and bodily identity significantly drove Europeans' responses to novel foods in the Americas [26]. After a first phase characterized by a notable adaptation to local foods, Europeans started importing European foods for a number of reasons related with their notion of health, race, cultural identity and religion. In the often-repeated story of the accidental discovery, by the black conquistador Juan Garrido, of three wheat grains in a sack of rice, one of which sprouted giving rise to all the wheat cultivated in Mexico, is hard not to see an allegory of the diffusion of Christianism and of its wheat-based liturgy. It is clear that Spaniards highly esteemed the cultivation of European plants in America even beyond their specific symbolic or religious meanings: Bernal Díaz del Castillo, quite prone to boast his own (and doubtful) participation in key events of the conquest of Mexico, mentioned orange sowing – and a curious Indigenous collaboration – among his feats:

> …I went to sleep in a high idols' house and there I sowed seven or eight orange seeds which I had brought from Cuba; they sprouted very well because it seems that the priests of those idols defended them from ants, and gave them water and kept them clean, because they noticed that they were plants so different from those they had. I have recorded this here so that it will be remembered that those plants were the first oranges ever sowed in New Spain: and after the conquest of Mexico […] I went in search of my oranges and transplanted them; and they were very good… [27].

With effects ranging from the wider environmental transformations to the intimacies of bodily nurture – passing through the realm of religious beliefs – the food-related dimension of the Columbian Exchange has been a veritable gold mine for recent scientific research, whose results were often disseminated outside the academy intersecting the veritable food-craze which is currently affecting popular literature and television programs.

The Invisible Exchange: Epidemics and Colonial Violence

Sadly, the Columbian Exchange had not only the mouth-watering effects often explored by food historians. An intense debate has centered on the role that infectious diseases such as smallpox, malaria, measles, typhoid or paratyphoid fevers and cholera had in causing the demographic collapse experienced by the Indigenous American populations after the Conquest.[3] Such debate is made difficult by the lack of secure data on the population of the Americas at contact, so that the evaluation of the rate of the subsequent demographic collapse depends strongly on the often-speculative estimates advanced in that regard. A well-established tendency to propose very high estimates for pre-contact Indigenous populations (e.g [28–30].) led

[3] Among the many works on these themes, see [28–36].

to the calculation of high rates of demographic decline – at times described as a veritable catastrophic one, with a loss of 90% of the Indigenous American population in the span of a century – that could only be explained by the devastating impact of epidemics. According to more cautious scholars (who often propose a decline around the 30% of the initial population), this monocausal explanation incurs serious pitfalls [36]. It prevents, for example, an accurate analysis of the diversity of demographic effects induced by the socio-cultural contexts in which the epidemics spread, so ignoring both those cases in which the extermination of the indigenous population preceded the first epidemics and those where the infectious diseases had very limited demographic effects due to proper care and prevention practices. This leads to an even more pernicious interpretative bias: if perceived as an involuntary effect of the movements of infected persons, the spread of epidemics in the Americas becomes an explanation which in some way "absolves" the European intruders, seen as mere unconscious carriers of pathogens whose existence they ignored. Much to the contrary – as the current COVID-19 pandemic is sadly showing us all – it is well known that the effects of epidemics are strongly influenced by social factors, so that the structural violence often pervading colonized indigenous societies (surely not an "unintended" factor) acted as a primary cause for the demographic collapse, thus blurring the distinction between involuntary and voluntary effects. According to Massimo Livi Bacci [36], the catastrophe was not the inevitable outcome of contact with Europeans, but rather the outcome of the interaction between natural factors and human behaviors. Apparently, this result of recent academic studies was very clear to witnesses, such as the Franciscan friar Toribio de Benavente (Motolinía) who, describing the ten "plagues" which had fallen upon the indigenous peoples of Mexico, started his list with smallpox and measles but immediately went on listing war, war-induced famines, labor organization, excessive tribute and taxation, forced work in mines, etc., thus sketching a lucid description of the violence-laden colonial society [37].

As already seen in the case of food circulation, the Columbian Exchange thus emerges as an extremely complex interplay between biological and social facts, to the degree that we could even perceive it as one of the best historical phenomena to radically question the very epistemological bases on which the biological vs. cultural divide is built.

Human-Animal Exchanges: Dogs and Humans in the Americas

The topic of the present volume – an infectious disease which affects both humans and other animals – seems especially apt to explore the complexity of the causes and effects of the Columbian Exchange. The topic becomes even more inspiring when we consider that the animals that mostly acted as infective agents for humans were dogs, that is, animals whose life had been deeply entangled with that of human actors for millennia. The effects of the Columbian Exchange in the context of human-animal relationships, indeed, is a topic which only recently has caught the attention of scholars (e.g [38, 39].) and, as far as dogs are specifically concerned, seems to be still in its infancy, despite promising approaches (e.g [40].).

According to current scholarly understandings, dogs repeatedly entered the Americas together with human colonists at least since 10,000 BP, then differentiating into a series of different breeds which spread all over the continent [41, 42]. In ancient Mesoamerica, for example, existed at least four different breeds of *Canis familiaris*: the *itzcuintli*, or common Mesoamerican dog; the *xoloitzcuintli*, or Mexican hairless dog; the *tlalchichi*, or short-legged dog; and the so-called Maya dog (Fig. 1.2). Dog-wolf and dog-coyote hybrids were also identified in the archaeological record. Some of these Mesoamerican breeds were further diffused in other American regions, so that a Mexican-derived hairless dog was also known in the pre-Colonial Andean world.[4]

As also occurred in the Old World, in the Americas the lives of dogs intersected those of humans in multiple ways, being employed as pets, hunting, herding, and

Fig. 1.2 Ceramic image of a *xoloitzcuintli (Ángel M. Felicísimo, Merida, Mexico). (*https://upload.wikimedia.org/wikipedia/commons/a/ae/Xoloitzcuintle_%2821058489659%29.jpg)

[4] On dogs in pre-Columbian America see [43, 44]. The Mesoamerican and Andean cases had been thoroughly studied by Raúl Álvarez and colleagues in a dearth of publications; see, at least, [45, 46] and references therein.

transport aids, food resources, and as sacrificial victims buried as religious offerings. Dogs must have been a familiar presence in Indigenous daily life, living in the households' patios and roaming in the markets in search of food (and where they were also sold as food). The strict bond linking dogs and humans is reflected in the widespread practice of burying dogs together with humans, a custom that – at least in the Nahua world of Mesoamerica – was linked with the idea that dogs would have helped the dead person to cross the Apanohuayan, an underworldly mighty river. This funerary practice, together with repeated appearance of dogs in Mesoamerican mythology (where their creation was described as the outcome of the decapitation of humans whose heads were then stuck to their backs), as well as the existence of anthropozoomorphic deities, combining human and canine traits (such as the Nahua god Xolotl, the "canine twin" of Quetzalcoatl), attesting to the degree of perceived entanglement between dogs and humans (Fig. 1.3). This sympatric relationship could have even led to some kind of overlapping between human and canine bodily

Fig. 1.3 Pre-Columbian illustration of Xolotl, the canine twin of Quetzalcoatl, from Codex Borgia. (https://commons.wikimedia.org/wiki/File:X%C3%B3lotl_1.jpg)

features: it has been proposed that the dental modifications observed on a human skeleton excavated in the Aztec site of Huixtoco would have been aimed at mimicking the shape of canine dentition and probably of that very female specimen which – after having been sacrificed, skinned, decapitated, quartered, cooked, eaten and partially reconstructed – was finally buried besides the human corpse [47].

The long isolation from the rest of the world that American dogs and humans had lived for thousands of years was broken with the colonial encounter, when both Old World humans and dogs, together with a host of "new" animals, were introduced in the newly "discovered" continent. The events of the Conquest attest the relevance of the man-dog relationship among the Spaniards: according to Francisco López de Gómara, the crew of a Spanish ship lost on an island in the area of Campeche managed to survive only thanks to the offices of a greyhound which they found on the island (probably left there by one of the previous Spanish expeditions in the area) and which used to bring them rabbits, hares, and deer [48]. In terms of the consequences of the Columbian Exchange on the human-animal relationship, one can only imagine the emotional effects that large animals such as horses and cattle, or the havoc-causing pigs which flourished in the tropical climates, must have had on indigenous people, whose larger domestic animals were smaller dogs, turkeys, guinea pigs and (in a highly circumscribed region) Andean camelids. As our opening citation showed, even "familiar" animals like dogs looked strange and fearsome. Greyhounds, mastiffs and great Danes must have been perceived as terrifying beings, especially when used in battle or to kill someone through the practice of "dogging", as famously denounced by Bartolomé de las Casas in his *Brevísima relación de la destrucción de las Indias* and also attested by archival documents [49].[5] According to the Spanish Jesuit Bernabé Cobo, who wrote around the mid of the seventeenth century, the dogs employed in battle by the Spaniards during the first phases of the conquest had become as fierce as tigers and the Natives were so frightened that they fainted only knowing that they had to fight against them. Nevertheless, Cobo noticed that Indigenous people soon become accustomed to these new dogs, so that in his times there was no Native household in Peru without Old World dogs, which were highly cherished [52]. The adoption of the ever-increasing population of European dogs by Native groups would have eventually led to the diminution of the local breeds. The astounding increase of the population of European dogs, also noticed by Cobo himself, must have been common to most American regions and was especially problematic in New Spain, where the newly imported dogs became a common presence both in the countryside and in the streets of colonial cities. The number of free-ranging dogs was so high that extermination campaigns, later also aimed at containing the diffusion of rabies, were often organized.[6] At the same time, the Spaniards also become accustomed to local American dogs, to the extent that the Franciscan friar Bernardino de Sahagún, while describing around 1570 the various kinds of dogs found in Central Mexico, wrote "…there

[5] On the use of dogs in battles see [50, 51].

[6] On dogs in colonial New Spain see [40, 53–55].

are other dogs called *tlalchichi*, small and chubby. They are very good to eat…" [56], thus showing that he must have savored them while sharing food with his indigenous collaborators. Sure enough, Europeans went on reproducing their usual cultural practices, keeping dogs as pets and even indulging in surprising cultural practices such as dogs' baptisms, wedding and funerals [57, 58]. One wonders what indigenous people would have thought of the jeweled collars that wealthy Europeans used to put on their pets, since adorning dogs with precious jewels was not an unknown practice in pre-colonial indigenous world, as attested by a recent finding in the Aztec Templo Mayor of a sacrificed, richly attired dog (or wolf), buried as part of a ritual offering [59].

From the initial clash, complex systems of exchange and adaptation to new conditions were made, so that American and Old World dogs became "social participants in the fabric of daily life" [40] of Natives, Europeans and Mestizos alike. Within this complex social scenario, Old World rabies must have found multiple ways to diffuse by the eighteenth century, becoming itself a relevant, non-human social actor in broader American life.

Conclusions

As even a brief introduction like the present one has shown, the Columbian Exchange has been one of the fundamental historical phenomena of the modern age, which actually shaped the world as we know it today. Its relevance has been so pervasive that is even difficult to define its limits, both in a chronological and conceptual terms.

As for the former aspect, it is clear that exchange of biological entities which followed the "discovery" of America was but the initial step of a much wider and still ongoing phenomenon. Every phase of sustained globalization – and the discovery of America initiated the early modern one – had, and still has, its own "exchange", as the current COVID-19 pandemic is reminding us every day.

The problem of the conceptual limits of the Columbian Exchange is even more tricky: as we saw, it is probably foolish to try to distinguish biological phenomena from cultural ones; on the other hand, if we expand the notion of the Columbian Exchange to the extent of transforming it in a catch-all label for any transatlantic exchange of ideas and cultural practices, we would incur in the risk of "diluting" it in a limitless cultural history. To avoid such a risk, we should probably keep close to Crosby's initial attempt, carefully avoiding any biological determinism, but also reaffirming the centrality and unavoidability of biological facets in human life.

It is for these reasons that the case of rabies seems especially interesting to explore the multiple dimensions of the Columbian Exchange. The European transatlantic navigations of the early modern age put in contact an enormous web of humans, dogs, livestock, and bats, mutually feasting in an almost endless chain of interactions and infections, whose past and present effects are still being thoroughly felt and explored.

References

1. Restall M, Sousa L, Terraciano K (eds) (2005) Mesoamerican voices. Native language writings from colonial Mexico, Yucatán and Guatemala. Cambridge University Press, New York
2. Crosby A (1972) The Columbian exchange: biological and cultural consequences of 1492. Greenwood Publishing Company, Westport
3. Crosby A (1986) Ecological imperialism. The biological expansion of Europe, 900–1900. Cambridge University Press, New York
4. Melville EGK (1994) A plague of sheep: environmental consequences of the conquest of Mexico. Cambridge University Press, Cambridge
5. Harper KN, Zuckerman MK, Harper ML et al (2011) The origin and antiquity of syphilis revisited: an appraisal of Old World pre-Columbian evidence for treponemal infection. Am J Phys Anthropol 146:99–133
6. Vos A, Nunan C, Bolled D et al (2011) The occurrence of rabies in pre-Columbian Central America: an historical search. Epidemiol Infect 139(10):1445–1452
7. Diamond J (1997) Guns, germs and steel: the fates of human societies. W. W. Norton, New York
8. Nunn N, Qian N (2010) The Columbian exchange: a history of disease, food, and ideas. J Econ Perspect 24(2):163–188
9. Earle R (ed.) (2009) European cuisine and the Columbian exchange, special issue of Food and History 7(1).
10. Earle R (2012) The Columbian exchange. In: Pilcher J (ed) The Oxford handbook of food history. Oxford University Press, online edition. https://doi.org/10.1093/oxfordhb/9780199729937.013.0019
11. Gentilcore D (2016) Food and health in early modern Europe. Diet, medicine and society, 1450–1800. Bloomsbury Academic, New York
12. McNeill JR (2019) The Columbian exchange. Encyclopaedia Britannica online. https://www.britannica.com/event/Columbian-exchange
13. Long J (1986) Capsicum y cultura: la historia del chilli. Fondo de Cultura Economica, Mexico City
14. Long J (ed) (2003) Conquista y comida. Consecuencias del encuentro de dos mundos. Univerisdad Nacional Autónoma de México, Mexico City
15. Foster N, Cordell L (eds) (1992) Chilies to chocolate: food the Americas gave the world. University of Arizona Press, Tucson
16. Kiple KF (2003) A movable feast: ten millennia of food globalization. Cambridge University Press, New York
17. Gentilcore D (2010) Pomodoro! A history of the tomato in Italy. Columbia University Press, New York
18. McCann J (2005) Maize and grace: Africa's encounter with a new world crop: 1500–2000. Harvard University Press, Cambridge
19. Carney J, Rosomoff R (2009) In the shadow of slavery: Africa's botanical legacy in the Atlantic world. University of California Press, Berkeley
20. Mintz SW (1985) Sweetness and power: the place of sugar in modern history. Penguin Books, New York
21. Schwartz SB (2004) Tropical Babylons: sugar and the making of the Atlantic world, 1450–1680. University of North Carolina Press, Chapel Hill
22. Earle R (2020) Feeding the people: the politics of the potato. Cambridge University Press, Cambridge
23. Domenici D (2020) Dei cibi che usavano i signori. Lessico gastronomico e progettualità politica nel mondo coloniale della Nuova Spagna (XVI sec.). In: Montanari M (ed) Cucina politica, Il linguaggio del cibo fra pratiche sociali e rappresentazioni ideologiche. Laterza, Roma-Bari, pp 59–74
24. Norton M (2006) Tasting empire: chocolate and the internalization of Mesoamerican aesthetics. Am Hist Rev 111(3):660–691

25. Norton M (2008) Sacred gifts, profane pleasures: a history of tobacco and chocolate in the Atlantic world. Cornell University Press, Ithaca and London
26. Earle R (2012b) The body of the conquistador. Food, race and the colonial Experiencw in Spanish America, 1492–1700. Cambdridge University Press, New York
27. Diaz del Castillo B (1991) Historia verdadera de la Conquista de la Nueva España. Alianza Editorial, Madrid
28. Dobyns HF (1966) Estimating aboriginal populations. An appraisal of techniques with a new hemispheric estimate. Curr Anthropol 7:395–449
29. Dobyns HF (1993) Disease transfer at contact. Annu Rev Anthropol 22:273–291
30. Cook SF, Borah W (1971-1979) Essays in population history, vol 1-3. University of California Press, Berkeley
31. Verano JW, Ubelaker DH (1992) Diseases and demography in the Americas. Smithsonian Institution Press, Washington, DC
32. Cook ND (1998) Born to die: disease and new world conquest, 1492–1650. Cambridge University Press, Cambridge
33. Henige DP (1998) Numbers from nowhere: the American Indian contact population debate. University of Oklahoma Press, Norman
34. Newson L (2001) Pathogens, places and peoples. In: Raudzens G (ed) Technology, disease and colonial conquests, sixteenth to eighteenth centuries. Brill, Boston, pp 167–210
35. Alchon SA (2003) A Pest in the land: new world epidemics in global perspective. University of New Mexico Press, Albuquerque
36. Livi Bacci M (2008) Conquest. The destruction of the American Indios. Polity Press, Cambridge
37. Motolinia T (1995) Historia de los Indios de la Nueva España. Porrúa, Mexico City
38. Alves AA (2011) The animals of Spain. An introduction to imperial perceptions and human interaction with other animals. Brill, Leiden
39. Norton M (2015) The chicken or the *Iegue*: human-animal relationships and the Columbian exchange. Am Hist Rev 120(1):28–60
40. Lipsett-Rivera S (2015) A new challenge: social history and dogs in the era of post-humanism, Sociedad indiana. http://socindiana.hypotheses.org/320
41. van Asch B et al (2013) Pre-Columbian origins of native American dog breeds, with only limited replacement by European dogs, confirmed by mtDNA analysis. Proc R Soc Lond B Biol Sci 280(1766):20131142
42. Wang GD, Zhai W, Yang HC et al (2016) Out of southern East Asia: the natural history of domestic dogs across the world. Cell Res 26:21–33
43. Allen GM (1920) Dogs of the American aborigines. Museum of Comparative Zoology, Cambridge
44. Schwartz M (1997) A history of dogs in the early Americas. Yale University Press, New Haven
45. Valadez Azúa R, Mendoza España V (2005) El perro como legado cultural. Nuevos aportes 2:15–35
46. Valadez Azúa R, Rodríguez Galicia B (2013) El perro y la civilización mesoamericana: análisis y contraste entre pensamiento antropológico tradicional y conocimiento científico actual. Boletín del Instituto de Investigaciones Antropológicas 16:1–17
47. Valadez Azúa R, Gamboa L, Vélez N et al (2004) Perros y prácticas rituales en una Antigua aldea de la cuenca de México. AMMVEPE 15(5):158–171
48. López de Gómara F (2007) Historia de la Conquista de México. Biblioteca Ayacucho, Caracas
49. Boornazian DL (2011) Manuscrito del aperreamiento (manuscript of the dogging): a 'dogging' and its implications for early colonial Cholula. Ethnohistory 58(5):585–611
50. Varner JG, Johnson Varner J (1983) Dogs of the conquest. University of Oklahoma Press, Norman
51. Mastromarino M (1984) Cry havoc and let loose the dogs of war: canines and the colonial American military experience, MA Thesis, College of William and Mary, Virginia
52. Cobo B (1964) Historia del Nuevo Mundo. In: Bernabé Cobo P (ed) Obras del. Atlas, Madrid

53. Ávila González JS (2007) Voces y ladridos. Ensayo sobre los perros de la ciudad de México, siglos XVIII y XIX, PhD thesis, Universidad Iberoamericana, Mexico City
54. Rojas Hernández L (2011), Muerto el perro, se acabó la rabia. Perros callejeros, vacuna antirrábica y salud pública en la ciudad de México, 1880–1915. MA thesis, Universidad Nacional Autónoma de México, Mexico City
55. Exbalin Oberto A (2014) Perros asesinos y matanzas de perros en la ciudad de México (siglos XXI-XVIII). Relaciones: Estudios de Historia y Sociedad 137:91–111
56. Sahagún B (1998) Historia General de las Cosas de Nueva España, vol III. Conaculta, Mexico City
57. Tortorici Z (2013) In the name of the father and the mother of all dogs`: canine baptisms, weddings, and funerals in Bourbon Mexico. In: Few M, Tortorici Z (eds) Centering animals in Latin American history. Duke University Press, Durham, pp 93–119
58. Proctor FT III (2014) Amores perritos: puppies, laughter and popular Catholicism in Bourbon Mexico City. J Lat Am Stud 46(1):1–28
59. López Luján L, Chávez Balderas X, Zúñiga Arellano B et al (2012) Un portal al inframundo. Ofrendas de animales sepultadas al pie del Templo Mayor de Tenochtitlan. Estudios de Cultura Náhuatl 44:9–40

Chapter 2
Glimpses into the Past: New World Contributions Towards Understanding the Basic Etiology, Pathobiology and Treatment of Rabies

Drishya Kurup, Charles E. Rupprecht, Stephen Scholand, Catherine Yankowski, and Matthias Schnell

Abstract Historical reports of rabies first appeared during the 1700s in colonial descriptions from throughout the Americas, although the disease was present much earlier. Until the 1900s, European beliefs and enlightenment held clear sway over the region, as to the possible nature of the disease, its pathobiology, and its treatment, in both theory and practice. Gradually, throughout the twentieth century, virologists within North and South America revolutionized the field through substantial progress in electron microscopy, biochemistry, immunobiology, pathophysiology, vaccinology, and anti-viral therapeutics. Now, during the start of the twenty-first century, New World scientists continue to make major basic and applied contributions into the biology of rabies virus, its pathogenesis, and potential therapy.

Keywords Etiology · Pathogenesis · Spontaneous generation · Therapy · Virion

"Authors Drishya Kurup, Charles E. Rupprecht, Stephen Scholand, Catherine Yankowski" have equally contributed to this chapter

D. Kurup · C. Yankowski · M. Schnell (✉)
Thomas Jefferson University, Philadelphia, PA, USA
e-mail: Matthias.Schnell@jefferson.edu

C. E. Rupprecht
Auburn University, Auburn, AL, USA

S. Scholand
Rabies Free World, Meriden, CT, USA

College of Medicine, University of Arizona, Tucson, AZ, USA

C. E. Rupprecht (ed.), *History of Rabies in the Americas: From the Pre-Columbian to the Present, Volume I*, Fascinating Life Sciences,
https://doi.org/10.1007/978-3-031-25052-1_2

On the Nature of Rabies

Introduction

> …Would it not be extraordinary, amazing, incredible, if, at this late day, it were proved that the thousands and hundreds of thousands of observations recorded from the birth of history to the present day, by the trained physician or veterinarian as well as by the layman, were misconceptions, that the authors were deceived, and that the disease was a myth? Where can a parallel be found to such a sudden and complete overthrow of an ancient and almost universally accepted conclusion concerning a phenomenon so accessible to observation and investigation?…. D.E. Salmon, Chief of the U.S. Bureau of Animal Industry, 1900

No one knows where, when, or how rabies originated. Nevertheless, a combination of logic, inference, anecdote, and phylogenetic data support the concept that rabies and other diseases perpetuated in the New World long before European colonization [1]. Inarguably, however, the 'Columbian Exchange' altered these disease dynamics in the Americas [2]. The resulting viral ensembles were forever changed by the post-fifteenth century invasions, including any available indigenous insights by the purposeful destruction and suppression of native codices, languages and traditions [3]. Considering incubation periods of weeks to months, as transit times improved, European colonizers introduced domestic animals (e.g., dogs) that were unknowingly infected with rabies virus from multiple endemic foci, principally within England, France, Portugal, and Spain [4]. Except for the eventual establishment of urban rabies some two centuries after first contact, the disease in the West would later appear differently within new reservoirs, skewing towards more sylvatic manifestations (e.g., among bats, coyotes, gray foxes, raccoons, skunks, etc.). Besides the need for successful cross-species transmission and adaptation, this pause and gradual recognition may have been because, objectively, compared to the Old World, the New World landscapes and cultures were quite novel, with human and domesticated animal populations that were less dense and less intertwined across the continents, and with unique faunal community compositions.

At home, Europeans were well aware of this zoonosis for centuries, particularly associated with dogs [4]. Nevertheless, approximately two centuries would pass before the colonists appreciated the somewhat 'sudden' appearance of rabies, emanating not only from dogs, but also among other mammalian carnivores, throughout their newly founded colonies [5]. Such long experience with this acute, progressive encephalitis and European thought during the 'Age of Discovery and Enlightenment' colored the perception of rabies in full measure up until the nineteenth century, particularly related to the basic nature of the disease, its detailed pathobiology, and potential treatment [6]. Earlier insights on most diseases were likely influenced initially by Christian missions and religious schools during the early conquests of the 1500s, rather than by pure academics. For example, often clerics believed these new lands were 'blest by their Creator and free from this scourge'.

If rabies was present, why did recognition take so long? Originally, regional communications were nonexistent, as the various localities were quite poor and isolated from each other, as various European powers competed for dominance. Eventually, foreign technological impositions would hybridize with local traditions of the Maya, Aztec, Inca and other indigenous peoples by necessity, as reflected in certain beliefs and cultural practices, such as related to animal bites and the use of 'madstones' [3]. Only much later would science and medicine figure more importantly than religiosity and superstitions, by the notable influences of the University establishments during the later 16th and 17th centuries, such as in 'New France' (e.g., the Collège des Jésuites, the Séminaire de Québec, etc.), through support for scholarly endeavors by the Spanish empire in Bolivia, Colombia, the Dominican Republic, Ecuador, Mexico, Peru, etc., and by the British higher educational systems within North America (e.g., Brown, the College of William and Mary, Columbia, Dartmouth, Harvard, Kings College, Princeton, Rutgers, the University of Pennsylvania, Yale, etc.) [7, 8]. Thus, a primary local understanding of rabies was greatly overshadowed by pervasive European thought for several hundred years, because the structured order was Old World by origin. Not until the eighteenth century would inhabitants of the New World themselves come to recognize clinically this disease and other zoonoses in which they were awash [9]. By the twentieth century a combination of factors, including colonial wars of independence, improved infrastructures, newfound wealth, abundant natural resources, economic growth, immigration, the development of academic institutions and organizations, and greater connectivity, allowed broader creative freedom. Such progress began to favor greater scientific contributions and originality, as reflected directly from the Americas.

Peculiar Denials, Professional Misunderstandings, and Public Ignorance

In general, unlike many human communicable diseases, rabies initially received rather short shrift, even in basic surveillance. For example, one public health communication published during 1891 by Joseph J. Kinyoun (an American physician, post-doctoral visiting scientist under Robert Koch and later founder of the U.S. Hygienic Laboratory), lamented the lack of investigation on rabies, reporting, "…Occasionally an isolated case is reported by a medical man, but this appears to be the exception, not the rule. The most reliable information that can be had at present is from the daily press: although this is not an infallible source of information, especially on medical subjects, some heed must be given to its reports…" [10]. While an association with dog bite was understood by many in the medical profession, causation in the animal was not appreciated by all to infection with a specific 'germ' but rather ran the gamut from spontaneous generation, 'putrid ailments', water deficiency and atmospheric interventions.

In line with this general lack of reporting on rabies (as it was not yet a notifiable disease), some members of the biomedical community still regarded rabies as a cross between myth and hysteria. According to Harold C. Ernst (a prominent physician and Harvard-trained bacteriologist), he hoped to convince people that "…there is distinctly such a disease as rabies, that it is a specific disease, and that it depends upon a special 'virus' which, whilst not isolated as to its identity, has certainly been traced to its place of most common occurrence…", yet he noted the persistence of "…a very great diversity of opinion among men, perfectly competent to judge, in regard to the existence even of such a disease…" [11]. Others, such as the Harvard-trained veterinary pathologist Langdon Frothingham, also alluded to such biomedical denial, commenting that "…no treatise on medicine has omitted this disease from its covers, and yet there are people today who deny its existence…" [12]. Austin Peters, from the Massachusetts State Board of Agriculture, gave additional insight on disease deniers, stating that "…many persons can be found in the community who do not believe in the existence of this disease, or if they do, are so fond of dogs that they are opposed to muzzling and restraining orders for its suppression. Many of these persons are people of ordinary intelligence and with fair education and are apparently sane in other respects…" [13]. Such opinions were not limited to the eastern U.S. seaboard. According to the authors of an early history of rabies in Southern California, "…There is no disease, perhaps, in which the public mind is so beclouded as with that of hydrophobia. The general public often denies the existence of the disease and even among the profession we have seen doctors who say they never saw a case of rabies, and who state therefore it does not exist…" [14].

Beyond scientific skepticism on the nature of infectious diseases, expressed by some members of professional societies (not only regarding rabies), others made public rejections of rabies as an old superstitious relic, from a time 'when devils were thought to run pell-mell to people and animals carrying disease' [15–17]. Such rejections grew out of an intertwining of philosophical, political, religious, social, and animal welfare attitudes. For example, Millicent Morden, a Canadian physician testifying before a U.S. Congressional Committee stated that "…rabies as known today was sponsored by an erratic chemist, who had a genius for business the likes of which the world has never known…" [15]. Moreover, Eleanor McBean, an anti-vaccination activist, claimed curiously "…germs are not the cause of disease in the first place and are useful wherever they are found in nature. Disease is not something to be cured, it is a cure. These diseases are not contagious but are merely cleansing measures of the body…" [17]. Despite such attitudes of deniers (some of which persist today!), by the end of the nineteenth century, the overall standards of science education had improved, the biomedical professions had grown, and the funding of research had increased throughout the region, leading 'Americans' to make original contributions to the field, in line with their European counterparts [18].

Applications From the Advent of Virology

By the start of the twentieth century, discoveries about the nature of rabies were intertwined with the classical birth of virology from other microbiological disciplines. Contributions to this discovery came from both hemispheres. Louis Pasteur and his laboratory team were credited undeniably with the historical development of the first rabies vaccine during the 1880s, although many basic questions about the agent still remained to be answered [6]. These inquiries included: What was the specific cause of the disease? Was it uni-modal or multi-causal? Would it arise spontaneously? What was the influence of the environment, such as the 'dog days' of summer? Was it a virulent toxin or deleterious enzyme? Could it simply be a bacterial 'microbe of infinite smallness'?

Scientists of this era posed such questions for multiple maladies affecting plants, animals, and humans, long before the coinage of the 'One Health' concept. Once again, European scientists were at the forefront. For example, between 1882–1898, the Russian botanist Dmitri Ivanovsky and the Dutch scientist Martinus Beijerinck, working independently, built upon the work of the German chemist Adolf Meyer, to advance an understanding of tobacco mosaic disease. Collectively, they knew the disease was infectious but were unable to observe the 'bacteria' microscopically or propagate the agents in artificial media [18]. At first, Ivanosky felt that this was a bacterial toxin, while Beijerinck created a reproducible infectious filtered juice (i.e., *contagium vivum fluidum*) and called it a "virus", a non-specific term that had been coined for centuries in reference to the "slimy liquid" or "poison" of biological origins, such as the associated frothy or stringy saliva from rabid dogs [6]. He also suggested that these dissolved substances could only grow within living cells.

As Ivanovsky and Beijerinck found with tobacco mosaic disease, Paul Frosch and Friedrich Loffler, working in Germany during 1898, found that the agent of foot-and-mouth disease could be transmitted from animal to animal, by fluid passed through porcelain filters [19]. Meanwhile, within the Americas during 1881, the Cuban physician Carlos Finlay theorized that *Aedes* mosquitoes were vectors of a "microscopic germ", causing yellow fever [20]. During 1900, James Carroll, Aristides Agramonte, and Walter Reed demonstrated that this same agent similarly passed through extremely fine filters [21]. During 1926, the prominent American physician Thomas Rivers defined viruses conceptually as 'obligate parasites whose reproduction depended upon living cells' [22]. From 1928–1937, the South African-American physician Max Theiler succeeded in isolation of yellow fever virus to develop a vaccine, and his distinguished research career at the Harvard University School of Tropical Medicine, the Rockefeller Foundation Virus Laboratory, and the Yale School of Medicine and Public Health, culminated in a Nobel Prize in Physiology or Medicine during 1951 [23].

The etiological agent of rabies would join the ranks of these "small filterable substances", although the precise nature of such pathogens (i.e., liquids, or particulates?) remained unresolved for years to come, until additional technical advances were realized by the 1930s. These advances included: tissue culture methods (such as by American zoologist Ross Granville Harrison); viral plaquing (such as by French-Canadian microbiologist Félix d'Herelle, who in 1917 used bacteriophages to enumerate "invisible antagonists"); and dual microscopy and crystallization, such as by the American biochemist Wendell Stanley and his colleagues, who suggested that the infectious crystals of tobacco mosaic disease were protein/nucleic acid-based – work which was awarded the 1946 Nobel Prize in Chemistry [24–26].

Until mid-century, animal viral diseases were typically classified based upon their hosts (e.g., hydrophobia in humans vs. furious rabies in dogs), transmission mode (e.g., respiratory, gastro-intestinal, etc.) or pathology (e.g., peritonitis, encephalitis, etc.), as little was known regarding morphology, chemical properties, or means of replication. In addition, not all microbiologists agreed with the emerging "virus" concept, including its applications to rabies. In 1903, the Italian Adelchi Negri observed that rabid animals had formations within neurons and concluded that these objects were caused by classical parasites [27, 28]. Following these descriptions by Negri, many prominent American scientists agreed, including esteemed scientists such as Watson, Williams, Lowden, Calkins, Poor, Stimson, Frothingham, Hadley, and Harris, who considered these neural inclusions to be intracellular parasites, associated with the Rhizopods or Microsporidia [29–31]. However, following Paul Remlinger's passage of infectious material through filters that would have precluded classical cells, other scientists asserted that Negri bodies were not protozoa, but rather degenerative changes wrought by the virus within neurons [32]. Decades would pass before Negri bodies were recognized essentially as "viral replication factories" rather than intracellular parasites, debris, or singular virions [33]. Their eventual recognition as associated with this viral encephalitis was a boon to basic laboratory diagnosis (Fig. 2.1). Microscopic methods for detecting inclusions improved, as did the propagation of *rabies virus* (RABV) in both avian embryos and *in vitro*, amplifying material for further research [34, 35].

The 1900s saw major scientific advances that improved the basic understanding of rabies (Table 2.1). For example, during the 1920s–1930s, research advances in physics (e.g., by Busch, Szilárd, Ruska, Knoll, Rudenberg, Lubcke, and Burton, among many others) resulted in the invention of the electron microscope and related techniques that enhanced the study of microbes, including RABV [18, 49, 51, 53, 54, 60, 61, 71, 82–85]. New tissue culture methods, fixatives, and methodologies improved visualization of virions [36, 37, 86, 87]. By the 1960s, elongated rod-like particles were observed in infected tissue cultures and murine neurons by negative contrast and thin-section electron microscopy [38, 40–42, 46]. Based upon these structural studies, RABV was classified taxonomically as a rhabdovirus [47, 50]. Besides classical RABV, 'rabies-related' viruses were also identified via serological and ultrastructural research [67, 88]. The advent of monoclonal antibodies provided additional support for antigenic relationships among these newly appreciated lyssaviruses [68].

By the mid-twentieth century, electron microscopy defined the basic particulate nature of the RABV virion (Fig. 2.2). Such laboratory introspection clarified the relationships of virions to the Negri body, the appreciation of distinct 'rabies-related

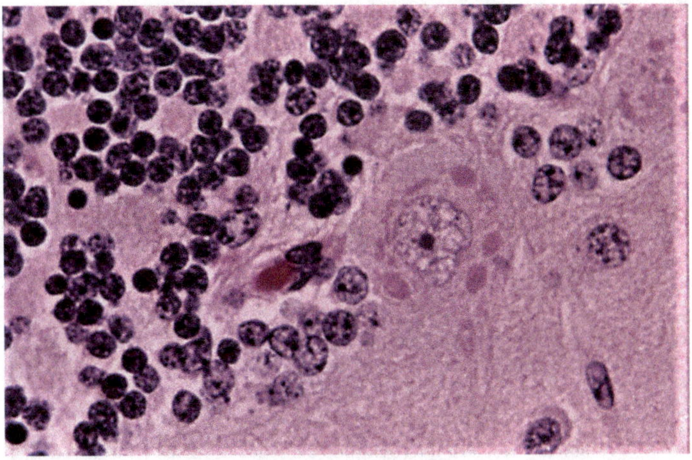

Fig. 2.1 From the time that Negri first described the pathognomonic intracytoplasmic inclusions that would bear his name, these structures were variously thought to be intracellular parasites, degenerative changes, or viral factories, as seen here in this micrograph of an infected Purkinje cell, in the cerebellum of a human rabies patient. (M. Fekadu, US HSUS Public Health Image Library)

viruses', and their affinity with other rhabdoviruses. Besides such technical progress in morphology, morphogenesis, and taxonomy, essential advances in biochemistry further delineated the functional nature of viruses [18, 83]. For example, during 1952, American researchers Hershey and Chase found evidence (while studying bacteriophages) that nucleic acids, not proteins, were the fundamental blueprints for life, advancing the basic understanding of the structure and replication of viruses [89, 90]. Additional early inferences and technical advances during the 1960s to date characterized the etiological model of rabies as a negative-stranded, non-segmented RNA virus, comprised of five monocistronic genes encoding for five structural proteins [43–45, 52, 55, 57, 58, 65, 69, 72–74, 76–79, 91]. These applied research findings on viral structure and function resolved the basic etiological nature of rabies, laid the groundwork for improved diagnostics and modern vaccines by technology transfer, and provided critical insights into pathobiology as a foundation for the design of future anti-viral drugs and experimental therapy. Building upon prior insights from Old World colleagues, the applied multi-disciplinary research contributions from the twentieth century to date, by numerous investigators within academia and government, including the Canadian Food Inspection Agency, the U.S. CDC, the Wistar Institute, Thomas Jefferson University, the Pan American Health Organization (PAHO) and at the Pasteur institutes throughout the Americas, have been instrumental to modern rabies laboratory-based surveillance, prevention and control.

Table 2.1 Selected twentieth Century Contributions on the Nature of Lyssaviruses by New World Scientists

Authors	Findings	Citation
Reagan and Bruckner (1950)	Early description of an apparent inclusion within a neuron of an infected mouse brain by electron microscopy, but no viral particles or internal structure were discernable	[30]
Hottle et al. (1951)	One of the earliest attempted electron micrographs of inclusion bodies from the hippocampus of a mouse inoculated with rabies virus, but no viral particles were recognized in association with inclusions	[31]
Kissling (1958)	Use of a tissue culture system for viral amplification studies	[36]
Almeida et al. (1962)	Original description of disrupted particles and ribbon-like internal material from the internal helix observed of fixed rabies virus in hamster kidney cells by negative contrast electron microscopy, mistakenly believed initially to be somewhat similar to myxoviruses	[37]
Matsumoto (1962)	Preliminary electron micrographic evidence of rabies virus as rod-like particles within murine nerve cells	[38]
Constantine (1962)	Rabies virus transmission by the nonbite route in a bat cave	[39]
Matsumoto (1963)	Description of elongated viral particles found within a matrix in the cytoplasm of nerve cells and of astrocytes of mice by thin section electron microscopy and inclusions appearing as dense rounded bodies containing various coarse structures, in association with viral particles	[40]
Pinteric et al. (1963)	Described membranous sacs covered with projections and containing wavy filaments in infected mouse brain by negative contrast electron microscopy, but no intact particles	[41]
Davies et al. (1963)	Sections of infected chicken embryo cells revealed elongated, bullet-shaped viral particles associated with masses of homogeneous material within the cytoplasm, budding from the cell membranes, and free in extracellular spaces, while rounded forms by negative staining showed a fringed outer surface	[42]
Johnson and Mercer (1964)	Rabies virus appeared to develop in the cytoplasm without any demonstrable nuclear involvement and staining reactions with acridine orange suggested that the agent was an RNA virus	[43]
Wiktor et al. (1964)	Adaptation and propagation of rabies virus in cell culture amplified material for later morphologic and biochemical analyses	[44]
Love et al. (1964)	Demonstration that Negri bodies contained RNA	[45]
Pinteric and Fenje (1966)	Association of bullet-shaped viral particles with infectivity, and description of the surface and internal details, with a membranous bleb of excess host cell membrane during budding, as observed by negative contrast electron microscopy	[46]
Melnick and McCombs (1966)	Suggested inclusion of rabies virus within the rhabdoviruses	[47]
Constantine (1966)	Nonbite transmission experiments in animals with bat rabies virus isolates	[48]

(continued)

Table 2.1 (continued)

Authors	Findings	Citation
Hummeler et al. (1967)	Bullet-shaped particles observed by electron microscopy, budding at the cell surface, consisting of an inner core of filamentous material surrounded by membranes of different densities, with the surface showing a honeycomb arrangement with protrusions 60 to 70 Å long, having a knoblike structure at their distal end	[49]
Jenson et al. (1967)	Based upon morphological similarities, support for rabies virus taxonomic inclusion as a rhabdovirus	[50]
Hummeler et al. (1968)	Morphology of rabies virus nucleoprotein	[51]
Sokol et al. (1968)	Purification of extra-cellular rabies virus	[52]
Dierks et al. (1969)	Electron microscopic observations revealed widespread involvement of neurons with accumulations of electron-dense intracytoplasmic matrices and moderate numbers of rabies virions associated with inclusions in the salivary glands, with extracellular budding of virions from the surface of marginal membranes of cells, exclusively in areas apical to nuclei and large accumulations of particles in dilated spaces lined with mucogenic cells	[53]
Jenson et al. (1969)	By electron microscopy, development of rabies virus in the trigeminal nerve and ganglion was identical to the brain, and virions noted within the endoplasmic reticulum of axons, discrete from viral matrices, representative of intraneural spread away from the infected perikaryon in the direction of axoplasmic flow	[54]
Sokol et al. (1969)	Isolation of nucleocapsids containing RNA from disrupted rabies virions	[55]
Correa-Giron et al. (1970)	Orally administered rabies virus leads to injection through the mucosa	[56]
Sokol et al. (1971)	Molecular characterization of rabies virus structural proteins	[57]
Vernon et al. (1972)	Model of rabies virion, correlating morphological data by electron microscopy with the biochemical composition	[58]
Hattwick et al. (1972)	Human survivor of rabies post-vaccination, after exposure and onset of symptoms	[59]
Murphy et al. (1973)	Demonstration of local rabies virus infection and transit from the inoculation site to the central nervous system via electron microscopy	[60]
Murphy et al. (1973)	Use of electron microscopy to define rabies virus infection of the central nervous system and centrifugal spread of virus to peripheral tissues	[61]
Murphy and Bauer (1974)	Early street rabies virus infection in striated muscle and later progression to the CNS	[62]
Baer et al. (1975)	Oral administration of a rabies vaccine to foxes provided antibody responses, as opposed to administration via the stomach	[63]
Porras et al. (1976)	Case report of survival after rabid dog bite and vaccination, with recovery after 1 year of symptoms	[64]
Dietzschold (1977)	Analysis of the oligosaccharide side chains on rabies virus glycoprotein	[65]

(continued)

Table 2.1 (continued)

Authors	Findings	Citation
Baer et al. (1977)	Experiments in mice supporting the possibility of survival from rabies	[66]
Harrison and Murphy (1978)	Documentation of lyssavirus infection of muscle spindles and motor end plates	[67]
Wiktor and Koprowski (1978)	Monoclonal antibodies for definition of specific antigens, detection of antigenic variants, and lyssavirus characterization	[68]
Dietzschold et al. (1978)	Purification of rabies virus glycoprotein	[69]
Clark (1978)	Rabies viruses increase in virulence when propagated in neuroblastoma cell culture	[70]
Charlton and Casey (1979)	In infected skunks, viral nucleocapsids, virions, and anomalous viral products found mainly in neuronal perikarya and dendrites, and less often in myelinated axons, while matrices, virions, and crystalloid structures were in muscle fibers at the inoculation site, with budding on the endoplasmic reticulum, neurotubules, and neuronal plasma membrane	[71]
Wunner et al. (1980)	Rabies virus-specific mRNA identified by analysis of encoded proteins	[72]
Anilionis et al. (1981)	Characterization of the cDNA sequence of rabies virus glycoprotein	[73]
Anilionis et al. (1982)	Deduction of rabies virus glycoprotein amino acid sequence	[74]
Fekadu et al. (1982)	Excretion of rabies virus in the salivia of dogs	[75]
Dietzschold et al. (1983)	Identification of an antigenic determinant of rabies virus glycoprotein that correlated with pathogenicity	[76]
Dietzschold et al. (1983)	Analysis of rabies virus soluble glycoprotein	[77]
Rayssiguier et al. (1986)	Cloning of rabies virus matrix protein mRNA	[78]
Larson and Wunner (1990)	Nucleotide and deduced amino acid sequences of rabies virus phosphoprotein	[79]
Smith et al. (1991)	Evidence of long incubation periods of rabies virus in humans	[80]
Hooper et al. (1998)	Humoral immune response, rather than CD8 T cells, considered critical for rabies virus clearance	[81]

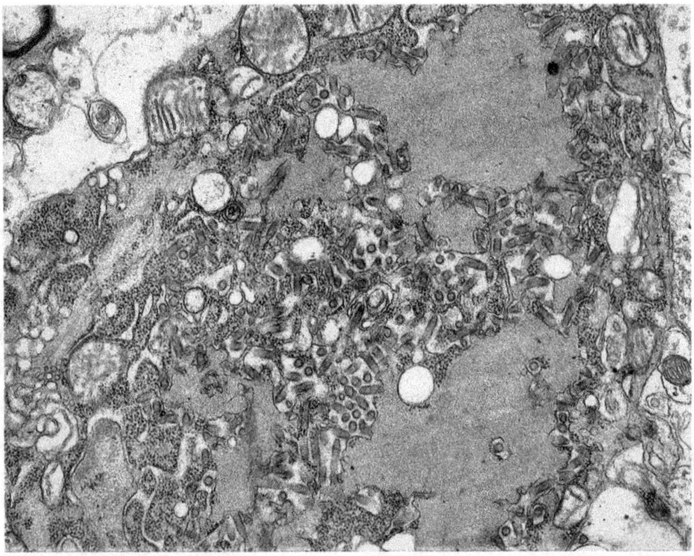

Fig. 2.2 Electron micrograph of bullet-shaped rabies virions and Negri bodies. (F. Murphy, US HSUS Public Health Image Library)

Rabies Virus Pathobiology

Researchers have long been puzzled by the variable and relatively long time for symptoms and signs to manifest after RABV spread to the central nervous system (CNS). Critical experiments performed in the 1960s and 1970s improved our understanding of its pathogenicity. For example, Baer et al. showed that RABV did not immediately travel to the CNS after infection, and amputation of the limb within hours after inoculation could rescue the animal from lethal encephalitis [92]. Similarly, removal of the sciatic nerve before footpad inoculation was a life-saving procedure, providing evidence for the neural spread of RABV infection [93]. But where did RABV reside until starting the journey to the CNS? Early infection seemed to occur almost exclusively in muscle tissue, followed by later progression to the CNS [62]. Researchers who injected mice in the footpad found that RABV may remain in the periphery for up to 18 days, matching closely the clinical experiences observed in humans and naturally infected animals [92].

This extended time in peripheral tissues before spreading to the CNS began an interesting area of research on immune evasion. Researchers performed an extensive study on the role of the muscle for twelve different isolates of street RABV in a hamster model. After intramuscular injections, investigators followed the infection course and initially detected RABV almost exclusively in striated muscle on day three. At day six, muscle infection was still pronounced, but at this point, neuromuscular spindle, nerve, and brain infections were also detected [62]. Progression of infection was continuous, with striking terminal accumulations of antigens, indicating that the initial infection targeted muscle cells. Researchers also showed further evidence of RABV infection of muscle spindles, and motor endplates in striated muscle of hamsters [67]. Harrison et al. documented that RABV was present in muscle spindles and motor endplates. Moreover, electron microscopy showed virus budding from intracellular and plasma membranes of muscle cells to the motor endplate of neurons.

Budding of RABV was also seen from the postsynaptic and adjacent membranes of dendrites and, to a lesser extent, from the plasma membrane of the perikaryon. Besides rodents, investigators used striped skunks (representative of a reservoir species) as an animal model, with a street RABV isolated from rabid skunks, mimicking natural infection [71]. Sequential studies indicated that after muscle infection, the neuromuscular spindles were a second site of viral antigen accumulation. Viral particles were found in the extracellular spaces of the neuromuscular junctions, likely due to the rapid movement of the genome centripetally across cellular junctions via membrane fusion allowing for subviral genomic moieties to pass. Some thought that these subviral moieties might be responsible for evading host defense mechanisms [94].

Charlton and Casey detected RABV antigens using immunofluorescence techniques in muscle cells at the inoculation site before observation in the CNS, supporting the idea that RABV initially infects muscle cells, followed by entry of the virus into neurons innervating the muscle and spreading centripetally to the brain via axonal transport at a fairly constant rate of 12–24 mm a day (46; 84). Initial entry into muscle cells was followed by the spread to a motoneuron using a neurotransmitter receptor, such as the acetylcholine receptor [95], or closely associated molecules [96]. *In vitro* studies have suggested two more putative receptors: the neural cell adhesion molecule, expressed at the neuromuscular junction, and the p75 neurotrophin receptor [97].

In contrast to the findings above emphasizing the initial infection of the muscle, Shankar et al. demonstrated direct entry of RABV into the central nervous system without prior local replication in the muscle cells [98]. However, this research used the RABV strain CVS24, a fixed mouse-adapted rabies virus that is highly neurotropic and probably lost some features of non-fixed street RABV isolates. The CVS-24 laboratory strain has a much shorter incubation time compared to street RABV. The shorter incubation times can likely be explained by the direct entry of such RABV into neurons. Taken together, these early research studies in large part indicate that the initial infection and replication of RABV in muscle cells is followed by the spread of the virus to the CNS via motor neurons.

While spread of RABV to the CNS is more well-studied, the virus spread within the CNS is complex and hypothesized to be as follows: (1) virus dissemination within extracellular spaces; (2) intra-axonal transport via fast axonal transport; and (3) cell-to-cell transmission between contiguous cells and their processes [97]. Electron microscopy has verified that RABV buds through axonal membranes with intra-axonal accumulation of virus particles concentrated at the nodes of Ranvier. This may be due to the presence of abundant organelle membranes for viral budding. After transit to the brain, RABV predominantly infects the hippocampus and cerebellum [99]. The olfactory bulbs in the nasal cavity are heavily infected in the late stages of infections and may be a site of aerosol viral entry, as might occur rarely under very unique ecological conditions, such as in a cave environment inhabited by millions of bats. The other late-stage sensory nerve endings involved are the fungiform and circumvallate papillae on the tongue that have exposed neuroepithelium (i.e., taste buds) which become heavily infected. Such organs may also contribute to viral shedding and transmission late in infection. Remarkably, the site of RABV maturation in the cells of the various target organs differs dramatically. In the nervous system, almost all virus buds upon membranes within cells that are not destroyed by the infection. However, in the salivary glands, nearly all virus buds upon plasma membranes of mucus-producing cells, immediately delivered into saliva, propagating viral spread. Late in infection, RABV infects the tonsils, adrenal gland, and the pancreas [94, 100]. Recent studies have provided some evidence for anterograde spread, highlighting that RABV axonal transport and spread of infectious virus in peripheral nerves is not exclusively retrograde [101].

As pathogenicity studies determined that RABV targets the CNS, it was logical to study viral occurrence in these cell types. Hence, one of the most commonly used cells for culturing RABV are neuroblastoma, or NA, cells. Minimal passaging of attenuated RABV in NA cells led to acquired virulence in strains of mice. In fact, RABV maintained virulence even after subsequent passage in non-NA cells [70]. This observation suggested that these obtained mutations increase RABV pathogenicity, specific to spread through the central nervous system.

Incubation Periods

The incubation periods of rabies are highly variable, ranging from <2 weeks to >8 years, depending in part on the concentration of the inoculum, the inoculation site, and the density of innervations. For example, severe bites on the hands, neck, face, and head mainly with bleeding lead to shorter incubation periods due to the decreased length to traverse to neurons and greater number of innervations in those areas [102]. Studies completed in the 1960s revealed that the dose of RABV infection is also an important factor for the progression and time to reach the CNS, with lower doses having longer incubation times [103]. It remains unclear where RABV resides during these long incubations times and if there is a certain threshold of needed before entering the CNS. Besides the demonstration in animal models, RABV can have very long incubation times in humans [80]. Although rare, several patients died of rabies years after the initial exposure. It remains unknown if RABV is in the CNS, causes a low level of persistent infection, or is dormant during that time.

The laboratory ERA strain and street variants of RABV have evidence of persistent infection of macrophages and primary cultures of murine bone marrow macrophages. Classically, investigators suggested that macrophages may serve as reservoirs of infection *in vivo*, and RABV may subsequently be activated from its persistent state. However, this has not been firmly established. Of note, several other cell lines can be persistently infected by RABV, so even muscle cells could serve as a potential local tissue depot to be later activated.

Rabies Virus Pathogenicity and Immune Responses

In addition to the dose, investigators found that the particular isolate of RABV also plays a role in pathogenicity, as seen by the relatively low dose of raccoon RABV required to infect animals compared to a bat isolate [104]. These raccoon isolates infected rodents easily, producing higher viral titers in the salivary gland, while a bat isolate did not.

Of course, the immune response to RABV plays an essential role in pathobiological outcomes. Researchers have shown that the humoral immune response is critical for controlling infection, whereas CD8+ cells do not play a significant role [81]. Suppression of the humoral and cellular responses with cyclophosphamide increased the overall mortality rate from 50 to 100% and delayed onset of signs and death by 1 to 2 weeks. This finding indicated that the immune system is essential to control RABV infection, but an immune response can accelerate paralysis and death. For example, when RAG2 knock-out mice are infected with RABV, it was 100% lethal in these immunodeficient mice, but the time of paralysis was delayed to 10 days. Experimental studies have shown strong evidence for upregulation of interferons, cytokines, and chemokines in response to RABV infection, driving an influx of immune cells into the CNS, particularly T cells. This immune cell influx cannot control RABV infection, and the host invariably dies [105]. Such findings indicate the importance of the immune response in the pathogenesis of RABV, but the immunopathology is still not well understood.

During 1983, a research group at the Wistar Institute in Philadelphia determined the antigenic region of the RABV glycoprotein responsible for pathogenicity. Nonpathogenic variants of RABV were selected after monoclonal antibody treatment and cloning of neutralization-resistant mutants unable to be neutralized by monoclonal antibodies specific for this region. The arginine amino acid residue at position 333 was determined critical for neurotropism through amino acid and nucleotide sequencing techniques [76]. This antigenic determination of pathogenicity has had practical implications for the generation of stable, nonpathogenic variants to be used for vaccination. In addition, this observation demonstrated how even a small genetic change could entail critical phenotypic outcomes, with the narrow conclusions on pathobiology resulting from limited sequence attempts.

Rabies Virus Transmission

Typically, infection with RABV occurs by the bite of an infected animal. If diagnosed early, it can be prevented successfully. In some cases, infections are not identified, as may occur with bat RABV transmission. In fact, most lethal RABV cases occurring in the USA during the last 10 years were bat-associated, including four of five cases during 2021 [106]. Bat exposures (bites and/or scratches) during the night to deep-sleepers (or in individuals affected by medication or dementia) may not be easily recognized based on the small size of the teeth and minimal injury. Pathobiological research is fundamental to understanding and managing potential exposures. For example, there have been several cases of non-traditional transfer of rabies virus by non-bite routes, such as through tissue or organ transplants, mucosal exposures, or aerosol inhalation. Experiments showed that mice who consumed brain tissue infected with a RABV of vampire bat origin acquired infection through buccal and lingual mucosae, the lung, and the intestine, with some animals recovering after disease [56]. Baer et al. showed that foxes developed serum-neutralizing antibodies to RABV after oral administration of an attenuated RABV but not when a similar dose was introduced via the stomach [63]. These pathobiological research findings emphasized the need for a vaccine bait that assured delivery of a vaccine dose to the oral mucosa. The oral route is now widely used for wildlife immunization with attenuated and recombinant viruses and has helped to eliminate RABV from parts of Europe and North America [107].

Domestic dogs, wild mesocarnivores, and bats are the major species responsible for RABV transmission, serving as disease reservoirs [108]. Infected dog bite is the most typical route of infection globally. Extensive dog vaccination programs in the North America helped eliminate circulating RABV in domestic dogs [109]. Throughout Latin America, enzootic areas remain due to deficiencies in dog vaccination, as well as conditions that bring people and especially children into contact with dogs. From 1993–2002, most (65%) human rabies cases were still transmitted by dogs, with the second most common transmitter being bats, according to the Regional Information System for Epidemiological Surveillance of Rabies in the Americas (SIRVERA) administrated by PAHO, a unique extension of WHO for the region [110]. Most of the attention for human rabies prevention was implicit from an understanding of local RABV perpetuation from indigenous fauna, predominantly after a bite, recognized or otherwise.

Although much less common, RABV transmission has also occurred through non-bite routes of infection. These non-bite exposures may include any contamination of open wounds, abrasions, or mucous membranes, including a scratch from a rabid animal that was contaminated with infectious material. In one tragic case, the lick from a rabid dog led to a child's death [111]. Exposure may also occur through airborne transmission. Two humans died in 1956 and 1958 after entering Frio Cave, a large limestone cavern near Uvalde, Texas. In response, Constantine conducted experiments with different wildlife to confirm the possibility of aerosol transmission of RABV [48]. Coyotes and foxes were kept in cages with varying levels of bat-proof and arthropod-proof measures, allowing variable exposures to bats, arthropods, or cave air only. After up to 30 days of exposure to varying degrees in the cave, all groups of coyotes and foxes died of rabies, even in the enclosures that only included exposure to cave air [39, 48]. Besides bites, RABV may be spread among bats via viral entry through the olfactory mucosa. Aerosolized RABV particles can be transmitted by resting bats that are vocalizing and flying bats that navigate through sonar, requiring bursts of ultrasound emanating from the nasopharynx, sites infected with RABV [112]. Laboratory accidents have caused human aerosol infection with RABV, after exposure to a "fixed" strain of RABV. Infection apparently resulted from inhalation of aerosols generated in a laboratory during the development and manufacture of animal rabies vaccines [113]. Lastly, there have been several transmissions after tissue and organ transplants from people who died with an unrecognized RABV infection [114]. By explanation, it is well-established that after the initial infection of the CNS, RABV spreads to multiple extraneural organs in human rabies [115].

Clearly, such non-bite exposures are exceptions to the rule, however, and are rare. The primary route of RABV transmission is through the bite of an infected animal, as accepted from decades of research on this classic transmission route. What we can learn from these unique and abnormal modes of transmission, is to not allow the conventional wisdom to limit our future understanding, prevention, and control. The potential of additional routes of transmission emphasizes that despite centuries of public health attention, research, and discovery, RABV remains a sinister and important pathogen.

Rabies Virus Pathology

Although RABV infection is characterized by severe neurological signs and a fatal outcome, gross pathological changes in the CNS are relatively less apparent, even absent, due to mild inflammatory reactions. Dysfunction rather than overt cell destruction is responsible for neural degeneration in rabies. Numerous autophagic compartments form early in infection of gangliocytes while several partially membrane-bound empty vacuoles form during the advanced stages of degeneration. The spinal cord is affected in the paralytic ascending form of rabies, with vacuolar lesions in the brainstem, hypothalamus, and limbic system. The severe clinical neurological disease that occurs in the late-stage results from severe beading of axons and dendrites of the cerebral cortex, brain stem, and cerebellum. Ultrastructural studies of the late stage of disease indicated that vacuoles can form in the cortical neurons, neuropil of the cerebral cortex, and axons and pre-synaptic nerve endings [102].

Infection is accompanied with mild histological changes in the brain, some recognized as spongiform lesions and Negri bodies. A spongiform lesion is best described as a vacuolar lesion in the neuropil of the gray matter, although infrequently found occurring in the basal ganglia, hippocampus, and hypothalamus. Rarely, rabies spongiform lesions can be as severe and widespread as seen in scrapie (prion)-infected brains. Careful examination of the distribution of the RABV antigen in skunk brains with rabies-induced spongiform change has revealed that membrane-bound vacuoles form in neuronal processes without the presence of RABV antigen in the same neuron, suggesting that RABV does not replicate in these vacuoles [116]. Classical Negri bodies—spherical, eosinophilic, cytoplasmic inclusion bodies in which viral RNAs (mRNAs as well as genomic and antigenomic RNAs) are synthesized—serve as histological demonstration of RABV infection of the brain (Figs. 2.1 and 2.2). These inclusions can fuse together and reversibly deform when encountered with a physical barrier. The RABV phosphoproteins (P) and nucleoproteins (N) have features similar to those of cellular proteins involved in liquid organelle formation, and their co-expression is sufficient to trigger the formation of Negri bodies. The N is a RNA-binding protein and the P protein contains intrinsically disordered domains. The RABV P protein plays a central role in the network of protein-protein interactions by providing a bridge between the polymerase L, N-RNA template, and cellular factors [117]. Experimental studies have verified the interaction of P with the dynein light chain LC8 to mediate the transport of RABV nucleocapsid in the CNS, although this has yet to be verified for natural infections. The P protein also plays an important role in IFN-induced antiviral responses [118]. In addition to the RABV proteins, Toll-like-Receptor-3 (TLR3) are also present in Negri bodies and play a critical role in their formation. Confirmation of this hypothesis was performed in TLR3 knock-out mice, wherein RABV-infected cells did not exhibit viral Negri body formation [119]. Classically, in human rabies cases, Negri bodies were found in only 50 to 90% of street RABV infections, demonstrating the need for alternative diagnostic techniques, such as the direct fluorescent antibody test during the 1940s, for sensitive and specific visualization of viral antigens.

Rabies Virus Excretion

Saliva is a primary and well-documented mode of RABV excretion, with studies documenting high viral loads. One study by Fekadu et al. described RABV excretion in the saliva 13 days before the appearance of disease onset, questioning the importance of monitoring suspect rabid dogs for longer periods than the currently accepted period of 10 days [75]. In addition, RABV was isolated from the tonsils of a dog that died 16 months after naturally 'recovering' from RABV infection. In this case, the dog was intermittently excreting RABV from the saliva, suggesting that viral sequestration and replication may have continued during the long clinical course or during apparent recovery phase [100]. To date, clinical observation after human exposure is only recommended from bites by dogs, cats, and ferrets, because of inadequate experimental and epidemiological data available for other species.

Rabies *Treatment*?

Clearer insights into the basic virology and pathophysiology of rabies enabled better-informed medical decisions in stark contrast to the prior century. Inarguably, applied research by New World investigators on tissue-culture vaccines and the advent of human rabies immune globulins modernized the approach to human post-exposure prophylaxis since the 1970s. However, a successful treatment for clinical rabies remains as elusive today as ever. In part, the tremendous successes of human prophylaxis and prevention via the mass vaccination of dogs, combined with low financial incentives for drug development curtailed progress on treatments for rabies. The devastating nature of the disease itself, with a case fatality of nearly 100% of victims succumbing, creates an atmosphere of despairing futility. How do we discover an effective treatment? Can we move beyond prophylaxis, which can effectively prevent, but not treat the disease after the manifestation of clinical signs? What happens to those in the throes of illness, especially children, when it is too late for prevention?

Treatments for rabies in the early history of the Americas derived from European methodologies that were imported along with colonization. Prophylaxis and treatment blurred somewhat at this stage, as the disease was not well understood in any meaningful way. For centuries, brutal wound cautery with glowing hot irons, boiling oil, or application of strong acids was a mainstay of intervention. Superstition and folk medicine affected other varied efforts, which included the application of poultices or 'madstones' to counteract the 'poison', forcing some patients into pools of water to 'counteract' the hydrophobia, or smothering between mattresses as an ultimate solution. Diverse potions and concoctions were tried, all without effect.

For any treatments to be useful would require a transition in basic understanding of rabies as a disease. In the Americas, even during the twentieth century, rabies pathobiology was far from being understood. Indeed, the classical field of virology was just emerging in the latter part of that century. Yellow fever virus, the first virus known to infect humans, was only *discovered* in 1901 [21–23]. In France, Pasteur and his colleagues intervened in saving the life of Joseph Meister, severely bitten by a presumed rabid dog, with the first historical rabies vaccination, using the dried spinal cords of infected rabbits, during 1885. While not even understood at that time that rabies was caused by a *virus*, that incredible moment rocked the world [3, 9, 11, 120]. Soon thereafter, exposed people everywhere flocked to France, including four Newark, New Jersey youths who were sent to Paris on the steamship *Canada* for immunization. After vaccination, their survival made Pasteur (and them!) 'American' heroes, amplifying the success story. Rabies could now be prevented. Soon, Pasteur institutes for vaccination sprang up around the globe, including throughout the Americas. However, the line remained blurred between such postexposure prophylaxis following an animal bite and actual treatment, after the onset of illness.

Insights into the potential medical management of clinical rabies may be distilled from reviewing extremely rare 'survival' cases. Inspiration that human survival might be possible was supported from animal studies. For example, during the 1970s, experiments using rodents with different RABV isolates, including from rabid foxes, skunks, bobcats, etc., yielded outcomes that included some survivors with neurologic sequelae (i.e., paralysis). Besides the effect of dose, route and severity, host attributes, such as individual immunologic responses were appreciated that might have allowed patients to escape death [66].

One of the more famous primary human survivors was Matthew Winkler, a 6-year-old Ohio boy who developed rabies in 1970 [59]. His remarkable story was summarized recently in a reminiscence from the local Lima News, June 14, 2022 [https://www.limaohio.com/top-stories/2022/06/14/reminisce-more-than-a-rifle-for-christmas-for-first-rabies-survivor/#:~:text=Winkler%20is%20believed%20to%20haveThe%20Lima%20News%20reported%20Dec]:

> "…Matt's unwanted odyssey into medical history began the night of Oct. 10, 1970, when a bat made its way from the attic of the Winkler family's century-old red brick farmhouse near Willshire in Van Wert County into his bedroom…At around 10 p.m., Matt screamed and his mother "flew up the dark staircase, flicking on lights. Matt lay there on the bed, in horror. On his left thumb, fastened by frenzied teeth, was a tiny brown creature," the AP wrote. "His father, Nick, tore the bat off and told his wife to find a jar. His mind held one thought that would have terrified any parent: rabies." Matt's mother cleansed the wound with soap, water and alcohol. The following morning, Nick took the bat to a veterinarian, "hoping that Matt would not have to endure the two weeks of rabies shots that he himself had endured after being bitten by a cat years earlier," the AP wrote. The veterinarian sent the bat by bus to Columbus for analysis by the Ohio Department of Health. Four days after Matt was bit, the report came back: The bat was rabid. The Winkler family doctor immediately began daily shots of the vaccine. But four days was too long when the diagnosis was rabies. "Rabies brings on a death so horrible that its victims once were closeted away when the symptoms appeared, both to protect others from infection by bite, and to spare the victims tortured nervous system from outside irritation," the AP wrote. The Parkway Elementary School first-grader began showing symptoms around Halloween, running a fever, and complaining of a stiff neck. On Nov. 3, Matt was taken to see Dr. Stechschulte at St. Rita's (now Mercy Health) in Lima, who admitted the boy for observation and prescribed antibiotics and other routine care. At first, the boy seemed to improve, though he tended to drift off to sleep. On Nov. 14, Dr. Thomas T. Weis, who was managing Matt's care while Stechschulte was out of town, found the boy's speech was garbled. Weis spent an hour examining Matt, at the end of which he called the doctors at the Communicable Disease Center (CDC) in Atlanta. "They concurred in his diagnosis," the AP wrote. "The boy could be flown to the National Institutes of Health in Bethesda, Maryland, by Air Force ambulance plane. There were experimental vaccines and serums available there which might be tried if nothing else worked." By the time Matt was diagnosed, experts were questioning if rabies need always be fatal since death usually came from complications, not from the disease itself. Could rabies be defeated by managing the complications, they wondered? "After two years of study, by late 1970, U.S. Public Health Service experts were looking for a case of rabies to give the new medical philosophy a try. Matthew Winkler became that case," the AP wrote. "Not in a decade had doctors been able to get to a rabies case so early." Matt's condition, however, had deteriorated and it was decided any attempt to move him would endanger his life. He would be treated at St. Rita's with the assistance of the CDC's top rabies expert, Dr. Michael Hattwick, who flew in from Atlanta. Over the next month, the team of doctors — Stechschulte, Weis, Hattwick and several others — dealt with symptoms as they arose. Matt was moved to intensive care and a nurse stood at his side. "His respiration was checked constantly," the AP noted. "The main areas of concern: heart, lungs and brain." On Nov. 16, the day Stechschulte returned, Matt was beginning to show signs of a lack of oxygen and a small airway was established in his throat. "It was a turning point," the AP wrote. "If there was a moment in Matthew Winkler's life when a yes or no decision was all important, it was then, the tracheostomy." Although there were more problems, the medical team was able to deal with them early and by Dec. 12, doctors noted Matt was able to walk on his own and felt less weakness in his left arm. He continued with speech and physical therapy. On Dec. 21, the news was announced at a press conference at St. Rita's, prompting a front-page headline in that day's edition of The Lima News: "Medical History in Lima! Doctors, Boy Beat Rabies." Stechschulte told The Lima News: "I don't feel any one thing we did has made the complete difference. It was a team effort by all. I happened to be the physician in charge and Matt is one child who survived." After

> another month of therapy at St. Rita's, Matt was released from the hospital on his seventh birthday, Jan. 27, 1971. "Being such a long-term hospital patient, Matt is aware of noise and there was plenty at his going-home," Hope Strong wrote in The Lima News. "'I'm used to it,' he bragged. 'I've had blood tests … they hurt sometimes.'" A year later, on his eighth birthday, he returned to St. Rita's to mark the anniversary. "There is no limit on his activities, and he had a great summer on the farm. Matt only has to see us periodically," Dr. Stechschulte told The Lima News on Jan. 21, 1972. In September 1975, five years after Matt made medical history, the AP visited the Winkler family on their Willshire farm and found Matt "a perfectly healthy farm boy who is learning to play the trumpet…".

Amazingly, after RABV exposure and vaccination, the prompt hospitalization post-illness with supportive care, his managing physicians worked actively to prevent the common complications of rabies (e.g., hypoxia, intracranial hypertension, cardiac arrhythmias, seizures, superinfection, etc.). Such efforts, combined with his overall health status and his young age, may have played a role in his recovery in toto.

Another survivor, also from the 1970s, involved a middle-aged woman from Argentina, who was severely bitten by a rabid dog [64]. She began vaccination 10 days later. Three weeks after the exposure, she developed severe encephalitic symptoms that persisted for several months. Just over a year later, she had almost completely recovered. This second outcome suggested that infection after rabid dog exposure, and not just bat RABV, could also be survivable with vaccination, medical care, and a great deal of luck, as realized for a handful of patients [64].

Reinvigorating the discussion, there is the celebrated story of Jeanna Giese, an unvaccinated rabies survivor from Wisconsin [121]. Like Matthew Winkler, young Jeanna was also bitten by a bat, but did not receive prophylaxis, because rabies was not suspected at that time, until she became ill a month later. As the *New England Journal of Medicine* reported in 2005, she was treated quite differently than prior vaccinated individuals, undergoing a new treatment known as the "Milwaukee protocol" (or MP). The MP was based not on 'pharmaceutically induced' immunity (i.e., use of human rabies immune globulin and rabies vaccine) in favor of allowing the native immune response to mature most specifically to the infecting viral antigens, while supporting the patient in an ICU setting under heavy sedation. Additionally, ketamine and amantadine – medications with *in vitro* activity against rabies virus – were added, and metabolic perturbations were closely managed. Supplementation with biopterin, a neurotransmitter precursor, was also thought fundamental in the management. The protocol appeared novel and elegant since it might allow a more targeted, effective CNS clearance. Jeanna survived after a protracted course with only mild sequelae, and even finished college, got married, and had children (Fig. 2.3).

Given these cases, it appeared that a rabies treatment was on the horizon. However, further iterations of the MP were disappointing [122]. Recently, clinicians in Brazil adjusted the MP, renaming it the 'Recife protocol', claiming that this adaptation saved the lives of two teenagers infected by bat bites [123]. Some ethicists questioned these efforts, given the poor outcomes the teenagers experienced – with neurological sequelae. The generic recommended approach for treating clinical rabies today remains palliative care.

Fig. 2.3 Rabies survivor, Ms. Jeanna Giese (center) flanked by her parents (and one of the chapter co-authors, Steve Scholand in the background)

Much has been gleaned over the years from the unsuccessful efforts to save rabies patients (as well as a few historical survivors from the Americas). Based on these results, Willoughby created a timeline for anticipated complications during treatment [124]. For example, salt wasting can be expected to develop at about day 5 of hospitalization, in addition to profoundly low biopterin levels during the disease course, with terminal vasospasm of cerebral arteries towards the end. Additionally, antibodies seem critical for RABV clearance from the CNS [125]. As such, additional research is examining the use of monoclonal antibodies or in situ elicitation of virus neutralizing antibodies to the CNS.

Besides the utility of medical management and specific use of novel biologics, the search for new anti-viral drugs has produced several candidates. For example, Favipiravir, a medication with antiviral activity against flaviviruses, influenza, and several other viruses, has found antiviral activity against RABV *in vitro*. Current opinion, however, suggests it may best be used in a multimodal approach, since *in vivo* experiments have not been very encouraging [126]. Another drug, IMT504, is an oligodeoxynucleotide, which strongly stimulates B-cells, T-cells, with cytokine production. Animal studies have shown promise in a preliminary model using rats [127].

To date, far too few rabid individuals are saved, despite heroic efforts. More thorough understanding of pathobiology, and investigation with relevant animal models, are anticipated to yield more promising results. For example, a 'Canine Rabies Treatment Initiative' is a recent attempt to create a coalition to combine current knowledge and available tools to treat naturally infected dogs for application to humans and other species [https://www.treatrabies.org/]. As of yet, no clinical trials with any such protocols have been conducted. This is partially because of the focus on prevention and control, as well as the status of rabies as a 'Neglected Tropical Disease' – an unfortunate distinction for an entity so catastrophic and deadly in nature. Still, a small, dedicated community of scientists, physicians, veterinarians, humanitarians, and others continue to collaborate in search of a viable treatment. Could a more advanced protocol be developed with additional novel therapies in the rabies armamentarium? Will the revolution in mRNA vaccine technology and antiviral strategies brought about by COVID-19 have trickle-down benefits for rabies interventions? Many questions remain, but there are solid foundations of science built over the past century into the nature of rabies, with glimpses of hope that an effective treatment may yet be found.

References

1. Rupprecht C, Kuzmin I, Meslin F (2017) Lyssaviruses and rabies: current conundrums, concerns, contradictions and controversies. F1000Res 6:184
2. Nunn NQN (2010) The Columbian exchange: a history of disease, food, and ideas. J Econ Perspect 24:163–188
3. Rupprecht CEFC, Mani RS, Palacios C, Sabeta CT, Ward M (2020) Rabies, 4th edn. Elsevier, San Diego, pp 1–42
4. Blancou J (2004) Historical perspective of rabies in Europe and the Mediterranean Basin. OIE, Paris, pp 15–24
5. Nadin-Davis SABJ (2004) Historical perspective of rabies in Europe and the Mediterranean Basin. OIE, Paris, pp 259–280
6. Wilkinson L (1988) Understanding the nature of rabies: an historical perspective. In: Rabies. Kluwer, Boston, pp 1–23
7. Ridder-Symoens eHd. (1996) A history of the university in Europe, vol II. Cambridge University Press
8. Stoeckel A (1976) Presidents, professors, and politics: the colonial colleges and the American revolution. Conspec Hist 1(3):45–56
9. Tarantola A (2017) Four thousand years of concepts relating to rabies in animals and humans, its prevention and its cure. Trop Med Infect Dis 2:5
10. Kinyoun JJ (1891) Rabies-its prevention and treatment. Public Health Pap Rep 17:162–168
11. Ernst HC (1891) Rabies: with an account of Pasteur's work and methods. J Mass Assoc Boards Health 1:92–101
12. Frothingham L (1905) Rabies. Am J Public Hygiene 15:446–462
13. Peters A (1907) Rabies in Massachusetts-symptoms, importance of early diagnosis and practical methods for its suppression. Am J Public Hygiene 17:94–115
14. Black SP, Powers LM (1910) History of rabies in southern California. Cal State J Med 8:369–372
15. Anonymous (1938) U.S. Congress. To prohibit experiments upon living dogs. Hearings before the Subcommittee on Public Health, Hospitals and Charities. HR 3890. US Government Printing Office,, 173 pp
16. Hotez PJ (2020) Anti-science extremism in America: escalating and globalizing. Microbes Infect 22:505–507
17. McBean E (1993) The poisoned needle. Health Research Books. 252 pp
18. Lustig A, Levine AJ (1992) One hundred years of virology. J Virol 66:4629–4631
19. Saldan JJ (2006) Science in Latin America. University of Texas Press, Austin. 264 pp

20. Finlay CJ (1903) New aspects of yellow fever etiology arising from experimental findings of the last three years. Public Health Pap Rep 29:68–70
21. Reed W, Carroll J, Agramonte A, Lazear JW (1900) The etiology of yellow fever-a preliminary note. Public Health Pap Rep 26:37–53
22. Horsfall FL (1965) Thomas Milton Rivers, September 3, 1888-May 12, 1962. Biogr Mem Natl Acad Sci 38:263–294
23. Frierson JG (2010) The yellow fever vaccine: a history. Yale J Biol Med 83:77–85
24. Stanley WM, Loring HS (1936) The isolation of crystalline tobacco mosaic virus protein from diseased tomato plants. Science 83:85
25. Stanley WM (1935) Isolation of a crystalline protein possessing the properties of tobacco-mosaic virus. Science 81:644–645
26. Loring HS (1942) The reversible inactivation of tobacco mosaic virus by crystalline ribonuclease. J Gen Physiol 25:497–505
27. Kristensson K, Dastur DK, Manghani DK, Tsiang H, Bentivoglio M (1996) Rabies: interactions between neurons and viruses. A review of the history of Negri inclusion bodies. Neuropathol Appl Neurobiol 22:179–187
28. Garbarino MC (2019) The origins of the fight against rabies: a journey between the history of neuroscience and university museum collections. Funct Neurol 34:196–204
29. Watson EM (1913) The Negri bodies in rabies. J Exp Med 17:29–42
30. Reagan RLBA (1950) Electron micrographs of Negri bodies found in rabies. J Infect Dis 87(3):213–216
31. Hottle GAMC, Peers JH, Wyckoff RW (1951) The electron microscopy of rabies inclusion (Negri) bodies. Proc Soc Exp Biol Med 77(4):721–723
32. Goodpasture EW (1925) A study of rabies, with reference to a neural transmission of the virus in rabbits, and the structure and significance of Negri bodies. Am J Pathol 1(547–582):3
33. Su JM, Wilson MZ, Samuel CE, Ma D (2021) Formation and function of liquid-like viral factories in negative-sense single-stranded RNA virus infections. Viruses 13:126
34. Dawson JR (1941) A study of chick-embryo-adapted rabies virus. Am J Pathol 17:177–188 7
35. Webster LT, Clow AD (1937) Propagation of rabies virus in tissue culture. J Exp Med 66:125–131
36. Kissling RE (1958) Growth of rabies virus in non-nervous tissue culture. Proc Soc Exp Biol Med 98:223–225
37. Almeida JD, Howatson AF, Pinteric L, Fenje P (1962) Electron microscope observations on rabies virus by negative staining. Virology 18:147–151
38. Matsumoto S (1962) Electron microscopy of nerve cells infected with street rabies virus. Virology 17:198–202
39. Constantine DG (1962) Rabies transmission by nonbite route. Public Health Rep 77:287–289
40. Matsumoto S (1963) Electron microscope studies of rabies virus in mouse brain. J Cell Biol 19:565–591
41. Pinteric L, Fenje P, Almeida JD (1963) The visualization of rabies virus in mouse brain. Virology 20:208–211
42. Davies MC, Englert ME, Sharpless GR, Cabasso VJ (1963) The electron microscopy of rabies virus in cultures of chicken embryo tissues. Virology 21:642–651
43. Johnson RT, Mercer EH (1964) The development of fixed rabies virus in mouse brain. Aust J Exp Biol Med Sci 42:449–456
44. Wiktor TJ, Fernandes MV, Koprowski H (1964) Cultivation of rabies virus in human diploid cell strain Wi-38. J Immunol 93:353–366
45. Love R, Fernandes MV, Koprowski H (1964) Cytochemistry of inclusion bodies in tissue culture cells infected with rabies virus. Proc Soc Exp Biol Med 116:560–563
46. Pinteric L, Fenje P (1966) International symposium on rabies, vol Standard Vol. 1. Talloires
47. Melnick JL, McCombs RM (1966) Classification and nomenclature of animal viruses, 1966. Prog Med Virol 8:400–409
48. Constantine DG (1966) Transmission experiments with bat rabies isolates: responses of certain Carnivora to rabies virus isolated from animals infected by nonbite route. Am J Vet Res 27:13–15

49. Hummeler K, Koprowski H, Wiktor TJ (1967) Structure and development of rabies virus in tissue culture. J Virol 1:152–170
50. Jenson AB, Rabin ER, Wende RD, Melnick JL (1967) A comparative light and electron microscopic study of rabies and Hart Park virus encephalitis. Exp Mol Pathol 7:1–10
51. Hummeler K, Tomassini N, Sokol F, Kuwert E, Koprowski H (1968) Morphology of the nucleoprotein component of rabies virus. J Virol 2:1191–1199
52. Sokol F, Kuwert E, Wiktor TJ, Hummeler K, Koprowski H (1968) Purification of rabies virus grown in tissue culture. J Virol 2:836–849
53. Dierks RE, Murphy FA, Harrison AK (1969) Extraneural rabies virus infection. Virus development in fox salivary gland. Am J Pathol 54:251–273
54. Jenson AB, Rabin ER, Bentinck DC, Melnick JL (1969) Rabiesvirus neuronitis. J Virol 3:265–269
55. Sokol F, Schlumberger HD, Wiktor TJ, Koprowski H (1969) Biochemical and biophysical studies on the nucleocapsid and on the RNA of rabies virus. Virology 38:651–665
56. Correa-Giron EP, Allen R, Sulkin SE (1970) The infectivity and pathogenesis of rabiesvirus administered orally. Am J Epidemiol 91:203–215
57. Sokol F, Stancek D, Koprowski H (1971) Structural proteins of rabies virus. J Virol 7:241–249
58. Vernon SK, Neurath AR, Rubin BA (1972) Electron microscopic studies on the structure of rabies virus. J Ultrastruct Res 41:29–42
59. Hattwick MA, Weis TT, Stechschulte CJ, Baer GM, Gregg MB (1972) Recovery from rabies. A case report. Ann Intern Med 76:931–942
60. Murphy FA, Bauer SP, Harrison AK, Winn WC Jr (1973) Comparative pathogenesis of rabies and rabies-like viruses. Viral infection and transit from inoculation site to the central nervous system. Lab Investig 28:361–376
61. Murphy FA, Harrison AK, Winn WC, Bauer SP (1973) Comparative pathogenesis of rabies and rabies-like viruses: infection of the central nervous system and centrifugal spread of virus to peripheral tissues. Lab Investig 29:1–16
62. Murphy FA, Bauer SP (1974) Early street rabies virus infection in striated muscle and later progression to the central nervous system. Intervirology 3:256–268
63. Baer GM, Broderson JR, Yager PA (1975) Determination of the site of oral rabies vaccination. Am J Epidemiol 101:160–164
64. Porras C, Barboza JJ, Fuenzalida E, Adaros HL, Oviedo AM, Furst J (1976) Recovery from rabies in man. Ann Intern Med 85:44–48
65. Dietzschold B (1977) Oligosaccharides of the glycoprotein of rabies virus. J Virol 23:286–293
66. Baer GM, Cleary WF, Diaz AM, Perl DF (1977) Characteristics of 11 rabies virus isolates in mice: titers and relative invasiveness of virus, incubation period of infection, and survival of mice with sequelae. J Infect Dis 136:336–345
67. Harrison AK, Murphy FA (1978) Lyssavirus infection of muscle spindles and motor end plates in striated muscle of hamsters. Arch Virol 57:167–175
68. Wiktor TJ, Koprowski H (1978) Monoclonal antibodies against rabies virus produced by somatic cell hybridization: detection of antigenic variants. Proc Natl Acad Sci U S A 75:3938–3942
69. Dietzschold B, Cox JH, Schneider LG, Wiktor TJ, Koprowski H (1978) Isolation and purification of a polymeric form of the glycoprotein of rabies virus. J Gen Virol 40:131–139
70. Clark HF (1978) Rabies viruses increase in virulence when propagated in neuroblastoma cell culture. Science 199:1072–1075
71. Charlton KM, Casey GA (1979) Experimental rabies in skunks: immunofluorescence light and electron microscopic studies. Lab Investig 41:36–44
72. Wunner WH, Curtis PJ, Wiktor TJ (1980) Rabies mRNA translation in Xenopus laevis oocytes. J Virol 36:133–142
73. Anilionis A, Wunner WH, Curtis PJ (1981) Structure of the glycoprotein gene in rabies virus. Nature 294:275–278
74. Anilionis A, Wunner WH, Curtis PJ (1982) Amino acid sequence of the rabies virus glycoprotein deduced from its cloned gene. Comp Immunol Microbiol Infect Dis 5:27–32

75. Fekadu M, Shaddock JH, Baer GM (1982) Excretion of rabies virus in the saliva of dogs. J Infect Dis 145:715–719
76. Dietzschold B, Wunner WH, Wiktor TJ, Lopes AD, Lafon M, Smith CL, Koprowski H (1983) Characterization of an antigenic determinant of the glycoprotein that correlates with pathogenicity of rabies virus. Proc Natl Acad Sci U S A 80:70–74
77. Dietzschold B, Wiktor TJ, Wunner WH, Varrichio A (1983) Chemical and immunological analysis of the rabies soluble glycoprotein. Virology 124:330–337
78. Rayssiguier C, Cioe L, Withers E, Wunner WH, Curtis PJ (1986) Cloning of rabies virus matrix protein mRNA and determination of its amino acid sequence. Virus Res 5:177–190
79. Larson JK, Wunner WH (1990) Nucleotide and deduced amino acid sequences of the nominal nonstructural phosphoprotein of the ERA, PM and CVS-11 strains of rabies virus. Nucleic Acids Res 18:7172
80. Smith JS, Fishbein DB, Rupprecht CE, Clark K (1991) Unexplained rabies in three immigrants in the United States. A virologic investigation. N Engl J Med 324:205–211
81. Hooper DC, Morimoto K, Bette M, Weihe E, Koprowski H, Dietzschold B (1998) Collaboration of antibody and inflammation in clearance of rabies virus from the central nervous system. J Virol 72:3711–3719
82. Kruger DH, Schneck P, Gelderblom HR (2000) Helmut Ruska and the visualisation of viruses. Lancet 355:1713–1717
83. Mettenleiter TC (2017) The first "virus hunters". Adv Virus Res 99:1–16
84. Iwasaki Y, Wiktor TJ, Koprowski H (1973) Early events of rabies virus replicaton in tissue cultures. An electron microscopic study. Lab Invest 28:142–148
85. Wiktor TJ, Dietzschold B, Leamnson RN, Koprowski H (1977) Induction and biological properties of defective interfering particles of rabies virus. J Virol 21:626–635
86. Luse SA (1960) Fixation and embedding of mammalian brain and spinal cord for electron microscopy. J Ultrastruct Res 4:108–112
87. Fenje P (1960) Propagation of rabies virus in cultures of hamster kidney cells. Can J Microbiol 6:479–484. https://doi.org/10.1139/m60-055
88. Shope RE, Murphy FA, Harrison AK, Causey OR, Kemp GE, Simpson DI, Moore DL (1970) Two African viruses serologically and morphologically related to rabies virus. J Virol 6:690–692
89. Raju TN (1999) The Nobel chronicles. 1969: Max Delbruck (1906–81); Salvador Luria (1912–91); and Alfred Hershey (1908–97). Lancet 354:784
90. Hershey AD, Chase M (1952) Independent functions of viral protein and nucleic acid in growth of bacteriophage. J Gen Physiol 36:39–56
91. Halonen PE, Murphy FA, Fields BN, Reese DR (1968) Hemagglutinin of rabies and some other bullet-shaped viruses. Proc Soc Exp Biol Med 127:1037–1042
92. Baer GM, Cleary WF (1972) A model in mice for the pathogenesis and treatment of rabies. J Infect Dis 125:520–527
93. Baer GM, Shantha TR, Bourne GH (1968) The pathogenesis of street rabies virus in rats. Bull World Health Organ 38:119–125
94. Murphy FA (1977) Rabies pathogenesis. Arch Virol 54:279–297
95. Lentz TL, Burrage TG, Smith AL, Crick J, Tignor GH (1982) Is the acetylcholine receptor a rabies virus receptor? Science 215:182–184
96. Lewis P, Fu Y, Lentz TL (2000) Rabies virus entry at the neuromuscular junction in nerve-muscle cocultures. Muscle Nerve 23:720–730
97. Mahadevan A, Suja MS, Mani RS, Shankar SK (2016) Perspectives in diagnosis and treatment of rabies viral encephalitis: insights from pathogenesis. Neurotherapeutics 13:477–492
98. Shankar V, Dietzschold B, Koprowski H (1991) Direct entry of rabies virus into the central nervous system without prior local replication. J Virol 65:2736–2738
99. Jackson AC, Reimer DL (1989) Pathogenesis of experimental rabies in mice: an immunohistochemical study. Acta Neuropathol 78:159–165
100. Fekadu M, Shaddock JH, Chandler FW, Baer GM (1983) Rabies virus in the tonsils of a carrier dog. Arch Virol 78:37–47

101. Potratz M, Zaeck LM, Weigel C, Klein A, Freuling CM, Muller T, Finke S (2020) Neuroglia infection by rabies virus after anterograde virus spread in peripheral neurons. Acta Neuropathol Commun 8:199
102. Singh R, Singh KP, Cherian S, Saminathan M, Kapoor S, Manjunatha Reddy GB, Panda S, Dhama K (2017) Rabies – epidemiology, pathogenesis, public health concerns and advances in diagnosis and control: a comprehensive review. Vet Q 37:212–251
103. Sikes RK (1962) Pathogenesis of rabies in wildlife. I. Comparative effect of varying doses of rabies virus inoculated into foxes and skunks. Am J Vet Res 23:1041–1047
104. Winkler WG, Schneider NJ, Jennings WL (1972) Experimental rabies infection in wild rodents. J Wildl Dis 8:99–103
105. Fooks AR, Banyard AC, Horton DL, Johnson N, McElhinney LM, Jackson AC (2014) Current status of rabies and prospects for elimination. Lancet 384:1389–1399
106. Anonymous. Centers for Disease Control and Prevention. Rabies in the U.S. Accessed on 5/1/21. https://www.cdc.gov/rabies/location/usa/surveillance/human_rabies.html
107. Cliquet F, Aubert M (2004) Elimination of terrestrial rabies in Western European countries. Dev Biol (Basel) 119:185–204
108. Scheffer KC, Iamamoto K, Asano KM, Mori E, Estevez Garcia AI, Achkar SM, Fahl WO (2014) Hematophagous bats as reservoirs of rabies. Rev Peru Med Exp Salud Publica 31:302–309
109. Rupprecht CE, Smith JS, Krebs JW, Childs JE (1997) Molecular epidemiology of rabies in the United States: reemergence of a classical neurotropic agent. J Neurovirol 3(Suppl 1):S52–S53
110. Belotto A, Leanes LF, Schneider MC, Tamayo H, Correa E (2005) Overview of rabies in the Americas. Virus Res 111:5–12
111. Leach CN, Johnson HN (1940) Human rabies, with special reference to virus distribution and titer. Am J Trop Med Hyg s1–20:335–340
112. Constantine DG, Emmons RW, Woodie JD (1972) Rabies virus in nasal mucosa of naturally infected bats. Science 175:1255–1256
113. Winkler WG, Fashinell TR, Leffingwell L, Howard P, Conomy P (1973) Airborne rabies transmission in a laboratory worker. JAMA 226:1219–1221
114. Srinivasan A, Burton EC, Kuehnert MJ, Rupprecht C, Sutker WL, Ksiazek TG, Paddock CD, Guarner J, Shieh WJ, Goldsmith C, Hanlon CA, Zoretic J, Fischbach B, Niezgoda M, El-Feky WH, Orciari L, Sanchez EQ, Likos A, Klintmalm GB, Cardo D, LeDuc J, Chamberland ME, Jernigan DB, Zaki SR, Rabies in Transplant Recipients Investigation T. (2005) Transmission of rabies virus from an organ donor to four transplant recipients. N Engl J Med 352:1103–1111
115. Jackson AC, Ye H, Phelan CC, Ridaura-Sanz C, Zheng Q, Li Z, Wan X, Lopez-Corella E (1999) Extraneural organ involvement in human rabies. Lab Investig 79:945–951
116. Bundza A, Charlton KM (1988) Comparison of spongiform lesions in experimental scrapie and rabies in skunks. Acta Neuropathol 76:275–280
117. Nikolic J, Lagaudriere-Gesbert C, Scrima N, Blondel D, Gaudin Y (2019) Structure and function of Negri bodies. Adv Exp Med Biol 1215:111–127
118. Chelbi-Alix MK, Vidy A, El Bougrini J, Blondel D (2006) Rabies viral mechanisms to escape the IFN system: the viral protein P interferes with IRF-3, Stat1, and PML nuclear bodies. J Interf Cytokine Res 26:271–280
119. Menager P, Roux P, Megret F, Bourgeois JP, Le Sourd AM, Danckaert A, Lafage M, Prehaud C, Lafon M (2009) Toll-like receptor 3 (TLR3) plays a major role in the formation of rabies virus Negri bodies. PLoS Pathog 5:e1000315
120. Hoenig LJ (1986) Triumph and controversy. Pasteur's preventive treatment of rabies as reported in JAMA. Arch Neurol 43:397–399
121. Willoughby RE Jr, Tieves KS, Hoffman GM, Ghanayem NS, Amlie-Lefond CM, Schwabe MJ, Chusid MJ, Rupprecht CE (2005) Survival after treatment of rabies with induction of coma. N Engl J Med 352:2508–2514
122. Zeiler FA, Jackson AC (2016) Critical appraisal of the Milwaukee protocol for Rabies: this failed approach should be abandoned. Can J Neurol Sci 43:44–51

123. Ledesma LA, Lemos ERS, Horta MA (2020) Comparing clinical protocols for the treatment of human rabies: the Milwaukee protocol and the Brazilian protocol (Recife). Rev Soc Bras Med Trop 53:e20200352
124. Anonymous. https://www.mcw.edu/-/media/MCW/Departments/Pediatrics/Infectious-Diseases/Milwaukee_protocol.pdf?la=en. Accessed 5/1/21
125. Dietzschold B, Kao M, Zheng YM, Chen ZY, Maul G, Fu ZF, Rupprecht CE, Koprowski H (1992) Delineation of putative mechanisms involved in antibody-mediated clearance of rabies virus from the central nervous system. Proc Natl Acad Sci U S A 89:7252–7256
126. Banyard AC, Mansfield KL, Wu G, Selden D, Thorne L, Birch C, Koraka P, Osterhaus A, Fooks AR (2019) Re-evaluating the effect of Favipiravir treatment on rabies virus infection. Vaccine 37:4686–4693
127. AN AM, Rodriguez J, Hernando-Insua A, Fló J, Lopez R, Sierra V, Paolazzi C, Larghi O, Horn D, Zorzopulos J, Elias F (2012) IMT504: a new and potent adjuvant for rabies vaccines permitting significant dose sparing. World J Vaccin 2(4):182–188. https://doi.org/10.4236/wjv.2012.24025

Chapter 3
The Diversity, Evolution and Emergence of Rabies Virus in the Americas

Edward C. Holmes and Erin H. Harvey

Abstract The Americas represent a unique geographic region where rabies virus variants derive from a variety of species, comprising both bats and carnivores, with ongoing spill-over infections to other mammalian hosts (including humans and domestic animals). As such, historically and spatially, the Americas constitute a microcosm of viral cross-species transmission and emergence, and hence of disease ecology in general. Despite the great advances made in our understanding of rabies through the deployment of evolutionary genomics and phylogeographic studies, important questions remain on the exact origin and timescale of its evolution in the New World. Future inquiry should shed light on the nature of genetic and ecological barriers that might prevent rabies virus establishing productive infections in new hosts. Enhanced laboratory-based surveillance, combined with metagenomic next-generation sequencing, is essential to identify additional rabies virus variants and reservoirs of regional epidemiological importance, as well as other lyssavirus species and hosts globally.

Keywords Biodiversity · Emergence · Evolution · Lyssavirus · Phylodynamics

Introduction

COVID-19 and its causative agent, the coronavirus SARS-CoV-2, provide a stark reminder of the fundamental principles of virus emergence and evolution. Central among these is that zoonoses – the process by which animal viruses jump to and emerge in humans – are commonplace and underpin most infectious disease emergence events in human history [1]. This narrative of an animal pathogen making a

E. C. Holmes (✉) · E. H. Harvey
School of Life & Environmental Sciences and School of Medical Sciences, The University of Sydney, Sydney, NSW, Australia
e-mail: edward.holmes@sydney.edu.au

C. E. Rupprecht (ed.), *History of Rabies in the Americas: From the Pre-Columbian to the Present, Volume I*, Fascinating Life Sciences,
https://doi.org/10.1007/978-3-031-25052-1_3

regular appearance in human populations is also true of rabies virus (RABV; species *Rabies lyssavirus*, genus *Lyssavirus*, family *Rhabdoviridae*), and the study of RABV in recent decades has helped write much of the textbook of virus evolution that has become a constant reference point during the COVID-19 pandemic. As a case in point that will be discussed in more detail below, a number of the bioinformatic techniques and tools employed to track virus evolution, particularly the phylogeographic methods that reconstruct the pattern and speed of virus spread on a spatial scale, cut their teeth in studies of animal RABV in North America [2].

The aim of this chapter is to provide an overview of the diversity, evolution and emergence of RABV and closely related lyssaviruses that infect a range of animal hosts. Although the focus will necessarily be on rabies in the Americas, the rapidity and fluidity of virus evolution, particularly as RABV readily jumps both species and geographic boundaries, makes it impossible to consider the Americas in splendid isolation. Rather, RABV diversity and evolution only makes sense if it is considered on a global scale, and the evolutionary origins and history of rabies will not be revealed in full until we have sampled viruses from a broad set of geographic localities. For example, one of the most important questions concerning RABV evolution in the Americas, and of RABV in general, is explaining the complex evolutionary connections between those viruses found in the Americas and those from the Old World. It is also important to recall that aside from Europe, perhaps more is known about rabies in the Americas than in any other locality, with far less data on virus ecology and evolution in parts of Africa and Asia.

One of the most notorious aspects of rabies is that each human infection is the result of an individual virus cross-species transmission (i.e. spill-over) event from an animal host. While 99% of the human rabies cases worldwide are associated with exposure to dog rabies, in the Americas these spill-over events generally involve an array of bat species that can live close to human habitation [3]. For example, *Tadarida brasiliensis* (the Mexican free-tailed bat) is present at high abundance in the Americas. Human rabies is therefore unusual in that it is a virus that is continually exposed to a particular host, but where there is effectively no human-to-human transmission. This key aspect of RABV biology automatically tells us two things that are central to understanding its evolution: (i) that the exposure of humans to zoonotic pathogens like RABV is a regular occurrence, but that (ii) there are intrinsic barriers that prevent viruses like RABV from establishing onward transmission cycles in new hosts, and that these barriers are only likely to be overcome with adaptive evolution. As such, rabies is a perfect illustration of the central importance of the human-animal interface in disease emergence, as it is the spatial proximity between humans and other animal species that provides the conduit for virus emergence to take place. Understanding the barriers that prevent RABV from establishing itself as a human-transmitted pathogen is arguably the outstanding question in RABV evolution.

Although there has been an understandable focus on rabies in humans, because there is no human-to-human transmission of RABV (or other lyssaviruses) this also means that humans are *not* the host where virus evolution takes place. Rather, the diversity of RABV lineages present in humans are a sample of those that circulate in animal populations, and to understand the evolution of RABV necessarily requires us to understand what is happening in these non-human hosts [4]. Similarly, while RABV infection in humans is almost invariably fatal, such that all variants of the virus seemingly have the ability to cause serious disease, whether this is also true of the virus in animal populations is less clear. It may be that some mammalian species, particularly bats, are less susceptible to overt disease than others, although this can be challenging to study. More broadly, rabies highlights the value of thinking about humans and animals as components within a connected ecological space, a notion that underpins the One Health perspective [5].

Bats and the Diversity of Lyssaviruses

Another of the parallels between RABV and SARS-CoV-2 is the central role played by bats as virus reservoirs. It is undoubtedly the case that bats harbour a wide diversity of coronaviruses, a small subset of which have successfully evolved onward transmission in a number of other mammalian species, including humans [6]. Similarly, the lyssaviruses in general and RABV in particular, were some of the first viruses in which the role of bats as hosts for RNA viruses became apparent, and this pattern now appears to be true of RNA viruses in general [7, 8]. Bat rabies is an increasingly important public health problem in the Americas, with relatively frequent spill-over transmission events from bats to other species, including humans and livestock. Indeed, one of the defining features of lyssavirus evolution is that these viruses are often able to readily transmit among different bat species, in contrast to the major transmission barriers that seem to block the onward transmission of RABV and other lyssaviruses species in non-volant mammals [4].

Members of the genus *Lyssavirus*, of which there are currently at least 17 classified species including *Rabies lyssavirus* (ICTV 2019), have been found on most continents, even Australia, and very strongly associated with bat hosts (Fig. 3.1). It is inevitable that additional bat lyssaviruses and lineages of RABV will be identified with more expansive sampling, and there are major gaps and ascertainment biases in those locations and mammalian species that have been screened for these viruses. For example, four of the 17 currently documented lyssavirus species are from Russia or the western Russian steppe region, while RABV is the only lyssavirus species found in American bats (Fig. 3.1). A simple manifestation of the biased and under-sampling of host species that plagues virology is the presence of overly long branches on viral phylogenetic trees, and which is very apparent on a phylogeny of the *Rhabdoviridae* as a whole, within which the lyssaviruses appear as a very distinct cluster (Fig. 3.2). As discussed below, it also seems that an as yet unsampled bat virus is the most likely explanation for the apparent absence of RABV in Old World bats. More broadly, the phylogeny of the *Rhabdoviridae* shows that while these viruses are able to infect a very wide range of hosts, including plants, invertebrates, vertebrates via vectors and vertebrates without vectors, the lyssaviruses are unique in that they are currently the only group that infect mammals exclusively (Fig. 3.2).

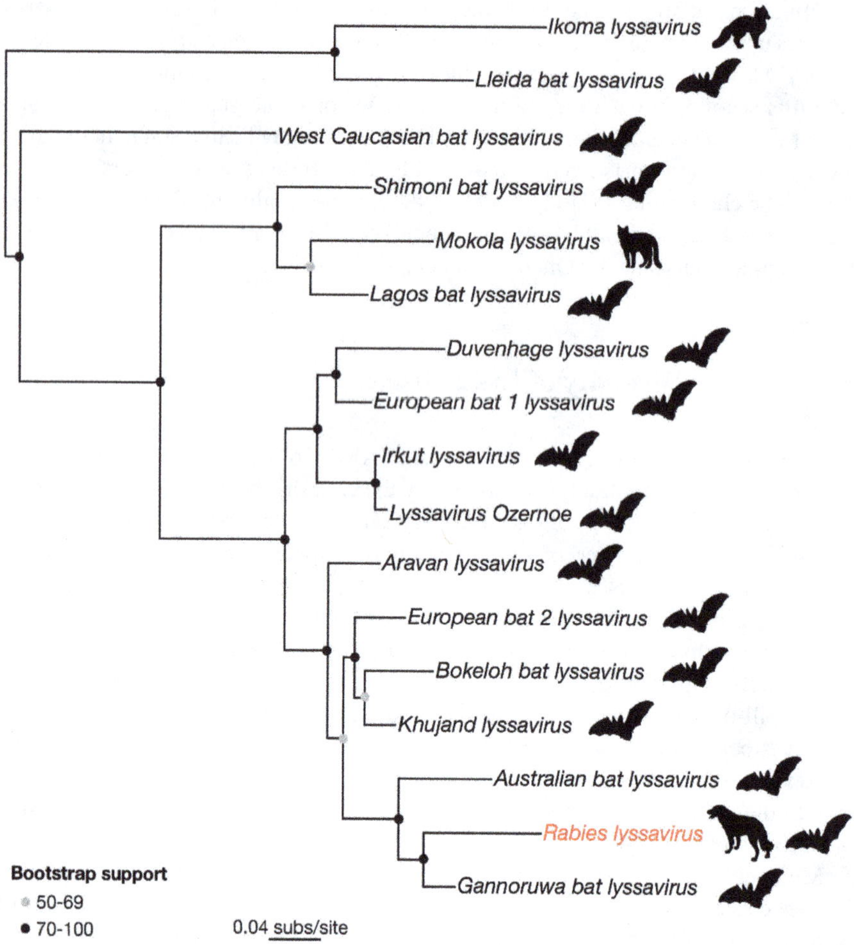

Fig. 3.1 Phylogenetic relationships of 17 currently known virus species within the genus *Lyssavirus* (*Rhabdoviridae*). Symbols represent the currently documented host of each virus species, although the natural reservoir hosts are unknown in a number of cases (for example, *Mokola lyssavirus*). The species *Rabies Lyssavirus* is shown in red. The phylogeny was estimated using an amino acid alignment of the conserved RNA-dependent RNA polymerase (RdRp) domain and the maximum likelihood method available within the IQ-TREE2 program [66] employing the JTTDCMut+F + I + Γ_4 substitution model. Bootstrap values are shown by node circles and the scale bar shows the number of amino acid substitutions per site. The tree was mid-pointed rooted for purposes of clarity only

Fortunately, modern metagenomic techniques are providing a powerful new way to characterise the virosphere and hence identify more novel lyssavirus species and lineages of RABV. Of particular importance is total RNA-sequencing (so-called "meta-transcriptomics"; 9) that has identified a growing number of novel lyssaviruses and related rhabdoviruses [9–12]. For example, a recent metagenomic survey of big brown bats (*Eptesicus fuscus*) in the USA identified two new rhabdoviruses [12],

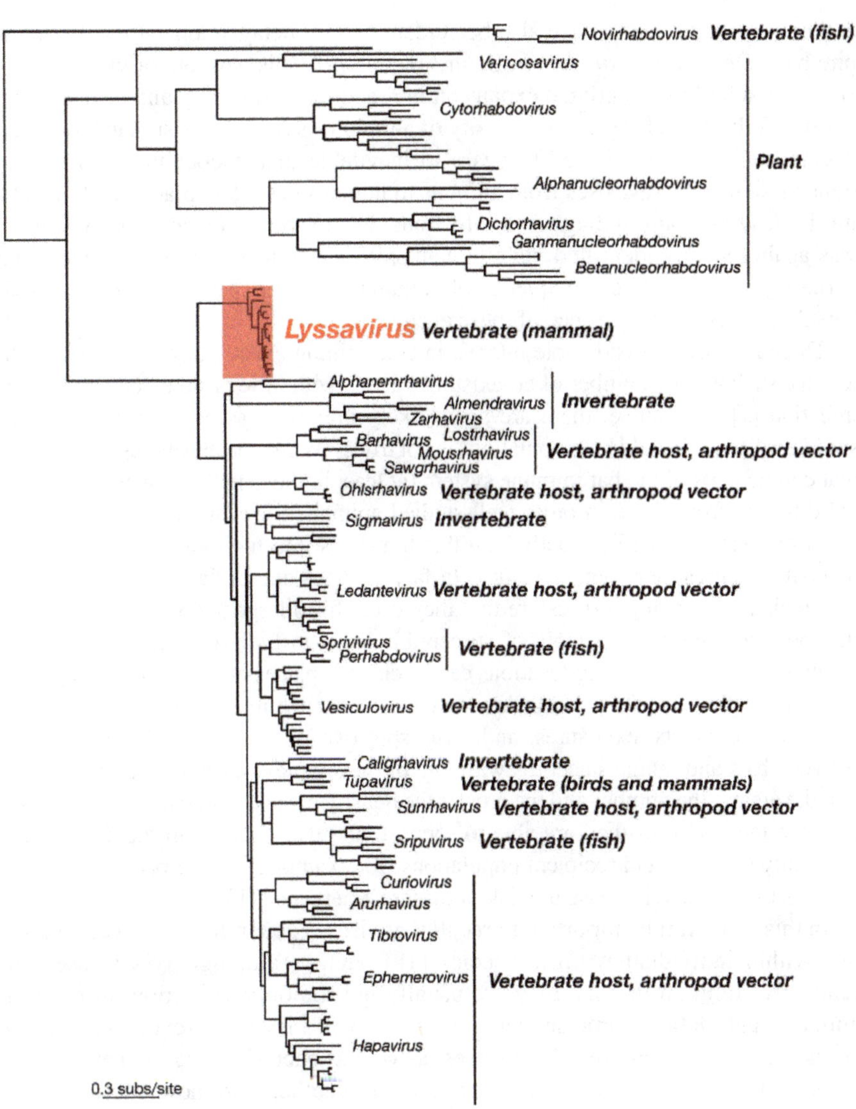

Fig. 3.2 Phylogenetic relationships of 191 virus species (30 genera) classified within the family *Rhabdoviridae* by the International Committee on Taxonomy of Viruses (ICTV; April 2021 release). The genus *Lyssavirus* is shown in red. The phylogeny was estimated using an amino acid alignment of the RdRp domain and the maximum likelihood method available within the IQ-TREE2 program [66] employing the LG + F + I + Γ_4 substitution model. The major taxonomic group of the hosts of each genus is marked. The scale bar shows the number of amino acid substitutions per site. The tree was mid-pointed rooted for purposes of clarity only, although the exact position of the root is unclear and could occur in a variety of places (for example, on the branch separating the fish from the other rhabdoviruses)

and novel rhabdoviruses have also been identified in metagenomic studies of vampire bats (*Desmodus rotundus*) in Latin America [9]. This new age of virus discovery has resulted in a continual expansion in rhabdovirus diversity and means that it is very likely that a far greater diversity of animal lyssaviruses exist than have been documented to date (Fig. 3.2). It is also inevitable that undocumented zoonotic transmission of lyssaviruses from animals to humans has taken place, likely resulting in dead-end and undiagnosed infections. For example, *Duvenhage lyssavirus* was again recently identified via a human spill-over infection [13]. It is therefore perhaps just the relative infrequency of human (and other species) interaction with bats that limits the occurrence of these events.

There has been considerable interest in determining exactly how bats are able to tolerate such a large number of co-existing viruses with little or no evidence of overt infection [7]. Of course, there are again likely important ascertainment biases, as studying disease in wild bat populations is not trivial. While recent studies have shown that components of the bat immune system (at least in those in those species studied) are different from those in other well-studied animals, for example involving interferon responses [14, 15], exactly how this translates into high virus abundance and diversity requires more detailed study. In fact, perhaps one of the main reasons why bats maintain so many viruses are that they often form large and dense populations that provide a continual supply of susceptible hosts to fuel virus transmission and evolution, and their roosting locations can be close to places of human habitation. For example, big brown bats in North America often roost in human structures, increasing the risk of contacts, exposures, and virus spill-over [16]. Hence, the relationship between bats and humans accords with two of the simplest rules of disease ecology: that the greater the number and the greater the density of susceptible hosts in a population, the more viruses they are able to support, and that changes in the density and proximity of donor and recipient populations (for example, the encroachment on bat habitats by humans) increase the risk of disease emergence [17].

In this context it is important to recall that viruses generally exist in communities within individual wildlife species [18]. Hence, although lyssaviruses are relatively common in bats, they are usually not the only virus present in these animals, and metagenomic sequencing studies will inevitably reveal that RABV co-exists with a variety of other viruses (as well as bacteria, fungi and eukaryotic parasites). This then raises the question of whether and how these myriad microbes interact with each other and whether these interactions impact host disease? Moreover, if such interactions exist, do they induce immunological cross-reaction and protection, or is there 'ecological interference' in which infection by one virus removes them from the susceptible class for another [19]? These are central questions in viral ecology that will require the generation of additional metagenomic sequence data.

The Genetic Diversity and Origins of RABV

Rabies is an almost global viral infection that circulates in a diverse set of animal reservoir hosts from the mammalian orders Chiroptera (bats) and Carnivora, with occasional spill-over infections to other animal species including humans and domestic animals. In addition, modelling studies suggest that there are a relatively high number of currently unrecognised potential bat and carnivore host species [20]. As such, the evolution of RABV provides a powerful example of a successful multi-host pathogen, but in which virus-host relationships can range from sustained transmission cycles (for example, bats) to transient dead-end spill-over events (for example, humans). Although the focus of this chapter and most studies of RABV evolution has been the virus side of this equation, it is also of crucial importance to consider the role played by the host. For example, an increase in the population genetic diversity of individual mammalian host species has been associated with an increase in the likelihood of that species being a donor in RABV cross-species transmission events [21]. It is also important to determine if and how virus susceptibility differs among species.

The most informative way to measure and visualise viral genetic diversity is through phylogenetic analysis. In the case of RABV, such studies initially involved gene fragments covered by individual PCR amplicons [22], and then whole genes, particularly the nucleoprotein (N) and glycoprotein (G) (see Ref. [23] for an illustrative example), until the rise of next-generation sequencing led comparisons of complete genome sequences to become the industry standard. At the time of writing, approximately 2000 complete genome sequences of RABV are available, representing well over 50 different host species. Most of these sequences are from animals sampled in North America.

The phylogenetic analyses of RABV undertaken to date have shown that virus genetic diversity can be classified into two main phylogenetic groups: (i) one that contains all the viruses sampled from bats, as well as from carnivores from North America, and which has been called the 'bat-related' virus group, and (ii) another that contains predominantly 'dog-related' viruses sampled from carnivores, to the exclusion of bats [24] (Fig. 3.3). Importantly, both groups are associated with a wide range of carnivore hosts and both have the ability to spill-over to human populations. At present, the 'bat-related' RABV group is confined to viruses that circulate among a diverse set of New World bat species (including the genera *Desmodus, Eptesicus, Lasionycteris, Lasiurus, Myotis, Tadarida* among others) as well as in some mesocarnivores such as skunks and raccoons that form a distinct phylogenetic cluster within the bat-related group [4, 23, 24]. In Latin America, there has been considerable recent attention on vampire bats as hosts, particularly as they act as reservoir species for RABV to infect other mammals, including humans and livestock [25, 26]. Similarly, RABV has also been reported in a number of non-human primate species in Latin America, including marmosets in Brazil, where they have also been associated with subsequent human infections [27].

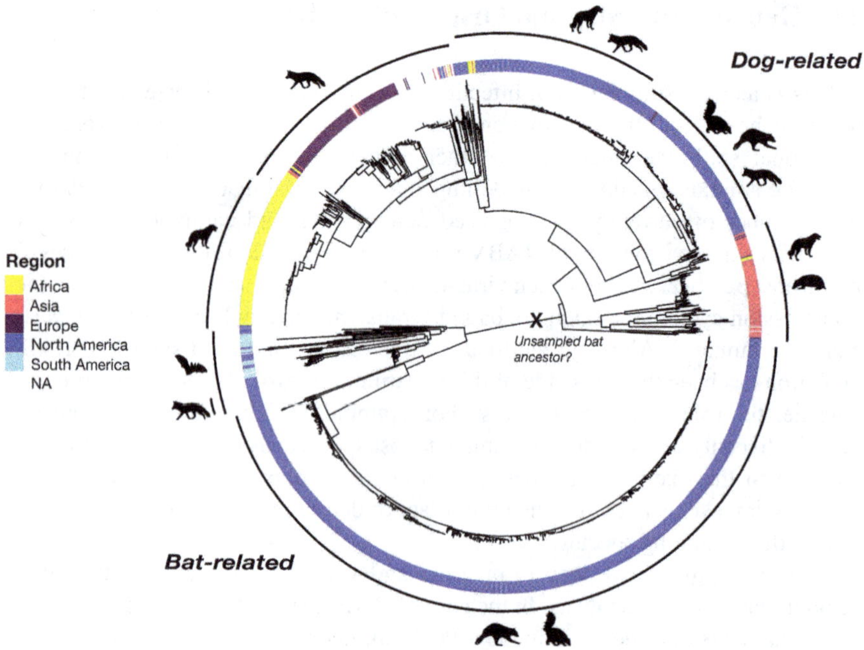

Fig. 3.3 Phylogenetic relationships of all available (1520) full genome nucleotide sequences of RABV (downloaded from NCBI/GenBank in March 2021). The phylogeny was estimated using the maximum likelihood method available within the IQ-TREE2 program [66] under the GTR + F + I + Γ_4 substitution model. Branches are coloured according to the geographic location of sampling and animal silhouettes depict the major animal host species in each component clade. Viruses from the major bat-related group of RABV are shown to the bottom-left, while those from the dog bat-related group are shown to top-right. The phylogenetic position of a putative unsampled bat ancestor for the dog-related cluster (that dominate in the Old World) is marked by a X symbol. (See Ref. [24] for additional phylogenetic analysis)

As its name implies, the 'dog-related' RABV group contains viruses that circulate effectively worldwide in dogs, as well as in various wildlife carnivore species that differ according to the geographic region of surveillance: for example, skunks (various species), foxes (various species), coyotes, etc., as carnivore hosts in the Americas, raccoon dogs and ferret-badgers in Asia, foxes and raccoon dogs in Europe, and mongooses and jackals in Africa. The presence of RABV in carnivore species such as ferret badgers and raccoon dogs is of particular interest, as these as animals are associated with widespread fur farming. Indeed, the translocation of raccoon dogs from Asia to north-eastern Europe for fur farming helped rabies virus to emerge and spread on that continent [28].

The 'dog-related' group can itself be further subdivided into a number of component clades, that generally reflect a combination of host species and the geographical area of sampling, and hence are the outcome of local *in situ* evolution [24]. For example, two clades are found in different parts of Africa (which have been denoted 'Africa 2' and 'Africa 3'), while another describes a large group of viruses of Asian origin, with a separate clade incorporating viruses from the Indian subcontinent. Importantly, many of the wild carnivore viruses from North America are associated with the so-called "Arctic-related clade" that contains viruses from such species as skunk, red fox and Arctic fox [29]. Interestingly, the Arctic-related clade has now been documented in a broad range of geographic locations, including Bangladesh [30]. The largest of the clades of dog-related viruses, at least in terms of phylogenetic diversity, is referred to as the "Cosmopolitan" clade and as its name implies it encompasses a relatively wide geographical distribution. Indeed, this clade has also been divided into a series of small subclades that again reflect local geographic distributions, such as Africa, Europe, the Middle East and Central Asia [24]. For instance, four subclades have been documented in the Americas, each associated with a particular set of host species and/or geographic region: (i) AM1 (i.e. "America 1"), that comprises viruses from skunks in the USA, (ii) AM2, containing viruses in dogs and wildlife from Mexico, (iii) AM3 – associated with dogs and foxes in Brazil, and (iv) AM4 – associated with skunks in California [24]. Hence, as is true globally, the genetic diversity of RABV in the Americas is structured by both the host species and geographic location of sampling.

What explains this phylogenetic diversity, both globally and in the Americas? Given that nearly all other species from the genus *Lyssavirus* have bat hosts, and that it is possible to root the RABV phylogeny with bat viruses (Figs. 3.1 and 3.3), the most parsimonious explanation is that the emergence of rabies virus as an important human pathogen was associated with the cross-species transmission from bats to carnivores [31]. Indeed, the key difference between RABV and other lyssaviruses is that RABV is the only one that has established itself in such a wide range of animal hosts (although the reasons are unclear). Although poor sampling means that there are likely undetected lyssaviruses in other mammalian species, the regularity with which novel bat lyssaviruses are identified and the propensity of bat RABV to spillover into humans makes it unlikely that there is an alternate animal reservoir yet to be identified. While this evolutionary trajectory is straightforward and mirrors the emergence of RNA viruses from bats to other mammalian hosts, with coronaviruses presenting a good example [32], a major phylogenetic puzzle remains: that the basal lineage of American bat RABV clusters with some carnivore RABV also from the Americas, with the dog-related RABV group sampled from other hosts and other locations (for example, Africa, Asia, Europe) forming a distinct lineage. This, at face value, suggests that the formative bat-to-carnivore cross-species transmission event took place in the Americas, with subsequent spread to the Old World, and therefore sits in direct contradiction to other anecdotal and historical document evidence: that RABV was present in parts of the Old World during the pre-Christian era [33], that the genetic diversity of RABV is greatest in the Old World, and that the Old World experiences by far the highest burden of rabies disease. The most likely solution to this phylogenetic knot is that there were *independent* cross-species transmission events from bats to carnivores in the New World and the Old World, but that the direct bat ancestor of the latter is yet to be sampled (Fig. 3.3). Given the very limited sampling of bat populations in the Old World undertaken to date, the hypothesis of an unsampled bat ancestor for RABV in this region seems highly plausible.

The Timescale of RABV Emergence and Evolution

The timing of key events in RABV evolution has also been the subject of considerable debate. Although it is obvious that RABV has been impacting human populations for centuries if not millennia, the exact timescale of the most important evolutionary events, including when the virus jumped from bats to carnivores, is uncertain. Due to its distinctive signs and symptoms, rabies is one of the oldest documented viral infections in humans [33]. It is therefore paradoxical that the studies of the timing of RABV evolution undertaken to date, all of which utilise the so-called 'molecular clock' of evolutionary change in gene sequences, depict an evolutionary process that has occurred on scale of less than 1000 years [23, 24, 34], whereas dogs have likely lived with humans for over 10,000 years [35]. For example, a molecular clock study of dog-related viruses suggested that these viruses shared a common ancestor between the years 1308–1510 and hence long after supposed documentary historical evidence for the presence of rabid dogs in humans, with a rapid global spread of these viruses associated with the rise of the intercontinental trade network during the fifteenth century [24]. Similarly recent divergent times have been estimated for bat-related RABV in the Americas [36]. Under this evolutionary timescale, rabies entered the Americas at least twice: once from the ancestral bat reservoir from which it established a stable non-volant animal reservoir, and likely occurring at some point prior to Columbian colonisation, and second a more recent post-Columbian importation event likely initially involving rabid dogs before spreading to a variety of local carnivore species.

Although the molecular clock can be a powerful tool when there is strong evidence for a constancy of evolutionary rate and when genetic distances can be estimated accurately, it is not without pitfalls. It is obviously essential to ensure that the data under consideration really do contain evidence that virus mutations occur at an approximately constant rate, without so much variation in the tick of the clock that they cannot easily be incorporated into a variety of statistical distributions [37]. Second, virus evolution is characterised by a strong "time-dependency", such that rate estimates tend to be higher toward the present than they are in the past [38]. The former is caused by the presence of transient deleterious (particularly nonsynonymous) mutations that have yet to be removed by purifying (i.e., negative) selection because they have adverse effects on some aspect of virus fitness, while the latter reflects the impact of multiple substitution at individual nucleotide sites that tends to progressively underestimate genetic distances as time passes (and even in the short-term; 40). Hence, while it is relatively straightforward to estimate the rate and timescale of RABV evolution in recent decades, the timing of the ultimate origin of RABV may be underestimated. It is therefore likely that the ancestry of rabies, including the Old World lineages of dog-related RABV, are older than suggested from molecular clock studies. This may explain the apparent discrepancy between ancient depictions of rabies in the historical literature and more recent molecular clock estimates of divergence times. It is also possible that any earlier lineages of carnivore RABV died-out, left no descendants, and hence cannot be sampled.

Molecular clock studies of RABV evolution are also complicated by the variation in tick rate between lineages. For example, one study reported strong molecular clock signal in dog-related RABV, but not bat-related RABV, indicative of markedly different evolutionary and epidemiological conditions depending on the dominant host species [24]. Another study showed that evolutionary rates in different lineages of New World bat RABV varied by over an order of magnitude [39], and in a manner that reflected the geographic region from which the bats were sampled, itself a function of bat seasonality. Although an interesting idea that clearly needs to be tested with additional data (as rate variation is not normally thought to be seasonal), it is also the case that the sample sizes of some of the bat virus lineages analysed were small and with only relatively weak temporal structure. Finally, it is noteworthy that all studies undertaken to date suggest that RABV is a relatively slowly evolving virus, with mean evolutionary rates of approximately 2×10^{-4} nucleotide substitutions per site, per year, and hence lower than seen in many other RNA viruses [40].

Ultimately, and more speculatively, in the future it might be possible to use the techniques of 'ancient DNA' to recover the virus from archival human and animal samples to help determine the antiquity of rabies. Although such techniques have been increasingly used to reveal the evolutionary history of DNA viruses, including hepadnaviruses (such as hepatitis B virus) and variola virus (the agent of smallpox), the problem of RNA degradation means that they have been used less often in the study of RNA viruses [41].

Host Adaptation of RABV

Rabies in humans is a transient, dead-end and usually fatal infection. That there is no onward transmission in *Homo sapiens* means that, by definition, the virus is not under long-term selection pressure in humans, such that all the traits observed reflect its adaptation to animal hosts. Despite this, the extent to which natural selection occurs as the virus spreads to other species is of considerable interest, not least because it is the cornerstone of disease emergence. The key question is whether RABV has to actively adapt to new host species as it moves between them, or whether it is such a host generalist that in many cases direct host adaptation is not necessary, in which case cross-species transmission can be effectively considered a neutral evolutionary process [4]. If the latter is true, disease emergence is more dependent on background ecological rather than evolutionary processes and revealing the competing roles of ecology and adaptive evolution in the emergence of RABV have been the source of long-term debate [42].

The traditional way the analyses of selection pressures have been performed are through comparisons of the relative numbers of nonsynonymous and synonymous nucleotide substitutions per site (as reflected in the ratio d_N/d_S). Most such studies on RABV have only revealed weak, or no, evidence for adaptive evolution concurrent with cross-species transmission events, with RABV evolution instead characterized by strong purifying selection in comparison to other RNA viruses [43, 44], including in association with host-switching events [45]. This is compatible with the idea that RABV has been subject to strong selective constraints because it is a 'generalist' virus that has to replicate in a broad range of host cell types and hence lacks evolutionary elbow room [43]. For example, comparisons of evolutionary rates and selection pressures revealed that the host-jumping events to different carnivore species did not involve apparent adaptive evolution [24]. Similarly, a skunk-to-fox cross-species transmission event in California was associated with only a few amino acid changes and hence a lack of clear evidence for adaptive evolution [46]. Although it has been suggested that host shifts of RABV in New World bats routinely involve episodic bursts of adaptive evolution [47], the very large number of positively selected sites detected, especially in relatively conserved genes like the polymerase that are not expected to assist host adaptation, suggests that many of the adaptive sites detected may be false-positives due to a poorly fitting analytical model. Experimental infection models using cell culture models of dogs and foxes also provided little direct evidence for molecular adaptation, and only a small number of the mutations that appeared in experimental passage also appeared in nature [44]. Interestingly, these studies also revealed that dog-adapted RABV was able to replicate in fox and fox cells relatively easily, while dogs and dog neuronal cells were not easily susceptible to fox adapted-RABV. Hence, there may be important "asymmetries" in the barriers to RABV infection in different hosts [44].

It is also the case that RABV lineages in different species are often associated with specific amino acid changes. For example, a particular suite of amino acid changes was associated with the emergence of RABV in ferret-badgers in Asia, and some host jumps (to ferret-badger and mongoose) were associated with increases in evolutionary rate, implying that jumps to some host species can entail localised adaptive evolution [24]. Positive selection has also been proposed in the case of an Arctic fox variant of RABV that emerged in skunk populations in Ontario, Canada [48], and evolutionary analysis of carnivore outbreaks in Arizona revealed that each was caused by an independent cross-species transmission event from bats [49]. Although, there was no statistically significant evidence for positive selection, convergent mutations were seen in the N, P, G and L proteins that may been associated with a fitness advantage [49].

The emergence of RABV in new hosts is therefore likely to be a complex, multifactorial adaptive process, involving only a small number of amino acid changes that, because of strong background genetic constraints, occur at a limited number of amino acid sites. These sites may be undetectable by standard methods of gene sequence analysis based on d_N/d_S comparisons that may perform poorly if natural selection is restricted to a single amino acid change appearing once. That RABV does not readily develop sustained transmission cycles in new hosts at the very least indicates that there must be major adaptive barriers to successful cross-species transmission such that host-jumping is unlikely to be a purely 'neutral' evolutionary process. Interestingly, analyses of North American bats have shown that the likelihood of successful cross-species transmission events declines with increasing phylogenetic distance between the host species, again as expected if there are major constraints on adaptive evolution [50] and which might be true of RNA viruses in general [51]. Clearly, the more and stronger constraints there are, the more likely that adaptive evolution in the virus is needed to overcome them. Accordingly, a combination of comparative and experimental analyses may be needed to fully tease apart the competing contributions of adaptation and ecology to explain complex host range of RABV.

The Phylogeography of RABV in the Americas

RABV has been one of the mainstays of virus phylogeography. In this case the phylogenetic analysis is used to track the spread of the virus within and between geographical regions. To date, RABV has proven especially useful in this respect because it usually spreads at a sufficiently low rate and in a sufficiently simple manner in animal populations (often through a simple spatial wave) that it is relatively easily to track through the analysis of viral genome sequences. In particular, there have been detailed studies of the spatial patterns of spread of RABV among different wildlife species in Ontario [52–54], while studies of the spatial dynamics of raccoon RABV in the north-eastern United States has been one of the key data sets used in the development of phylogeographic methods [2, 55, 56]. Detailed phylogeographic studies documented the demographic and spatial spread of raccoon RABV in the eastern states of the USA, setting the benchmark for studies in this area [57], while the sophisticated Bayesian analyses of New World bat populations has shed important new light on the patterns and determinants of spread in these species in different localities [58]. For example, detailed phylogeographic analyses of vampire bat-associated RABV in Costa Rica demonstrated that the viruses from the animals sampled in both North and South America share a single common ancestor, such that they represent a single defined epizootic outbreak [59]. This international linkage fuels an ongoing process of RABV lineage extinction and reinvasion [59]. While at first glance rabies might appear as a static disease, in reality it is anything but uniform.

More recently, a long-distance translocation virus event from southern New York State, USA, to Hamilton, Ontario, Canada was identified as responsible for an outbreak of raccoon rabies in the latter [60]. This highlights the importance of interactions between humans and wildlife, and how humans have changed the infectious disease dynamics of some animal pathogens. In fact, there is increasing evidence for the role of humans in the spread of rabies among animal populations. Although rabies is primarily a disease of wildlife, because humans also move animal populations (deliberately and inadvertently) they may also contribute to the phylogeography of RABV in wildlife. For example, in North Africa it has been suggested that the movement of dogs by humans has had a major impact of the geographic structure of the virus in dogs in this region [61], although lower rates of spatial spread have been documented in China [62]. Similarly, in the Americas it has been suggested that human-mediated movement of raccoons was responsible for the major spatial shift of the virus from the southern to the northern states [63].

Concluding Remarks

Much of the textbook on RNA virus evolution and emergence has been written through the lens of rabies, providing important lessons that are currently being used to our benefit in tracking the ongoing evolution of SARS-CoV-2. The Americas are a geographic region of particular interest because the variants of RABV present in this locality derive from a variety of hosts, comprising both bats and carnivores, with ongoing spill-over infections to other hosts (including humans). As such, the Americas represent a microcosm of cross-species transmission and emergence, and hence of disease ecology in general. It is also inevitable that increased animal sampling combined with metagenomic next-generation sequencing will identify more animal variants of RABV and other lyssaviruses, as well as additional animal reservoirs [64].

Going forward, it will be important to put the extent of cross-species transmission of RABV into a more quantitative framework, from which we will be able to better understand the threat posed by these major zoonotic viruses. In particular, it will be important to sample (perhaps through non-invasive faecal means) lyssaviruses in all the mammalian hosts in an entire ecosystem at a single time-point. With these data in hand, it should be possible to estimate the frequency with which lyssaviruses are found in different host species, if only transiently, and hence the overall rate of virus exposure, successful host jumping and lineage extinction. As well as giving RABV ecology and evolution a firmer, more quantitative basis, the data obtained will also shed light on the nature of the barriers, both genetic and ecological, that might prevent RABV establishing productive infections in new hosts, and that remains perhaps the most pressing question in RABV evolution. Are more host species exposed to RABV infection than we currently know? Do any of these hosts experience undocumented disease? It is also possible that different levels of virus virulence between donor and recipient species impact the likelihood of successful cross-species transmission [65]. These are all questions that need be addressed.

Despite the great advances made in our understanding of rabies through the deployment of evolutionary genomics, important questions remain on the exact origin and timescale of its evolution in the Americas. Although the answers to some may require analytical advances, such as the more precise dissection of selection pressures and the incorporation of complex patterns of lineage-specific rate variation in molecular clocks, it is inevitable that the metagenomic survey of a greater sample of animal taxa will help also provide data of generic importance. Ongoing studies of the ecology and evolution of RABV in the Americas will be central to this endeavour.

References

1. Woolhouse MEJ, Taylor LH, Haydon DT (2001) Population biology of multihost pathogens. Science 292:1109–1112
2. Lemey P, Rambaut A, Drummond AJ, Suchard MA (2009) Bayesian phylogeography finds its roots. PLoS Comp Biol 5:e1000520

3. Noah DL, Drenzek CL, Smith JS, Krebs JW, Orciari L, Shaddock J et al (1998) Epidemiology of human rabies in the United States, 1980 to 1996. Ann Intern Med 128:922–930
4. Fisher CR, Streicker DG, Schnell MJ (2018) The spread and evolution of rabies virus: conquering new frontiers. Nat Rev Microbiol 16:241–255
5. Cunningham AA, Daszak P, Wood JLN (2017) One health, emerging infectious diseases and wildlife: two decades of progress? Philos Trans R Soc Lond Ser B Biol Sci 372:20160167
6. Cui J, Li F, Shi ZL (2019) Origin and evolution of pathogenic coronaviruses. Nat Rev Microbiol 17:181–192
7. Wang L-F, Walker PJ, Poon LLM (2011) Mass extinctions, biodiversity and mitochondrial function: are bats "special" as reservoirs for emerging viruses? Curr Opin Virol 1:649–657
8. Luis AD, Hayman DT, O'Shea TJ, Cryan PM, Gilbert AT, Pulliam JR et al (2013) A comparison of bats and rodents as reservoirs of zoonotic viruses: are bats special? Proc Biol Sci 280:20122753
9. Bergner LM, Mollentze N, Orton RJ, Tello C, Broos A, Biek R et al (2021) Characterizing and evaluating the zoonotic potential of novel viruses discovered in vampire bats. Viruses 13:252
10. Bergner LM, Orton RJ, Benavides JA, Becker DJ, Tello C, Biek R et al (2020) Demographic and environmental drivers of metagenomic viral diversity in vampire bats. Mol Ecol 29:26–39
11. Coffey LL, Page BL, Greninger AL, Herring BL, Russell RC, Doggett SL et al (2014) Enhanced arbovirus surveillance with deep sequencing: identification of novel rhabdoviruses and bunyaviruses in Australian mosquitoes. Virology 448:146–158
12. Hause BM, Nelson E, Christopher-Hennings J (2020) Novel and diverse non-rabies rhabdoviruses identified in bats with human exposure, South Dakota, USA. Viruses 12:1408
13. Paweska JT, Blumberg LH, Liebenberg C, Hewlett RH, Grobbelaar AA, Leman P et al (2006) Fatal human infection with rabies-related Duvenhage virus, South Africa. Emerg Infect Dis 12:1965–1967
14. Schountz T, Baker ML, Butler J, Munster V (2017) Immunological control of viral infections in bats and the emergence of viruses highly pathogenic to humans. Front Immunol 8:1098
15. Zhou P, Tachedjian M, Wynne JW, Boyd V, Cui J, Smith I et al (2016) Contraction of the type I IFN locus and unusual constitutive expression of IFN-α in bats. Proc Natl Acad Sci U S A 113:2696–2701
16. Osborne C, Cryan PM, O'Shea TJ, Oko LM, Ndaluka C, Calisher CH et al (2011) Alphacoronaviruses in New World bats: prevalence, persistence, phylogeny, and potential for interaction with humans. PLoS One 6:e19156
17. Plowright RK, Parrish CR, McCallum H, Hudson PJ, Ko AI, Graham AL et al (2017) Pathways to zoonotic spillover. Nat Rev Microbiol 15:502–510
18. Wille M, Shi M, Eden J-S, Klaassen M, Hurt AC, Holmes EC (2019) Virome heterogeneity and connectivity in waterfowl and shorebird communities. ISME J 13:2603–2616
19. Rohani P, Green CJ, Mantilla-Beniers NB, Grenfell BT (2003) Ecological interference between fatal diseases. Nature 422:885–888
20. Worsley-Tonks KEL, Escobar LE, Biek R, Castaneda-Guzman M, Craft ME, Streicker DG et al (2020) Using host traits to predict reservoir host species of rabies virus. PLoS Negl Trop Dis 14:e0008940
21. Rodríguez-Nevado C, Lam TT-Y, Holmes EC, Pagán I (2018) The impact of host genetic diversity on virus evolution and emergence. Ecol Lett 21:253–263
22. Sacramento D, Badrane H, Bourhy H, Tordo N (1992) Molecular epidemiology of rabies virus in France: comparison with vaccine strains. J Gen Virol 73:1149–1158
23. Bourhy H, Reynes J-M, Dunham EJ, Dacheux L, Larouss F, Hong VTQ et al (2008) The origin and phylogeography of dog rabies virus. J Gen Virol 89:2673–2681
24. Troupin C, Dacheux L, Tanguy M, Sabeta C, Blanc H, Bouchier C et al (2016) Large-scale phylogenomic analysis reveals the complex evolutionary history of rabies virus in multiple carnivore hosts. PLoS Pathog 12:e1006041
25. Benavides JA, Rojas Paniagua E, Hampson K, Valderrama W, Streicker DG (2017) Quantifying the burden of vampire bat rabies in Peruvian livestock. PLoS Negl Trop Dis 11:e0006105

26. Johnson N, Aréchiga-Ceballos N, Aguilar-Setien A (2014) Vampire bat rabies: ecology, epidemiology and control. Viruses 6:1911–1928
27. Kotait I, Oliveira RN, Carrieri ML, Castilho JG, Macedo CI, Pereira PMC et al (2019) Non-human primates as a reservoir for rabies virus in Brazil. Zoonoses Public Health 66:47–59
28. Bourhy H, Kissi B, Audry L, Smreczak M, Sadkowska-Todys M, Kulonen K et al (1999) Ecology and evolution of rabies virus in Europe. J Gen Virol 80:2545–2558
29. Kuzmin IV, Hughes GJ, Botvinkin AD, Gribencha SG, Rupprecht CE (2008) Arctic and Arctic-like rabies viruses: distribution, phylogeny and evolutionary history. Epidemiol Infect 136:509–519
30. Jamil KM, Ahmed K, Hossain M, Matsumoto T, Ali MA, Hossain S et al (2012) Arctic-like rabies virus, Bangladesh. Emerg Infect Dis 18:2021–2024
31. Badrane H, Tordo N (2001) Host switching in *Lyssavirus* history from the Chiroptera to the Carnivora orders. J Virol 75:8096–8104
32. Corman VM, Muth D, Niemeyer D, Drosten C (2018) Hosts and sources of endemic human coronaviruses. Adv Virus Res 100:163–188
33. Neville J (2004) Rabies in the ancient world. In: King AA, Fooks AR, Aubert M, Wandeler AI (eds) Historical perspective of rabies in Europe and the Mediterranean basin. World Organisation for Animal Health (OIE), Paris, pp 1–12
34. Hughes GJ, Orciari LA, Rupprecht CE (2005) Evolutionary timescale of rabies virus adaptation to North American bats inferred from the substitution rate of the nucleoprotein gene. J Gen Virol 86:1467–1474
35. Bergström A, Frantz L, Schmidt R, Ersmark E, Lebrasseur O, Girdland-Flink L et al (2020) Origins and genetic legacy of prehistoric dogs. Science 370:557–564
36. Kuzmina NA, Kuzmin IV, Ellison JA, Taylor ST, Bergman DL, Dew B et al (2013) A reassessment of the evolutionary timescale of bat rabies viruses based upon glycoprotein gene sequences. Virus Genes 47:305–310
37. Ho SY, Duchêne S (2014) Molecular-clock methods for estimating evolutionary rates and timescales. Mol Ecol 23:5947–5965
38. Duchêne S, Holmes EC, Ho SYW (2014) Analyses of evolutionary dynamics in viruses are hindered by a time-dependent bias in rate estimates. Proc R Soc B 281:20140732
39. Streicker DG, Lemey P, Velasco-Villa A, Rupprecht CE (2012) Rates of viral evolution are linked to host geography in bat rabies. PLoS Pathog 8:e1002720
40. Holmes EC, Dudas G, Rambaut A, Andersen KG (2016) The evolution of Ebola virus: insights from the 2013-2016 epidemic. Nature 358:193–200
41. Duchêne S, Ho SYW, Carmichael AG, Holmes EC, Poinar H (2020) The recovery, interpretation and use of ancient pathogen genomes. Curr Biol 30:R1215–R1231
42. Mollentze N, Biek R, Streicker DG (2014) The role of viral evolution in rabies host shifts and emergence. Curr Opin Virol 8:68–72
43. Holmes EC, Woelk CH, Kassis R, Bourhy H (2002) Genetic constraints and the adaptive evolution of rabies virus. Virology 292:247–257
44. Bonnaud EM, Troupin C, Dacheux L, Holmes EC, Monchatre-Leroy E, Tanguy M et al (2019) Comparison of intra- and inter-host genetic diversity in rabies virus during experimental cross-species transmission. PLoS Pathog 15:e1007799
45. Marston DA, Horton DL, Nunez J, Ellis RJ, Orton RJ, Johnson N et al (2017) Genetic analysis of a rabies virus host shift event reveals within-host viral dynamics in a new host. Virus Evol 3:vex038
46. Borucki MK, Chen-Harris H, Lao V, Vanier G, Wadford DA, Messenger S, Allen JE (2013) Ultra-deep sequencing of intra-host rabies virus populations during cross-species transmission. PLoS Negl Trop Dis 7:e2555
47. Streicker DG, Altizer SM, Velasco-Villa A, Rupprecht CE (2012) Variable evolutionary routes to host establishment across repeated rabies virus host shifts among bats. Proc Natl Acad Sci U S A 109:19715–19720
48. Nadin-Davis SA, Fehlner-Gardiner C (2019) Origins of the arctic fox variant rabies viruses responsible for recent cases of the disease in southern Ontario. PLoS Negl Trop Dis 13:e0007699

49. Kuzmin IV, Shi M, Orciari LA, Yager PA, Velasco-Villa A, Kuzmina NA et al (2012) Molecular inferences suggest multiple host shifts of rabies viruses from bats to mesocarnivores in Arizona during 2001-2009. PLoS Pathog 8:e1002786
50. Streicker DG, Turmelle AS, Vonhof MJ, Kuzmin IV, McCracken GF, Rupprecht CE (2010) Host phylogeny constrains cross-species emergence and establishment of rabies virus in bats. Science 329:676–679
51. Longdon B, Hadfield JD, Webster CL, Obbard DJ, Jiggins FM (2011) Host phylogeny determines viral persistence and replication in novel hosts. PLoS Pathog 7:e1002260
52. Nadin-Davis SA, Casey GA, Wandeler AI (1994) A molecular epidemiological study of rabies virus in Central Ontario and western Quebec. J Gen Virol 75:2575–2583
53. Nadin-Davis SA, Feng Y, Mousse D, Wandeler AI, Aris-Brosou S (2010) Spatial and temporal dynamics of rabies virus variants in big brown bat populations across Canada: footprints of an emerging zoonosis. Mol Ecol 19:2120–2136
54. Nadin-Davis SA, Sampath MI, Casey GA, Tinline RR, Wandeler A (1999) Phylogeographic patterns exhibited by Ontario rabies virus variants. Epidemiol Infect 123:325–336
55. Dellicour S, Rose R, Pybus OG (2016) Explaining the geographic spread of emerging epidemics: a framework for comparing viral phylogenies and environmental landscape data. BMC Bioinformatics 17:82
56. Real LA, Henderson JC, Biek R, Snaman J, Jack TL, Childs JE et al (2005) Unifying the spatial population dynamics and molecular evolution of epidemic rabies virus. Proc Natl Acad Sci U S A 102:12107–12111
57. Biek R, Henderson JC, Waller LA, Rupprecht CE, Real LA (2007) A high-resolution genetic signature of demographic and spatial expansion in epizootic rabies virus. Proc Natl Acad Sci U S A 104:7993–7998
58. Seetahal JFR, Velasco-Villa A, Allicock OM, Adesiyun AA, Bissessar J, Amour K et al (2013) Evolutionary history and phylogeography of rabies viruses associated with outbreaks in Trinidad. PLoS Negl Trop Dis 7:e2365
59. Streicker DG, Fallas González SL, Luconi G, Barrientos RG, Leon B (2019) Phylodynamics reveals extinction-recolonization dynamics underpin apparently endemic vampire bat rabies in Costa Rica. Proc Biol Sci 286:20191527
60. Nadin-Davis S, Buchanan T, Nituch L, Fehlner-Gardiner C (2020) A long-distance translocation initiated an outbreak of raccoon rabies in Hamilton, Ontario, Canada. PLoS Negl Trop Dis 14:e0008113
61. Talbi C, Lemey P, Suchard MA, Abdelatif E, Elharrak M, Jalal N et al (2010) Phylodynamics and human-mediated dispersal of a zoonotic virus. PLoS Pathog 6:e1001166
62. Tian H, Feng Y, Vrancken B, Cazelles B, Tan H, Gill MS et al (2018) Transmission dynamics of re-emerging rabies in domestic dogs of rural China. PLoS Pathog 14:e1007392
63. Dobson A (2000) Raccoon rabies in time and space. Proc Natl Acad Sci U S A 97:14041–14043
64. Franka R, Constantine DG, Kuzmin I, Velasco-Villa A, Reeder SA, Streicker D et al (2006) A new phylogenetic lineage of rabies virus associated with western pipistrelle bats (*Pipistrellus hesperus*). J Gen Virol 87:2309–2321
65. Mollentze N, Streicker DG, Murcia PR, Hampson K, Biek R (2020) Virulence mismatches in index hosts shape the outcomes of cross-species transmission. Proc Natl Acad Sci U S A 117:28859–28866
66. Minh BQ, Schmidt HA, Chernomor O, Schrempf D, Woodhams MD, von Haeseler A et al (2020) IQ-TREE 2: new models and efficient methods for phylogenetic inference in the genomic era. Mol Biol Evol 37:1530–1534

Chapter 4
The Ecological Range and Principles of Wildlife Rabies Virus Perpetuation in the Americas

Amy T. Gilbert

Abstract Rabies virus is one of the most ancient and lethal zoonoses and has circulated for centuries in the Americas. Rabies virus infections in humans and animals are notifiable and laboratory-based surveillance including the characterization of positive cases has been critical to advancing knowledge regarding the description of species-specific reservoirs among various bat and carnivore wildlife. With notable progress in the regional goal of eliminating human rabies deaths due to canine rabies virus, continuation of surveillance targeting domestic animals and wildlife supports accurate risk communication and patient management across temperate to tropical regions. The process of viral emergence continues across the Americas, through expansion into novel hosts or host populations, and provides opportunities to measure in real-time the ecological and evolutionary factors associated with the phenomenon. As a result, our understanding of cross-species transmission and host shift processes is increasing rapidly, which can be used adaptively to guide surveillance and ecological studies among known and suspected wildlife reservoir hosts of importance to human, animal, and ecosystem health.

Keywords Community ecology · Epizootiology · Host range · Host switch · Population ecology · Rabies virus

A. T. Gilbert (✉)
US Department of Agriculture, Animal and Plant Health Inspection Service, Wildlife Services, National Wildlife Research Center, Fort Collins, CO, USA
e-mail: amy.t.gilbert@usda.gov

C. E. Rupprecht (ed.), *History of Rabies in the Americas: From the Pre-Columbian to the Present, Volume I*, Fascinating Life Sciences,
https://doi.org/10.1007/978-3-031-25052-1_4

Historical Context

Biogeography

Among lyssaviruses, the Americas are unique in that only rabies lyssavirus infects wild and domestic animals in this region [1, 2]. The biogeographic patterns of lyssavirus diversity [2], and comparisons with other bat associated viruses or pathogens (e.g., SARS-like beta-coronaviruses [3] or *Pseudogymnoascus destructans* which causes the disease White Nose Syndrome [4]), support the idea of more recent evolutionary timescales of pathogen diversity and host range among the New World bat fauna. Evidence suggests that rabies virus invaded and diversified among New World bats in recent time, relative to the more ancient global evolutionary history and radiations of bats and carnivores. Although the original split and directionality of bat and carnivore rabies virus host switch remain controversial and unresolved based on estimates from the genetic data, these two lineages have been evolving independently and may have variable ecology, epidemiology and transmission risk to humans and animals [5–8].

The Chiroptera evolved from a Palearctic common ancestor and is classified into two major suborders [9–11]. Neotropical bats show evidence of a single invasion from Africa during the Eocene ~40 mya, with later arrival of carnivores to South America, including the North American procyonids, and even later still the artiodacytls and perissodactyls from North America via the Panamanian land bridge ~3 mya [12]. The Yinpterochiropera is only present in the Old World and is comprised of 6 families with ~354 described species. In contrast, the cosmopolitan Yangochiroptera comprises at least 12 families with ~762 described species [13]. At least 407 species have been described from the Vespertilionidae (Yangochiroptera), with two extant subfamilies (i.e., Myotinae, Vespertilioninae; [14]) and the greatest number of recognized rabies virus reservoirs in the Americas [15, 16]. Recent evidence based on host genomic comparisons supports a basal ancestry of bats with a group of mammals known as the Fereuungulata (Cetartiodactyla + Perissodactyla + Pholidota + Carnivora), which when including bats is referred to as the Scrotifera (Chiroptera + Fereuungulata) [10]. Perhaps, given these basal relationships and geographic restriction of Pholidota to the Old World, it is possible rabies virus underwent one or more host switch(es) to the Carnivora, considering the exceptional diversity of carnivores (especially canids) in the New World [16]. These ancestral relationships may also serve as a baseline from which to examine host range and emergence phenomena among other bat-associated viruses [17, 18]. Within the Americas, bat rabies virus reservoirs are mostly from the Vespertilionidae, with fewer associations described from the Molossidae and Phyllostomidae. Among these three families, only phyllostomid bats are endemic to the Americas.

Within the Carnivora, two major suborders include 15 families: Caniformia (9 families, 165 species) and Feliformia (6 families, 121 species). The diversification history of the Carnivora is geographically complex, with asymmetrical historic intercontinental exchanges of major families (e.g., Canidae, Felidae; [19, 20]) within this order of mammals. Evolutionary signatures suggest that some families (e.g., Canidae, Ursidae, Mustelidae) migrated from North America and underwent explosive radiations in Eurasia. In contrast, other families (e.g., Felidae, Procyonidae, Mephitidae) invaded North America from Eurasia and with more uniform diversification patterns (e.g., greater extinction rates with rising diversity) in North America [19]. These patterns lend support to an idea that rabies virus switched into North American canids first and then later invaded Eurasia, but could also explain the infrequent nature of rabies virus infections reported among several North American felids (e.g., bobcats, *Lynx rufus*) [21]. Among caniforms, the Canidae (35 species, e.g., dogs [*Canis lupus familiaris*], coyotes [*Canis latrans*], foxes [*Vulpes spp.*]) are represented by several rabies virus reservoirs and the greatest richness occurs in North America, despite exceptional diversity of the Mustelidae (59 species). There are fewer reported rabies virus reservoirs from the Procyonidae (14 species), Mephitidae (12 species), and feliform Herpestidae (33 species) in the Americas [15, 16, 22, 23], patterns which may be expected because these families reportedly arrived later to the New World from Eurasia [19]. The Procyonidae and nearly all Mephitidae are currently restricted to the Americas, whereas the Canidae and Herpestidae have much broader distributions, occasionally due to human-mediated translocation [24–28]. In summary, the complex patterns of phylogenetic exchange and diversification of carnivores globally is consistent with a North American origin of the Canidae and rabies virus shift into carnivores from bats. These patterns may also explain the differences in diversity of carnivore rabies virus reservoirs, given some families arrived to the New World much later in evolutionary time and may be underrepresented in terms of taxonomic and reservoir richness in the Americas.

Rabies virus infections of bats in the Americas were present before modern-day settlers arrived [29]. The history of infection in bats is largely independent of the history of infection in carnivores within the Americas and the Arctic, where the virus was likely introduced and spread by settler movement of infected companion animals and emergent cycles between dogs and wildlife. This led to a historic divergence in the phylogeny of rabies viruses circulating in the Americas, where the bat-associated lineages form a monophyletic clade which is distinct from spillover and host shift events associated with the Arctic and the Cosmopolitan canine lineages of rabies virus [5], with greater influence of the Arctic variant for temperate regions of North America [30, 31]. The raccoon (*Procyon lotor*) variant of rabies virus shares a common ancestor with the south-central skunk variant, and both share closest ancestry to bat rabies viruses [32, 33], with reports of a rare recombination mixed ancestry event [34, 35].

Host Shifts

The evidence regarding the host range of rabies virus in the Americas demonstrates that despite being coded by a small (~12 kb) and conserved RNA genome comprised of five genes under strong purifying selection, the virus can adapt to taxonomically diverse wildlife reservoirs within variable landscapes and ecological communities [33, 36, 37]. The potential for adaptation stems in part from a lack of RNA proofreading by the virus-dependent polymerase during replication within host cells, which is present across the Order Mononegavirales and contributes to generation of quasi-species during replication [38, 39]. Transmission of rabies virus from infected to susceptible hosts is under selective pressure during neuronal travel with limited replication in and excretion from the salivary glands into the saliva and transferred to the new host through bite or mucous membrane contact. Iteration of this transmission process may fix mostly synonymous nucleotide polymorphisms with or without functional significance amongst the backdrop of quasi-species generated through replication, resulting in a bottleneck or restricted transmission of virus genotypic diversity passed on to the next host [8, 40, 41]. Most of the spontaneous polymorphisms may be deleterious, yet the processes of lineage sorting and host adaptation cumulatively can result in a host switch with independent onward transmission, which contributes to our understanding of rabies virus epizootiology among domestic and wild mammal communities [39, 41–45].

Some have reported that certain rabies virus variants may be pre-adapted to switch from bat to carnivore hosts [44] or between carnivore hosts [45]. Another study documented periodic bouts of selection pressure post-invasion across novel hosts and by variable molecular pathways across New World bat reservoirs [43]. One clear driver for a host shift may involve repeated spillover infection of the novel host [31], as most low-frequency spillover infections are destined to result in dead-end infections [8, 46]. However, with regular spillover infections, it is possible that even rare virus genotypes may have an opportunity to become adapted and fixed [45]. These observations likely underly a multi-factor spectrum, rather than an exclusive single factor, governing rabies virus host range, ecology, and evolution in the Americas.

Analysis of the host range and switching phenomena has also revealed unifying evolutionary constraints, whereby bats and carnivores share common patterns and factors influencing rabies virus infections and host range [16], consistent with factors governing broader patterns of virus richness and host range across mammals [18, 47, 48]. Phylogenetic relatedness and geographic range overlap have been implicated as drivers of rabies virus host range and switching. Although all mammals are susceptible to rabies virus infection, the ability of the virus to perpetuate in novel hosts outside of bats and carnivores appears restricted [8]. Notable exceptions have been documented and may point to ecological opportunity [49]. Even within the bats and carnivores, only a small fraction of the described species are known or potential rabies virus reservoirs [16]. Population and community ecology will continue to play an important role in the wildlife rabies virus emergence process, with emphasis on factors that bring animal hosts into greater contact such as geographic range overlap and ecological similarity. Factors may also include multi-species roosting aggregations or swarming behaviors among bats and increased dietary flexibility among carnivores (e.g. scavenging) [50].

Host and Population Range

While several wildlife species are known reservoirs of rabies virus variants with broad distribution in the Americas (Fig. 4.1), many species are not recognized reservoirs. Several host species are not recognized to be reservoirs throughout their natural (e.g. red fox, *Vulpes vulpes*) [31] or introduced geographic range (e.g. small Indian mongoose; *Urva auropunctata*) [51]. The landscape level environmental conditions under which the virus can perpetuate in reservoir host populations can be studied with climate or ecological niche modeling to reveal the importance of intra- and inter-species interactions in processes of maintenance [52–55]. Furthermore, management has restricted the circulation of specific rabies viruses (e.g., dog-coyote, Arctic fox [*Vulpes lagopus*] and raccoon variants) in North America among targeted wildlife populations through oral rabies vaccination programs, yet other carnivore and bat rabies virus variants represent stable long-term foci in the Americas [15, 33, 56, 57]. In contrast, certain rabies viruses, such as south-central skunk and common vampire bat (*Desmodus rotundus*) variants continue to emerge and re-emerge among novel populations of the reservoir species, with implications for spillover to domestic animals and exposure to humans [58–60] and impacts on wildlife conservation [61, 62].

Novel variants have emerged from bat and canine rabies virus lineages with spillover to other wildlife in the Americas [23, 63–67]. Hampering the understanding rabies virus host range among bats and other wildlife is a lack of routine species identification of specimens submitted for diagnosis and virus characterization of positive cases across the Americas [68–70], where approximately one third of positive cases in terrestrial mammals may be typed to determine the rabies virus

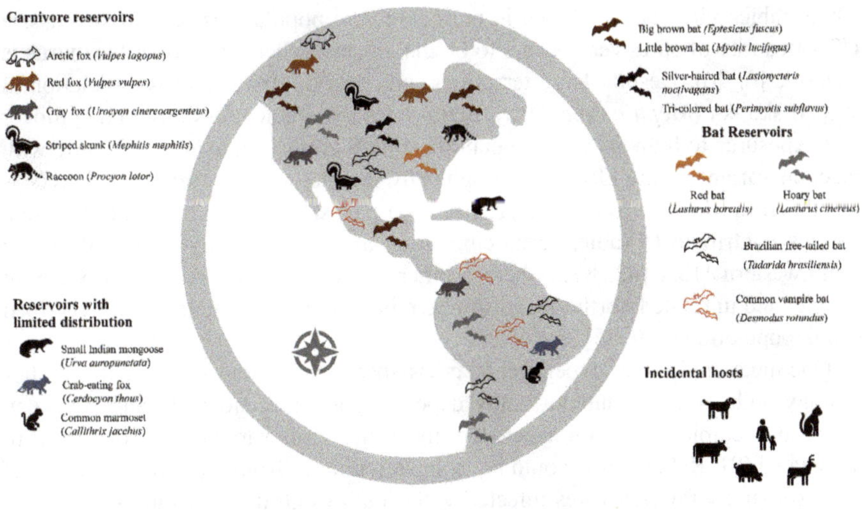

Fig. 4.1 Examples of common reservoirs of rabies virus throughout the Americas

variant in the United States during 2010–2015 [71]. Improving the specificity of the data collected through the national and regional passive public health surveillance systems could benefit our knowledge of host range, and intra-host population range, of rabies virus infections among wildlife for effective risk communication to public and animal health authorities [72, 73].

Ecology and Epizootiology of Virus Circulation

Despite imperfect surveillance, there are recognizable similarities and differences between the processes governing circulation in the temperate versus tropical rabies reservoir host populations [16, 74], including the evidence of abortive infections among reservoir hosts, and vector host species [75]. Among temperate bat and carnivore populations, circulation appears to be driven largely by interactions of a synchronized annual birth pulse of susceptible young as well as winter host heterothermy with overwintering of infections. This contrasts with less synchronized birth processes and greater importance of spatial connectivity among host populations inhabiting tropical areas [59, 76–80]. Mega-carnivores do not appear to be routinely reported with rabies virus infections across temperate to Arctic areas despite serving as potential vectors of exposure to humans, livestock and/or pets [59]. This phenomenon is likely related to the population fecundity and dietary flexibility of smaller bodied mesocarnivores [16].

From a community perspective, differences among bat and mesocarnivore rabies virus circulation principally involve different levels of ecological interaction with and spillover infections to other mammals, with implications for human exposures and infection [6, 7, 67]. Spillover infections are much more frequently reported where rabies virus is circulating in mesocarnivore populations, and in the eastern USA this rate of spillover appears to be the greatest where raccoon rabies virus is enzootic [6, 46]. Despite large territories with enzootic rabies virus circulation in striped skunks (*Mephitis mephitis*), these areas experience less reported spillover and exposures to humans and companion animals compared to areas enzootic with raccoon variant in the USA. Geographically distinct skunk reservoirs are recognized across Mexico [81], and there has been past speculation that multiple skunk species contribute to rabies virus circulation as a "community reservoir" in the Southwestern USA [82, 83], whereas other areas of the mid-western USA have distinct and unrelated northern and southern infection cycles maintained by striped skunk populations [59, 84].

One meta-analysis of experimental cross-species infections revealed that pathobiology and infection outcome may depend upon the genetic distance between donor and recipients as well as whether the rabies virus variant is bat or carnivore associated [8]. Rabies virus could be isolated regularly from the salivary glands of domestic and wild carnivores infected with South Central skunk rabies virus suggesting that related vector hosts may be competent for onward transmission [85]. As such, social contacts may drive patterns of conspecific transmission among wildlife.

Areas where rabies virus circulates in insectivorous bats generally have much lower levels of spillover infection to other mammals and exposures to humans and pets, yet insectivorous bat rabies viruses continue to be the leading cause of human rabies in temperate regions of the Americas [6, 55, 86]. In contrast, the circulation of rabies virus in common vampire bats leads to a fundamentally different exposure profile for domestic animals and humans due to their obligate behavior of feeding on mammalian blood [7, 67]. Insufficient information has been documented regarding fruit bat reservoirs in the Americas, yet certain fruit bats species can be locally abundant across disturbed areas and may commonly roost in homes or buildings in the Neotropics [87].

Species richness among mammals is primarily driven by environmental and biological factors, including elevation, home range size and trophic level in the continental USA [88]. Anthropogenic land use negatively impacted total mammal species richness in the USA, but rabies virus emergence (i.e., outbreaks) associated with wildlife reservoirs may be more prone to occur during anthropogenic land use change or conversion [89]. Greater dietary breadth has been linked to a reservoir status among 23 carnivore species globally, in addition to faster age to sexual maturity and production of greater litter size(s), helping to explain common intrinsic host traits likely to be important for rabies virus transmission and perpetuation among wild carnivore populations in the Americas [16]. This finding also helps to explain why only mesocarnivores have been reported as rabies virus reservoirs, in contrast to true carnivores. Certain bat and mesocarnivore reservoirs are ecological generalists, and may be locally abundant across rural and developed areas (e.g., raccoons, coyotes, big brown bats (*Eptesicus fuscus*), Brazilian free-tailed bats (*Tadarida brasiliensis*) [90–95]).

Multiple species of insectivorous bats and the common vampire bat have long been recognized as reservoirs of rabies virus in the Americas [2, 73, 96, 97]. Among 29 bat reservoirs of rabies virus, spatial distribution in North America with greater geographic range area and litter size were associated with reservoir status [16]. Having a greater geographic range may present greater opportunities for inter- and intra-specific viral exchange and perpetuation, with reduced virus lineage extinction that may be more common among bat species with limited geographic distribution(s) [42, 69]. Most of the taxonomic diversity of bats inhabiting temperate North America are insectivorous in diet. However, among the mostly frugivorous and tropically distributed Phyllostomidae, only the Jamaican fruit-eating bat (*Artibeus jamaicensis*) and great fruit-eating bat (*A. lituratus)* have commonly been reported to be infected with rabies virus, yet are infrequently implicated as reservoirs of independent circulation [73, 98–102].

In North America, raccoons, striped skunks, coyotes, arctic foxes, red foxes and gray foxes (*Urocyon cinereoargenteus*), as well as invasive small Indian mongooses in the Caribbean region, all are historically recognized wildlife rabies reservoirs [30, 31, 58, 103–109]. At least one focus of gray fox rabies virus may have naturally went extinct in eastern North America [110], potentially due to competition via an emerging raccoon rabies virus epizootic in the mid-Atlantic US starting in the late 1970s. Other unmanaged regional and circumpolar foci have remained relatively stable through time [15]. In the southcentral region of the USA and northern Mexico, a canine variant of rabies virus shifted into coyote populations in Mexico and spread to the USA but was intensively managed in the USA and possibly may have gone extinct in Northern Mexico, as supported by public health surveillance regarding the elimination of human rabies deaths mediated by dogs in Mexico. Gray foxes have also been long-standing primary targets for rabies virus management in the southcentral and southwestern USA [111–115]. In South America, there have been a number of host shifts of canine rabies virus into wild carnivore populations, including gray fox populations in Colombia [116], and crab-eating foxes (*Cerdocyon thous*) and hoary foxes (*Lycalopex vetulus*) in Brazil [117, 118]. The maintenance of robust surveillance during efforts to eliminate canine rabies virus in the Americas will be crucial to detect other emergent host shifts of this lineage into wildlife populations [67, 70].

The small Indian mongoose was introduced to and became established on many Caribbean islands [119]. Populations of small Indian mongooses can be locally abundant in protected areas and mongooses tend to have small home ranges where abundant [120–122]. The emergence of rabies in the small Indian mongoose within a subset of Caribbean islands has been described as a series of host shifts resulting from local spillover of canine rabies virus variants from domestic dogs that were introduced to different islands of the region, including Cuba, Hispaniola, Puerto Rico and Grenada [51, 107]. Several Caribbean islands with small Indian mongoose are apparently free of rabies (e.g. US Virgin Islands, [123]). The variable emergence of rabies virus in small Indian mongoose populations across the Caribbean may be due to the absence of canine rabies introduction and establishment among certain islands and/or main biogeographic invasion pathways associated with oceanic exploration [24, 119].

Other Mammals and Vertebrates

Both bats and rodents are speciose orders of mammals, which begs the question of why rabies virus does not perpetuate among rodents. Outside of infrequent spillover cases and rarer cases of oral rabies vaccine spillover, rodents are generally considered a low-risk wildlife host for rabies virus infections [124, 125]. This may partly stem from the fact that rodents are not closely related to bats or carnivores [10]. Moreover, the movements and vagility of rodents is reduced on average compared to most bats, which may limit opportunities for virus spread and perpetuation across rodent communities [47]. Wild rodents that share ecological similarity or overlap with mesocarnivore reservoir hosts or populations may be more likely to reported with spillover infections (e.g. woodchuck, *Marmota monax*; [125]), as are wild rodents subject to vampire bat predation (e.g. capybara *Hydrochoeris hidrochaeris*) [126]. Despite deep evolutionarily divergence from bats and carnivores, marsupials such as the Virginia opossum (*Didelphis virginiana*) may exhibit ecological strategies similar to mesocarnivores in the North America, which make them important nontarget competitors for oral rabies vaccine baits and hosts to spillover infections of rabies virus (Slate et al. 2020). Nevertheless, the Virginia opossum reportedly has reduced susceptibility to and is considered a dead-end host of rabies virus infection [127]. Another marsupial with strong evidence of rabies virus exposure is the big-eared opossum (*Didelphis aurita*) in Brazil [128]. In general, wild mammals other than carnivores or bats are infrequently reported as infected with rabies virus in the Americas [15].

One recent study reported a lower viral richness among birds compared to mammals [18]. There is unlikely to be a substantial number of spillover cases and pressure to establish a reservoir among birds, even raptors. Birds may scavenge on rabies virus-infected carcasses and can be infected with rabies virus, but the likelihood of being exposed to a sufficient infectious dose in nature may be low by this route [129, 130]. Most vertebrates possess salivary glands but only mammals have true compound and compact salivary glands among vertebrates [131], which may be important for rabies virus dissemination through the saliva among infected vertebrates. While the evolution of teeth and jaw structures among vertebrates remains contested [132], it is noteworthy that modern birds lack teeth capable of percutaneous bite contact which is the primary mechanism of rabies virus transmission.

Future Ecology of Rabies Viruses in the Americas

Emergence phenomena have been described regularly throughout the history of rabies in the Americas. In some instances, the occurrence of traveling wave phenomena and host biology or community ecology have been used to predict where infections will occur. Many other aspects of emergence have been understudied (i.e., regional host population maintenance) and thus not easily predicted [133–136]. Enhanced rabies surveillance attention may prioritize taxa with ecological and evolutionary similar traits to known rabies virus reservoirs [16], for a broader understanding of the roles of mammalian community ecology and evolution in perpetuating this highly lethal zoonosis in the Americas.

Conclusions

Significant gains have been made in the elimination of human rabies deaths due to canine rabies virus circulation in the Americas. Further success will be realized by the continuity of robust laboratory-based surveillance efforts to confirm elimination from target domestic and wild animal populations. Throughout the Americas, there is incomplete and inconsistent surveillance effort focused on wildlife. Complete elimination of rabies virus from dog populations in the Americas may incidentally increase pressure for shifts of rabies virus to novel wildlife hosts in areas with heightened contact and opportunity for spillover infections. A shifting and dynamic landscape and community of susceptible wildlife populations will necessitate new strategies for the prevention and control of human and domestic animal rabies virus exposures associated with wildlife rabies circulation. Bats and carnivores will continue to serve as more common wildlife rabies reservoirs, but complex and varied intra- and inter-specific community interactions are also influential to explain the transmission and distribution of viral circulation across wildlife reservoir populations, species and communities of mammals in the Americas. Given the recognized dynamic host range of rabies virus, it is unlikely that true eradication in the Americas will ever be achieved. However, the costs of living with rabies (i.e., exposures, morbidity and mortality) can all be successfully reduced through control and management actions targeting domestic and wild carnivore reservoirs and public education bolstered in at-risk areas to prevent human-animal-wildlife contacts and exposures, particularly among diverse and broadly distributed bat rabies virus reservoirs.

References

1. Rupprecht CE, Turmelle A, Kuzmin IV (2011) A perspective on lyssavirus emergence and perpetuation. Curr Opin Virol 1(6):662–670
2. Banyard A et al (2020) Bat rabies. In: Fooks AR, Jackson A (eds) Rabies: Scientific basis of the disease and its management. Academic Press, pp 231–276
3. Olival KJ et al (2020) Possibility for reverse zoonotic transmission of SARS-CoV-2 to free-ranging wildlife: a case study of bats. PLoS Pathog 16(9):e1008758
4. Leopardi S, Blake D, Puechmaille SJ (2015) White-nose syndrome fungus introduced from Europe to North America. Curr Biol 25(6):R217–R219
5. Troupin C et al (2016) Large-scale Phylogenomic analysis reveals the complex evolutionary history of rabies virus in multiple carnivore hosts. PLoS Pathog 12(12):e1006041
6. Pieracci EG et al (2019) Vital signs: trends in human rabies deaths and exposures – United States, 1938-2018. MMWR Morb Mortal Wkly Rep 68(23):524–528
7. Schneider MC et al (2009) Rabies transmitted by vampire bats to humans: an emerging zoonotic disease in Latin America? Rev Panam Salud Publica 25(3):260–269
8. Mollentze N et al (2020) Virulence mismatches in index hosts shape the outcomes of cross-species transmission. Proc Natl Acad Sci U S A 117(46):28859–28866
9. Springer MS et al (2001) Integrated fossil and molecular data reconstruct bat echolocation. Proc Natl Acad Sci U S A 98(11):6241–6246
10. Jebb D et al (2020) Six reference-quality genomes reveal evolution of bat adaptations. Nature 583(7817):578–584
11. Amador LI, Almeida F, Giannini N (2020) Evolution of traditional aerodynamic variables in bats (Mammalia: Chiroptera) within a comprehensive phylogenetic framework. J Mamm Evol 27:549–561
12. Lim BK (2009) Review of the origins and biogeography of bats in South America. Chiroptera Neotropical 15:391–410
13. Wilson DE, Reeder DM (2005) Mammal species of the world. A taxonomic and geographic reference. Johns Hopkins University Press, p 2142
14. Hoofer SR, Van den Bussche RA (2003) Molecular phylogenetics of the chiropteran family Vespertilionidae. Acta Chiropterol 5:1–59
15. Gilbert A (2018) Rabies virus vectors and reservoir species. Rev Sci Tech 37(2):371–384
16. Worsley-Tonks KEL et al (2020) Using host traits to predict reservoir host species of rabies virus. PLoS Negl Trop Dis 14(12):e0008940

17. Streicker DG, Gilbert AT (2020) Contextualizing bats as viral reservoirs. Science 370(6513):172–173
18. Mollentze N, Streicker DG (2020) Viral zoonotic risk is homogenous among taxonomic orders of mammalian and avian reservoir hosts. Proc Natl Acad Sci U S A 117(17):9423–9430
19. Pires MM, Silvestro D, Quental TB (2015) Continental faunal exchange and the asymmetrical radiation of carnivores. Proc Biol Sci 282(1817):20151952
20. Pires MM, Silvestro D, Quental TB (2017) Interactions within and between clades shaped the diversification of terrestrial carnivores. Evolution 71(7):1855–1864
21. Krebs JW et al (2003) Rabies among infrequently reported mammalian carnivores in the United States, 1960–2000. J Wildl Dis 39(2):253–261
22. Nyakatura K, Bininda-Emonds OR (2012) Updating the evolutionary history of Carnivora (Mammalia): a new species-level supertree complete with divergence time estimates. BMC Biol 10:12
23. Dell'Armelina Rocha PR et al (2020) Unexpected rabies variant identified in kinkajou (Potos flavus), Mato Grosso, Brazil. Emerg Microbes Infect 9(1):851–854
24. Louppe V et al (2020) The globally invasive small Indian mongoose Urva auropunctata is likely to spread with climate change. Sci Rep 10(1):7461
25. Louppe V et al (2019) Current and future climatic regions favourable for a globally introduced wild carnivore, the raccoon Procyon lotor. Sci Rep 9(1):9174
26. Tsuchiya MTN et al (2020) Whole genome sequencing of procyonids reveals distinct demographic histories in kinkajou (Potos flavus) and northern raccoon (Procyon lotor). Genome Biol Evol 13(1):evaa255
27. Chipman R et al (2008) Downside risk of wildlife translocation. Dev Biol 131:223–232
28. Berentsen A, Pitt WC, Sugihara R (2018) Ecology of the small Indian mongoose (Herpestes auropunctatus) in North America. In: Pitt WC, Beasley JC, Witmer GC (eds) Ecology and Management of Terrestrial Invasive Species in the United States. CRC Press, Boca Raton, pp 251–267
29. Hayman DT et al (2016) The global Phylogeography of lyssaviruses – challenging the 'Out of Africa' hypothesis. PLoS Negl Trop Dis 10(12):e0005266
30. Hanke D et al (2016) Spatio-temporal analysis of the genetic diversity of Arctic rabies viruses and their reservoir hosts in Greenland. PLoS Negl Trop Dis 10(7):e0004779
31. Nadin-Davis SA, Fehlner-Gardiner C (2019) Origins of the arctic fox variant rabies viruses responsible for recent cases of the disease in southern Ontario. PLoS Negl Trop Dis 13(9):e0007699
32. Kuzmina NA et al (2013) A reassessment of the evolutionary timescale of bat rabies viruses based upon glycoprotein gene sequences. Virus Genes 47(2):305–310
33. Streicker D, Biek R (2020) Evolution of rabies virus. In: Fooks AR, Jackson A (eds) Rabies: Scientific basis of the disease and its management. Academic Press, pp 83–101
34. Ding NZ et al (2017) A permanent host shift of rabies virus from Chiroptera to Carnivora associated with recombination. Sci Rep 7(1):289
35. Deviatkin AA, Lukashev AN (2018) Recombination in the rabies virus and other lyssaviruses. Infect Genet Evol 60:97–102
36. Mollentze N, Biek R, Streicker DG (2014) The role of viral evolution in rabies host shifts and emergence. Curr Opin Virol 8:68–72
37. Fisher CR, Streicker DG, Schnell MJ (2018) The spread and evolution of rabies virus: conquering new frontiers. Nat Rev Microbiol 16(4):241–255
38. Kissi B et al (1999) Dynamics of rabies virus quasispecies during serial passages in heterologous hosts. J Gen Virol 80(Pt 8):2041–2050
39. Nadin-Davis S (2020) Molecular epidemiology. In: Fooks AR, Jackson A (eds) Rabies: Scientific basis of the disease and its management. Academic Press, pp 143–193
40. Holmes EC et al (2002) Genetic constraints and the adaptive evolution of rabies virus in nature. Virology 292(2):247–257
41. Marston DA et al (2017) Genetic analysis of a rabies virus host shift event reveals within-host viral dynamics in a new host. Virus Evol 3(2):vex038

42. Streicker DG et al (2010) Host phylogeny constrains cross-species emergence and establishment of rabies virus in bats. Science 329(5992):676–679
43. Streicker DG et al (2012) Variable evolutionary routes to host establishment across repeated rabies virus host shifts among bats. Proc Natl Acad Sci U S A 109(48):19715–19720
44. Kuzmin IV et al (2012) Molecular inferences suggest multiple host shifts of rabies viruses from bats to mesocarnivores in Arizona during 2001-2009. PLoS Pathog 8(6):e1002786
45. Borucki MK et al (2013) Ultra-deep sequencing of intra-host rabies virus populations during cross-species transmission. PLoS Negl Trop Dis 7(11):e2555
46. Wallace RM et al (2014) Right place, wrong species: a 20-year review of rabies virus cross species transmission among terrestrial mammals in the United States. PLoS One 9(10):e107539
47. Luis AD et al (2015) Network analysis of host-virus communities in bats and rodents reveals determinants of cross-species transmission. Ecol Lett 18(11):1153–1162
48. Albery GF et al (2020) Predicting the global mammalian viral sharing network using phylogeography. Nat Commun 11(1):2260
49. Kotait I et al (2019) Non-human primates as a reservoir for rabies virus in Brazil. Zoonoses Public Health 66(1):47–59
50. Theimer TC et al (2017) Ecological potential for rabies virus transmission via scavenging of dead bats by Mesocarnivores. J Wildl Dis 53(2):382–385
51. Seetahal JFR et al (2018) Rabies in the Caribbean: A situational analysis and historic review. Trop Med Infect Dis 3(3):89
52. Kim BI et al (2014) A conceptual model for the impact of climate change on fox rabies in Alaska, 1980–2010. Zoonoses Public Health 61(1):72–80
53. Huettmann F, Magnuson EE, Hueffer K (2017) Ecological niche modeling of rabies in the changing Arctic of Alaska. Acta Vet Scand 59(1):18
54. Hayes MA, Piaggio AJ (2018) Assessing the potential impacts of a changing climate on the distribution of a rabies virus vector. PLoS One 13(2):e0192887
55. Escobar LE et al (2013) Ecology and geography of transmission of two bat-borne rabies lineages in Chile. PLoS Negl Trop Dis 7(12):e2577
56. Gilbert A, Chipman R (2020) Rabies control in wild carnivores. In: Fooks AR, Jackson A (eds) Rabies: Scientific basis of the disease and its management. Academic Press, pp 605–654
57. Meske M et al (2021) Evolution of rabies in South America and inter-species dynamics (2009–2018). Trop Med Infect Dis 6:98
58. Kuzmina NA et al (2013) The phylogeography and spatiotemporal spread of south-central skunk rabies virus. PLoS One 8(12):e82348
59. Pepin KM et al (2017) Predicting spatial spread of rabies in skunk populations using surveillance data reported by the public. PLoS Negl Trop Dis 11(7):e0005822
60. Benavides JA et al (2020) Defining new pathways to manage the ongoing emergence of bat rabies in Latin America. Viruses 12(9):1002
61. O'Shea TJ et al (2016) Multiple mortality events in bats: a global review. Mammal Rev 46:175–190
62. Ellwanger JH, Chies JAB (2019) The triad "dogs, conservation and zoonotic diseases" – an old and still neglected problem in Brazil. Perspect Ecol Conserv 17(3):157–161
63. Garcés-Ayala F et al (2017) Molecular characterization of atypical antigenic variants of canine rabies virus reveals its reintroduction by wildlife vectors in southeastern Mexico. Arch Virol 162(12):3629–3637
64. Arechiga-Ceballos N et al (2010) New rabies virus variant found during an epizootic in white-nosed coatis from the Yucatan peninsula. Epidemiol Infect 138(11):1586–1589
65. Vargas-Linares E et al (2014) Rabia en Potos flavus identificados en el departamento de madre de dios, Perú. Rev Peru Med Exp Salud Publica 31:88–93
66. Kobayashi Y et al (2013) Isolation of a phylogenetically distinct rabies virus from a tufted capuchin monkey (Cebus apella) in Brazil. Virus Res 178(2):535–538
67. Rocha SM et al (2017) Epidemiological profile of wild rabies in Brazil (2002–2012). Transbound Emerg Dis 64(2):624–633

68. Pieracci EG et al (2020) Evaluation of species identification and rabies virus characterization among bat rabies cases in the United States. J Am Vet Med Assoc 256(1):77–84
69. Nadin-Davis S, Alnabelseya N, Knowles MK (2017) The phylogeography of Myotis bat-associated rabies viruses across Canada. PLoS Negl Trop Dis 11(5):e0005541
70. Freire de Carvalho M et al (2018) Rabies in the Americas: 1998–2014. PLoS Negl Trop Dis 12(3):e0006271
71. Pieracci EG et al (2020) Evaluation of rabies virus characterization to enhance early detection of important rabies epizootic events in the United States. J Am Vet Med Assoc 256(1):66–76
72. Escobar LE, Peterson AT (2013) Spatial epidemiology of bat-borne rabies in Colombia. Rev Panam Salud Publica 34(2):135–136
73. Constantine DG (2009) Bat rabies and other lyssavirus infections, in Circular 1329. US Geological Survey, Fort Collins, p 68
74. Streicker DG et al (2012) Rates of viral evolution are linked to host geography in bat rabies. PLoS Pathog 8(5):e1002720
75. Gold S et al (2020) Rabies virus-neutralising antibodies in healthy, unvaccinated individuals: what do they mean for rabies epidemiology? PLoS Negl Trop Dis 14(2):e0007933
76. George DB et al (2011) Host and viral ecology determine bat rabies seasonality and maintenance. Proc Natl Acad Sci U S A 108(25):10208–10213
77. Blackwood JC et al (2013) Resolving the roles of immunity, pathogenesis, and immigration for rabies persistence in vampire bats. Proc Natl Acad Sci U S A 110(51):20837–20842
78. McClure KM et al (2020) Variation in host home range size decreases rabies vaccination effectiveness by increasing the spatial spread of rabies virus. J Anim Ecol 89(6):1375–1386
79. Davis AD et al (2016) Overwintering of rabies virus in silver haired bats (Lasionycteris noctivagans). PLoS One 11(5):e0155542
80. Theimer TC et al (2017) Den use and heterothermy during winter in free-living, suburban, striped skunks. J Mammal 98(3):867–873
81. Müller T, Freuling C (2020) Rabies in terrestrial animals. In: Fooks AR, Jackson AC (eds) Rabies: Scientific Basis of the Disease and Its Management. Acadmic Press, pp 195–230
82. Dragoo JW et al (2004) Identification of skunk species submitted for rabies testing in the desert southwest. J Wildl Dis 40(2):371–376
83. Hass CC, Dragoo JW (2006) Rabies in hooded and striped skunks in Arizona. J Wildl Dis 42(4):825–829
84. Oertli EH et al (2009) Epidemiology of rabies in skunks in Texas. J Am Vet Med Assoc 234(5):616–620
85. Jimenez I et al (2019) Isolation of rabies virus from the salivary glands of wild and domestic carnivores during a skunk rabies epizootic. J Wildl Dis 55(2):473–476
86. Fenton MB, Jackson AC, Faure PA (2020) Bat bites and rabies: the Canadian scene. Facets 5(1):367–380
87. Dias RA et al (2019) Spatiotemporal distribution of a non-haematophagous bat community and rabies virus circulation: a proposal for urban rabies surveillance in Brazil. Epidemiol Infect 147:e130
88. Howard C, Flather CH, Stephens PA (2019) What drives at-risk species richness? Environmental factors are more influential than anthropogenic factors or biological traits. Conserv Lett 12(2):e12624
89. Stoner-Duncan B, Streicker DG, Tedeschi CM (2014) Vampire bats and rabies: toward an ecological solution to a public health problem. PLoS Negl Trop Dis 8(6):e2867
90. Hadidian J et al (2010) Raccoons (Procyon lotor). In: Gehrt SD, Riley SP, Cypher BL (eds) Urban carnivores: ecology, conflict, and conservation, vol 35-47. The Johns Hopkins University Press, Baltimore
91. Breck SW et al (2019) The intrepid urban coyote: a comparison of bold and exploratory behavior in coyotes from urban and rural environments. Sci Rep 9(1):2104
92. Turmelle AS et al (2010) Ecology of rabies virus exposure in colonies of Brazilian free-tailed bats (Tadarida brasiliensis) at natural and man-made roosts in Texas. Vector Borne Zoonotic Dis 10:165–175

93. Streicker DG et al (2013) Anthropogenic roost switching and rabies virus dynamics in house-roosting big brown bats. Vector Borne Zoonotic Dis 13(7):498–504
94. Leslie MJ et al (2006) Bat-associated rabies virus in skunks. Emerg Infect Dis 12(8):1274–1277
95. O'Shea TJ et al (2014) Variability in seroprevalence of rabies virus neutralizing antibodies and associated factors in a Colorado population of big brown bats (Eptesicus fuscus). PLoS One 9(1):e86261
96. Constantine DG (1967) Bat rabies in the southwestern United States. Public Health Rep 82(10):867–888
97. Seetahal JFR et al (2019) Of bats and livestock: the epidemiology of rabies in Trinidad, West Indies. Vet Microbiol 228:93–100
98. Escobar LE et al (2015) Bat-borne rabies in Latin America. Rev Inst Med Trop Sao Paulo 57(1):63–72
99. Calderon A et al (2019) Frugivorous bats in the Colombian Caribbean region are reservoirs of the rabies virus. Ann Clin Microbiol Antimicrob 18(1):11
100. Zieger U et al (2017) Natural exposure of bats in Grenada to rabies virus. Infect Ecol Epidemiol 7(1):1332935
101. Seetahal JFR et al (2020) The serological prevalence of rabies virus-neutralizing antibodies in the bat population on the Caribbean island of Trinidad. Viruses 12(2):178
102. Fahl WO et al (2012) Desmodus rotundus and Artibeus spp. bats might present distinct rabies virus lineages. Braz J Infect Dis 16(6):545–551
103. Rosatte RC et al (2007) Elimination of arctic variant rabies in red foxes, metropolitan Toronto. Emerg Infect Dis 13(1):25–27
104. Bigler WJ, McLean RG, Trevino HA (1973) Epizootiologic aspects of raccoon rabies in Florida. Am J Epidemiol 98(5):326–335
105. Everard CO, Everard JD (1988) Mongoose rabies. Rev Infect Dis 10(Suppl 4):S610–S614
106. Kuzmin IV et al (2008) Arctic and Arctic-like rabies viruses: distribution, phylogeny and evolutionary history. Epidemiol Infect 136(4):509–519
107. Zieger U et al (2014) The phylogeography of rabies in Grenada, West Indies, and implications for control. PLoS Negl Trop Dis 8(10):e3251
108. Davis R et al (2013) Genetic characterization and phylogenetic analysis of skunk-associated rabies viruses in North America with special emphasis on the central plains. Virus Res 174(1–2):27–36
109. Gremillion-Smith C, Woolf A (1988) Epizootiology of skunk rabies in North America. J Wildl Dis 24(4):620–626
110. Carey AB, Giles RH, McLean RG (1978) The landscape epidemiology of rabies in Virginia. Am J Trop Med Hyg 27(3):573–580
111. Maki J et al (2017) Oral vaccination of wildlife using a vaccinia-rabies-glycoprotein recombinant virus vaccine (RABORAL V-RG®): a global review. Vet Res 48(1):57
112. Sidwa TJ et al (2005) Evaluation of oral rabies vaccination programs for control of rabies epizootics in coyotes and gray foxes: 1995-2003. J Am Vet Med Assoc 227(5):785–792
113. Clark KA et al (1994) Epizootic canine rabies transmitted by coyotes in south Texas. J Am Vet Med Assoc 204(4):536–540
114. Fearneyhough MG et al (1998) Results of an oral rabies vaccination program for coyotes. J Am Vet Med Assoc 212(4):498–502
115. Velasco-Villa A et al (2008) Enzootic rabies elimination from dogs and reemergence in wild terrestrial carnivores, United States. Emerg Infect Dis 14(12):1849–1854
116. Paez A et al (2007) Molecular epidemiology of rabies in Colombia 1994-2005 based on partial nucleoprotein gene sequences. Virus Res 130(1–2):172–181
117. Carnieli P Jr et al (2008) Characterization of rabies virus isolated from canids and identification of the main wild canid host in northeastern Brazil. Virus Res 131(1):33–46
118. Favoretto SR et al (2013) The emergence of wildlife species as a source of human rabies infection in Brazil. Epidemiol Infect 141(7):1552–1561

119. Louppe V et al (2021) Dispersal history of a globally introduced carnivore, the small Indian mongoose Urva auropunctata, with an emphasis on the Caribbean region. Biol Invasions 23:2573–2590
120. Johnson SR et al (2016) Estimates of small Indian mongoose densities: implications for rabies management. J Wildl Manag 80(1):37–47
121. Berentsen AR et al (2020) Home range estimates for small Indian mongooses (Urva auropunctata) in southwestern Puerto Rico. Caribb J Sci 50(2):225–235
122. Pitt WC, Sugihara R, Berentsen A (2015) Effect of travel distance, home range, and bait on the management of small Indian mongooses, *Herpestes auropunctatus*. Biol. Invasions 17(6):1743–1759
123. Browne AS et al (2021) Determination of freedom-from-rabies for mongoose populations in the United States Virgin Islands, 2019–2020. PLoS Negl Trop Dis 15(7):e0009536
124. WHO (2018) WHO expert consultation on rabies, third report, in WHO Technical Report Series. World Health Organization, Geneva, p 195
125. Fitzpatrick JL et al (2014) Rabies in rodents and lagomorphs in the United States, 1995–2010. J Am Vet Med Assoc 245(3):333–337
126. Marcelo Azevedo de Paula Antunes J et al (2017) Rabies virus exposure of Brazilian free-ranging wildlife from municipalities without clinical cases in humans or in terrestrial wildlife. J Wildl Dis 53(3):662–666
127. Constantine DG (1966) Transmission experiments with bat rabies isolates: reaction of certain Carnivora, opossum, and bats to intramuscular inoculations of rabies virus isolated from free-tailed bats. Am J Vet Res 27(116):16–19
128. Araujo DB et al (2014) Antibodies to rabies virus in terrestrial wild mammals in native rainforest on the north coast of Sao Paulo State, Brazil. J Wildl Dis 50(3):469–477
129. Gough PM, Jorgenson RD (1976) Rabies antibodies in sera of wild birds. J Wildl Dis 12(3):392–395
130. Shannon LM et al (1988) Serological survey for rabies antibodies in raptors from California. J Wildl Dis 24(2):264–267
131. Tucker R (1958) Taxonomy of the salivary glands of vertebrates. Syst Zool 7(2):74–83
132. Rucklin M et al (2012) Development of teeth and jaws in the earliest jawed vertebrates. Nature 491(7426):748–751
133. Streicker DG et al (2016) Host-pathogen evolutionary signatures reveal dynamics and future invasions of vampire bat rabies. Proc Natl Acad Sci U S A 113(39):10926–10931
134. Childs JE et al (2000) Predicting the local dynamics of epizootic rabies among raccoons in the United States. Proc Natl Acad Sci U S A 97(25):13666–13671
135. Benavides JA, Valderrama W, Streicker DG (2016) Spatial expansions and travelling waves of rabies in vamipre bats. Proc R Soc Lond Ser B Biol Sci 283(1832):20160328
136. Trewby H et al (2017) Processes underlying rabies virus incursions across US-Canada border as revealed by whole-genome Phylogeography. Emerg Infect Dis 23(9):1454–1461

Chapter 5
Historical Laboratory Contributions Supporting Rabies Diagnosis and Disease Prevention and Control in the Americas

Susan A. Nadin-Davis, Lillian A. Orciari, Elaine R. Fernandes, and Pamela A. Yager

Abstract Rabies was one of the first documented diseases in history. However, rapid, sensitive and specific methods of detection of the viral etiological agent have become available only in relatively modern times. The historical account of the emergence of current laboratory techniques of rabies virus detection and characterization is outlined in this chapter, in relation to disease prevention and control in the Americas. Up until the start of the twentieth century, rabies was diagnosed by a set of non-specific clinical signs. The development of improved microscopy facilitated a more scientific approach with the histological identification of inclusions (Negri bodies) in the brains of infected animals, with later visualization by electron microscopy, which improved after the ability to passage the virus in experimental animals and cell culture. Development of the direct fluorescent antibody (DFA) method in the 1950s led to a major paradigm shift and enabled accurate and specific detection of rabies virus antigens. Subsequently, many advances in the life sciences furthered our capability to characterize rabies virus variants. Hybridoma technology enabled the generation of monoclonal antibodies, further improving the DFA test and facilitating discrimination of viral variants through antigenic typing. Additional progress occurred with the invention of the polymerase chain reaction (PCR) which permitted highly sensitive and specific detection of viral nucleic acids and opened the door to

---

Author Susan A. Nadin-Davis is retired.

S. A. Nadin-Davis (✉)
National Reference Centre for Rabies, Ottawa Laboratory Fallowfield, Canadian Food Inspection Agency, Ottawa, ON, Canada

L. A. Orciari · P. A. Yager
Poxvirus and Rabies Branch, Division of High Consequence Pathogens and Pathology, Centers for Disease Control and Prevention, Atlanta, GA, USA

E. R. Fernandes
Pasteur Institute, Division of Rabies Diagnosis, Laboratory of Immunohistochemistry and Laboratory of Serology, São Paulo, SP, Brazil

C. E. Rupprecht (ed.), *History of Rabies in the Americas: From the Pre-Columbian to the Present, Volume I*, Fascinating Life Sciences,
https://doi.org/10.1007/978-3-031-25052-1_5

extensive molecular epidemiological studies of rabies virus variants, thereby expanding our knowledge of the diversity of these viruses across the Americas and providing insight into viral spread to aid in prevention and control. A plethora of other laboratory techniques, including several serological methods to measure the immune status of subjects, aided such efforts. More recent tests, including the direct rapid immunohistochemistry test (DRIT) and other rapid point of care methods still under evaluation, extend diagnostic capability outside of central laboratories, enabling accurate and rapid assessment in more remote regions. Improvement of existing tests, as well as development of new diagnostic methods, improve enhanced laboratory-based surveillance and better support management of this disease from the Arctic to Amazonia.

Keywords Diagnosis · Laboratory · Microscopy · Molecular biology · Sequencing · Virology

Introduction

Rabies is one of the most feared of all diseases. Reference to dog bites in the Eshnunna Code (>2000 BCE) suggests this disease was recognized even in ancient times. However, rapid and reliable laboratory confirmation of rabies in humans and animals, essential to making post-exposure prophylaxis decisions, was not possible until the twentieth century (Fig. 5.1). This chapter reviews the history of laboratory diagnosis, with emphasis on the key contributions made by scientists in the Americas (Fig. 5.2). The WHO/OIE Collaborating Centers for Rabies in the Americas, regional state public health and veterinary diagnostic laboratories, and research institutions within the countries of the region, as well as international conferences for sharing of current developments and research, such as the Rabies in the Americas (RITA) annual meeting, played important roles in the advancement of knowledge within the rabies field. This chapter also includes brief historical background in each section, starting with clinical diagnosis, and is divided into parts according to methodologies in chronological order of discovery, with updates or current usages as applicable.

Diagnosis by Observation of Clinical Signs

Historically, diagnosis of rabies cases was made by observation of clinical signs in the suspect animal. Identification of the biting animal was essential. The most notable and easily recognized signs of rabies in animals were aggressive or erratic behavior, excessive salivation, paresis (itching) and paralysis, tremors, change in the tone of the bark or vocalization, loss of appetite (aversion to drink and eat) and death within 10 days of illness onset. The earliest attempts at post-mortem diagnosis of suspect

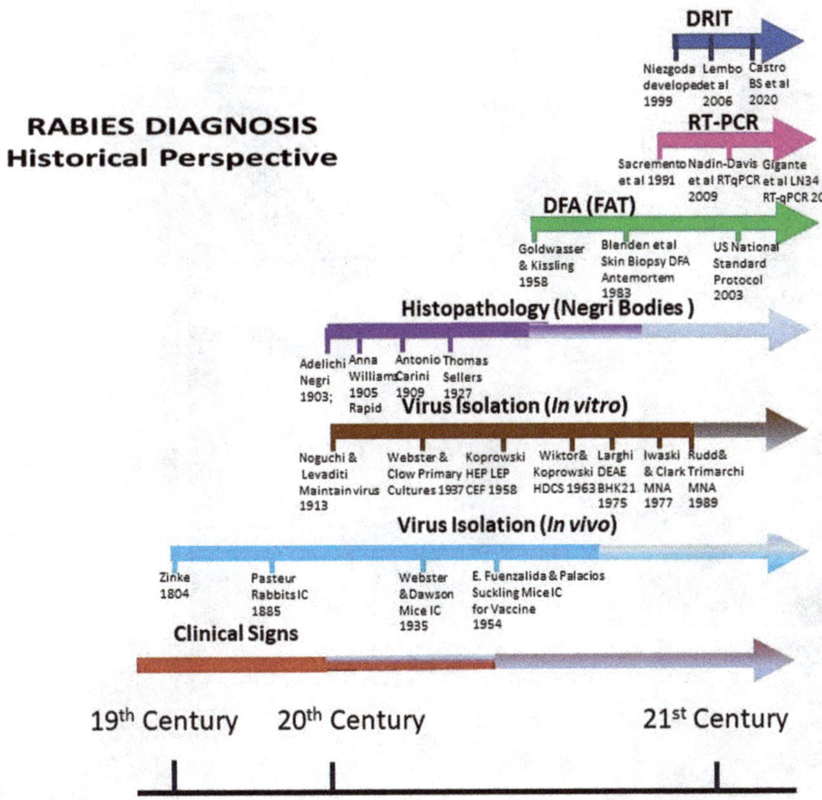

Fig. 5.1 Illustration of important methods for primary rabies diagnosis since the start of the nineteenth century. Multi-dimensional timelines were created for each of the tests presented. When the timeline is solid, the test is being used in the timeframe as a primary diagnostic method. As new or improved tests replace the primary test, the line shading is reduced. Key contributors to the development of the tests are placed on the timeline. However, due to space limitations all major contributors could not be listed. This was particularly true for the DRIT and RT-PCR timelines. In these cases, the first contributor to develop the test was listed together with two others

rabies cases in dogs was performed by examining the contents of the stomach postmortem for pica (unusual articles such as stones, sticks, wood, soil etc.) consumed in the furious phase of illness. Suspicion of rabies in the dog which bit Joseph Meister (first successful recipient of Pasteur's rabies vaccine) was strengthened by this method. Unfortunately, there are many countries in which reliable or timely laboratory testing is not available. Rabies suspicion and post-exposure prophylaxis are still based primarily on the clinical signs of rabies in a suspect animal, when observation is possible. Classically, with respect to human rabies, cases caused by rabid dogs may be based upon specific symptoms, such as hydrophobia, photophobia, excessive salivation, and aggressiveness, features characteristic of furious rabies.

Fig. 5.2 Selected photos of some of the individuals from the Americas who made significant historical contributions to rabies virus methodology

Top row: (left to right) Dr. Anna Wessel Williams (developed the first rapid 2-h test for rabies diagnosis by Negri body detection on brain smears with Giemsa stain); Dr. Antonio Carini, (Director of Pasteur Institute of São Paulo, Brazil from 1906 to 1914, and first to provide laboratory confirmation of rabies transmission from hematophagous bats to herbivores); Drs. Hilary Koprowski and Tadeusz Wiktor, (middle and right of this photo, responsible for the first licensed human diploid cell culture vaccine used in humans, and the first to develop monoclonal antibodies (MAbs) for rabies diagnosis and typing, as well as many other contributions too numerous to list here; SA Plotkin on left is administering vaccine to Dr. Koprowski);

(continued)

Laboratory Diagnostic Methods

Virus Isolation

In-Vivo Isolation

The first laboratory diagnoses of rabies were made by *in vivo* isolation of "the disease-causing agent in animals". Initial attempts at isolation of rabies virus were performed using dogs and rabbits by Zinke in 1804 [1] and by Galtier using rabbits

Fig. 5.2 (continued)
Second row (left to right): Dr. Thomas Forte Sellers, (Georgia Department of Health, developed the Sellers staining method for detection of Negri Bodies in brain impressions and smears that reduced the time to make a rabies diagnosis from a few hours to a few minutes.); Dr. Robert Kissling, (Investigator from CDC, USA, who with a Guest Research at CDC R. Goldwasser developed the direct fluorescent antibody test for rabies diagnosis.); Dr. Leslie T. Webster (from the Laboratories of the Rockefeller Institute for Medicine (LRIM), NY, NY, who with J.R. Dawson developed a sensitive and reproducible *in vivo* isolation method, the mouse inoculation test.); Dr. H. Fred Clark (who with Yuzo Iwasaki first used MNA cells for *in vitro* isolation of rabies virus);

Third row (left to right): Dr. Klaus Hummeler, (Division of Experimental Pathology, The Children's Hospital of Philadelphia, School of Medicine, University of Pennsylvania, who collaborated with colleagues at the Wistar Institute to use anti-rabies conjugates labeled with ferritin and thin section electron microscopy (TEM) to better understand the nature of inclusion bodies and the rabies virus intracellular infection cycle); Dr. Kenneth M. Charlton (Animal Diseases Research Institute, Ottawa, Canada, also explored the pathogenesis of rabies viruses using EM and peroxidase rabies conjugates for rabies virus antigen detection, investigated the ultrastructure of spongiform lesions found in skunk and fox rabies cases, studied oral rabies vaccination (ORV) with attenuated rabies vaccines, and examined the antigenic characteristics of rabies viruses); Dr. Oscar Larghi (formerly of Pan American Zoonoses Center, PAHO, WHO Ramos Mejia, Argentina, seen distributing certificates at the Colegio de Veterinarios de la provincia de Buenos Aires on World Rabies Day; Dr. Larghi was the first to demonstrate the increased sensitivity of DEAE treated BHK-21 for use in the isolation of rabies street viruses; he was a PAHO representative for the WHO sponsored meetings and consultation on MAbs in Diagnosis and Research in the 1980s and was interested in diagnostic techniques for rabies and simplification of applications); Dr. Bernhard Dietzschold, (formerly of the Department of Microbiology and Immunology, Thomas Jefferson University, Philadelphia, was extensively involved in the development and evaluation of rabies vaccines and in the use of MAbs, including for therapeutic purposes, he was especially renowned for his studies aimed at improving our understanding of rabies virus pathogenesis);

Fourth Row (left to right): Dr. Makonnen Fekadu, (CDC, Atlanta, GA, first reported on an immunohistochemical (IHC) test employing an avidin biotin peroxidase system, encouraged colleagues at CDC to develop an improved FFDFA test, studied rabies pathogenesis and ORV, and was the first to document the "carrier-state" in an Ethiopian rabid dog case); Dr. Amir Hamir (Pathologist, College of Veterinary Medicine, University of Pennsylvania, developed a reliable IHC test using highly reactive MAbs from the Wistar Institute); Dr. George M. Baer (Former Chief, Rabies Program, CDC Atlanta, GA, was known as the father of ORV, co-developer of the RFFIT for virus neutralizing antibodies, the founder of the annual Rabies in the Americas Meeting (RITA), and a proponent for sharing of resources and inclusiveness of technology in the Americas; he was well known as a mentor and encouraged new and innovative ideas for rabies diagnosis); Dr. Ivanete Kotait (Researcher at the Pasteur Institute in São Paulo, Brazil, participated in numerous studies on rabies in Latin America and embraced many technological initiatives up until her retirement in 2013) in 1879 [2]. This testing method involved making fresh cutaneous or subcutaneous wounds and inoculation with saliva from a rabid animal, mimicking a route of natural infection. These attempts gave inconsistent and unreliable results due to the intermittent rabies virus shedding in saliva. Pasteur, with knowledge of rabies pathogenesis gained from his and others research, used brain tissue suspensions rather than saliva from rabid animals and intracerebral inoculation of rabbits (optimal sample and route) to successfully isolate the disease-causing agent. Later, guinea pigs were used and were also found to be sensitive hosts. In 1935, Leslie T. Webster and J.R. Dawson, from the Laboratories of the Rockefeller Institute for Medicine (LRIM), New York, demonstrated increased sensitivity, with diagnosis as early as 7–15 days when Swiss albino mice "specifically bred for high susceptibility to neurotropic viruses" were used as the host. The method described by Webster, involving preparation of 10% brain suspensions and inoculation of 0.03 ml intracerebrally in mice, became the routine procedure for isolation in three-week-old mice, both as a biological confirmatory test for rabies diagnosis and the basis for a serum neutralization test in mice. Inoculated mice were observed for up to 30 days after inoculation, and those presenting clinical signs were euthanized and their brains tested for evidence of viral infection [3]. This same procedure for the mouse inoculation test was described in detail in the WHO Laboratory Techniques in Rabies by Hilary Koprowski from the Wistar Institute, Philadelphia, PA [4]. However, intracerebral inoculation of suckling mice (3 days old or less) with (0.01 ml of brain suspension) was found to be most sensitive, as a confirmatory test to isolate rabies "street viruses" or amplify virus. Neonatal mice (intracerebrally inoculated) were used by Fuenzalida and Palacios, at the Bacteriological Institute of Chile in 1954, to produce vaccine, first for dogs then later for humans, that was applied in South America and Mexico for decades. Such vaccines elicited less adverse affects than those produced by the sheep brain origin nerve tissue vaccines of the time [5]. Although previously essential to most national and regional rabies laboratories, the use of animals for routine diagnostic purposes as a "biological confirmatory test" is now discouraged. The large number of animals required raised significant animal welfare issues, and virus isolation can be performed in cell cultures using modern techniques with the same sensitivity as animal inoculation, but with faster results and without the maintenance requirements of laboratory animals or safety concerns. Ideally, *in vivo* testing should be reserved for efficacy and safety studies for biologics or virulence and pathogenesis studies.

In-Vitro Isolation

The first attempts at *in vitro* isolations of rabies virus were made in the beginning of the twentieth century. Earliest of these were separate experiments by Noguchi and Levaditi (1913) to maintain rabies virus in CNS tissue pieces from experimentally infected animals *in vitro*. Blood products, such as clotted monkey blood, were used to provide nutrients to maintain virus, but without virus amplification [6]. However, in 1937 Leslie T. Webster and Anna Clow, LRIM, were able to propagate rabies virus from rabid mouse brain suspensions inoculated in primary tissue cultures from embryonic brains of mice, chickens and rabbits with Tyrode plus 10% monkey serum [7]. The tissue cultures were incubated 3–4 days at 37 °C and serially passaged regularly with the virus titer checked by mouse inoculation. The isolation procedure of Webster and Clow demonstrated amplification of rabies virus. Prior to this protocol, K. Kanazawa, from Japan had attempted primary tissue culture with rabbit embryo cultures with some success, but virus titers did not increase or were not as high. Many rabies investigators during the first half of the twentieth century attempted inoculations of primary tissue cultures, as was done with other neurotropic viruses at the time. Such knowledge was useful for future vaccine production. In the early 1950s chicken embryo fibroblast cultures were used by Koprowski and colleagues at the Viral and Rickettsial Research, Lederle Laboratories, American Cyanamid Company, Pearl River, NY for the preparation of LEP (low egg passage) and HEP (high egg passage) Flury vaccines (a strain named from the isolate obtained from a young U.S. girl who died of rabies in 1930) used for immunizations of animals and humans, respectively [8–10]. In 1958, Robert Kissling and Dorothy R. Reese, Communicable Disease Center, Montgomery, AL, used non-nervous tissue, primary hamster kidney cultures to amplify rabies viruses [11]. Later, the baby hamster kidney (BHK)-21 (clone 13) fibroblast cell line was established in March of 1961 by I. Macpherson and M. Stoker [12]. Kissling and Reese, at the Center for Disease Control (CDC), Atlanta, GA, used hamster kidney cells to isolate rabies virus, demonstrated amplification, and the "potential usefulness" of such laboratory 'fixed' virus grown in cell culture for vaccine preparation and efficacy in animal challenges [13]. Rabies viruses for vaccine production were adapted to established cell lines, such as BHK-21 and Human Diploid cell strains (HDCS), producing high titered virus and high antigenicity, by T. Wiktor and H. Koprowski, at the Wistar Institute, Philadelphia, PA [14]. Later, BHK-21 cell cultures, treated with diethylaminoethyl-dextran (DEAE), to increase sensitivity to viral infection, were proposed for isolation of street rabies viruses for diagnosis by Oscar Larghi, Pan American Zoonoses Center, Pan American Health Organization (PAHO), WHO Ramos Mejia, Argentina. Larghi compared isolation of rabies virus from saliva of rabid dogs using DEAE-treated BHK-21 cells to isolation in sucking mouse brain and demonstrated greater sensitivity using the treated BHK-21 cells [15]. Robert J. Rudd and colleagues at the Division of Laboratories and Research, New York State Department of Health, Albany, NY compared isolation in BHK-21 (clone 13) cells treated with DEAE to isolation in 3-week-old mice to evaluate use of cell culture as a potential more rapid confirmatory test for the DFA test [16]. Indeed, it was a common practice for most rabies laboratories to test all negative samples by mouse inoculation. In 1977, Yuzo Iwasaki and H Fred Clark, from the Wistar Institute, described the use of mouse neuroblastoma (MNA) cells to isolate LEP Flury rabies virus, as part of a virulence study [17]. Abigail Smith, at the Department of Epidemiology and Public Health, College of Medicine, Yale University, New Haven, CT in collaboration with the Viral and Rickettsial Disease Laboratory, California State Department of Health Berkeley, CA, compared isolation of street virus in CER (a BHK related cell line) to MNA cells and demonstrated equal results to mouse inoculation [18]. Later, Rudd and Charles V. Trimarchi demonstrated the greater sensitivity of MNA cells compared to BHK-21. As such, MNA cells remained the cell line of choice for rabies virus isolation and a replacement for the mouse inoculation test [16, 19]. With the development of rapid and sensitive methods for antigen detection (e.g., DFA, DRIT, IHC) and nucleic acid detection (RT-PCR, real-time RT-PCR) during the second half of the twentieth and the twenty-first century, the usefulness of isolation as a primary diagnostic method or a confirmatory test for the antigen detection tests has diminished. Isolation methods are still useful in detecting the presence of infectious virus in samples. Cell cultures are used in propagation, amplification, and quantification of virus and antibodies, for vaccine production, for determination of the safety of vaccine lots and to study rabies virus pathogenesis.

Direct Detection Methods

Microscopic Detection of Inclusions (Negri Bodies)

With development of improved microscopes in the early 1900s, diagnostic contributions by scientists in the Americas accelerated. Initially, brain tissue sections were examined for non-specific evidence of histopathologic signs of encephalitis, such as inflammatory cells, perivascular cuffing, and Babes nodules. However, in 1903 an Italian, Adelchi Negri, assistant pathologist to Camillo Golgi, was encouraged by his mentor to examine both rabid dog and rabbit brain tissue sections with some of the "new" histologic stains. Using Mann's stain (containing methylene blue and eosin), Negri discovered eosinophilic intracytoplasmic inclusion bodies thought by him to be the parasite causing rabies, and he published the research in 1904 [20]. At the same time as Negri's discovery, a pathologist, Anna Wessel Williams, at the New York City Department of Health, made the same discovery using Giemsa stain (containing methylene blue, eosin and Azure B), but she was not the first to publish (and hence Negri lent his name forever after to the detected inclusions). The staining procedure used by Williams had advantages over that used by Negri, as it was the first truly rapid test for rabies diagnosis performed on brain smears [21]. Although the Giemsa stain could be applied for 30 min to 24 h, Williams found her procedure worked best with two-hour staining followed by decolorization and slide reading, looking for inclusions with typical internal structures. Later, Williams developed a more rapid staining technique (a modification of the Van Giesen stain) for detection of Negri bodies [22]. Brain smears after drying were fixed for 10 seconds in methyl alcohol with 10% picric acid, followed by addition of the stain solution (prepared with saturated alcoholic solution of fuchsin 1 drop, saturated alcoholic solution of methylene blue 5 drops and 10 cc of water) to the brain smear slide and heated on a flame until it steamed, then rinsed with tap water, reducing staining time to just minutes. In contrast, the procedure used by Negri was performed on tissue sections requiring two or more days to process, embed and section. Mouse inoculation was performed on all samples which were negative for detection of Negri bodies with these staining techniques.

William's modified procedure for the detection of Negri bodies was a method of choice in the Americas for more than 20 years, until Thomas Forte Sellers, Georgia Department of Public Health, Atlanta, GA, developed an improved staining technique. An important and historic application of Negri body detection was made in 1908 by Antonio Carini, Technical Director of the Pasteur Institute, São Paulo, Brazil. While investigating an outbreak of rabies in the State of Santa Catarina, Brazil, in which 3000 cattle and 1000 horses died, Carini hypothesized the disease was transmitted from hematophagous bats to livestock. He found evidence of bat attacks on herbivores during the day, and laboratory confirmation was made by examination of the CNS tissues of animals that died. He demonstrated histopathologic signs of encephalitis, including inflammation, perivascular cuffing, and numerous Negri bodies in the cytoplasm of nerve cells [23].

The Sellers staining method, developed in 1927, is a simplified procedure for Negri body detection prepared on fresh brain impressions or smears (without fixing or drying) [24]. The rapid test could be performed in a few minutes, allowing for faster administration of rabies vaccination in exposed humans, following a positive result. Impression slides were added to the Sellers stain (2 parts methylene blue to 1-part basic fuschin) for 1–5 s (depending on thickness) at room temperature, washed with tap water, then viewed with a light microscope for typical magenta intracytoplasmic like inclusions, or "Negri bodies" having internal basophilic granules. Like the methods used by Williams, the Sellers rapid test could distinguish inclusions by internal staining characteristics. By modern standards, like the other tests for detection of rabies virus inclusions non-specifically with histologic stains, it lacked sensitivity (50–80%) and specificity compared to later highly sensitive and specific antigen or molecular nucleic acid detection tests. Due to this limitation, especially since large inclusions were most easily recognised in the Purkinje cells of the cerebellum, in the hippocampus and in the midbrain at later stages of disease progression, diagnosis was usually supplemented with mouse inoculation as a confirmatory test. The Sellers stain was used around the world for decades, until its replacement with the DFA and other rabies virus-specific antigen detection tests, such as the DRIT.

Electron Microscopy (EM) for Form and Function

Observation of rabies virions by electron microscopy (EM) allowed for the examination of their ultrastructure, shape, and size. The technique provided supportive evidence of a rhabdovirus infection but required careful examination by expertly trained personnel of numerous observational fields of the sample. As such, it was impractical and too costly for routine diagnosis [25]. Clearly, EM, besides visualization of morphology and structure of rabies virus particles, has been an essential research tool to elucidate the intracellular (cytoplasmic) processes of the infection cycle, sites of replication and budding of complete virus. The technology for EM was invented by German scientists Max Knoll and Ernst Ruska in 1932 [26]. However, EM was not used in rabies research until 1951 by George Hottle and colleagues at the NIH, Bethesda, MD, who observed typical large and small inclusion bodies in sections of mouse brains (provided by Harry A. Rubin CDC, Montgomery, AL) from animals euthanized 7, 9 and 12 days after IC inoculation [27]. This rabies EM reference is rarely cited because no intact virus particles were observed, and the granulations described within the inclusions were not recognizable as the unwound coils of ribonucleoprotein (RNP). Bullet-shaped viral particles were observed in rabies virus-infected chick embryo cultures in 1963 by M. C. Davies and colleagues, Biochemical Research Section and Viral and Rickettsial Laboratories, Lederle Laboratory and in rabies infected BHK-21 cells by Pascu Atanasiu and researchers at the Pasteur Institute, Paris, France, with definitive association of these particles with infectivity established by collaborators [28, 29]. In the 1960s and 1970s several investigators, such as from the Pasteur Institute, Wistar Institute, CDC, Institute for Virus Research, Kyoto University, Kyoto, Japan, and Federal Research Institute for Animal Virus Diseases, Tübingen, West Germany generated high resolution images using a negative staining process and ultrathin sections in Transmission EM (TEM) to better understand rabies virus structure [30]. These TEM studies demonstrated that the rabies virus is cylindrical with one round or conical end and one planar or concave end, that all lyssaviruses are approximately 180 by 75 nm, and that these viruses are contained in an envelope bilayer derived from the host cell cytoplasmic membrane during viral budding. Peplomers, glycoprotein (G protein) trimeric spike inserts of approximate 10 nm, were found within the surface of the virus envelope. The inner surface lining of the envelope was formed by the matrix (M) protein, bound with the G protein-envelope and the viral RNP core. The RNP was composed of tightly wound viral RNA, encapsulated by the phosphorylated nucleoprotein (N protein) and associated with the phosphoprotein (P) and large polymerase (L) protein. The TEM use applied for rabies virion detection was described by Matsumoto and colleagues [31]. This helped to resolve issues with morphology, internal virion structure and the processes involved in "morphogenesis" from virion adsorption by the cell to the final budding of complete virus from the cell. Matsumoto and collaborators demonstrated the development of rabies virions in association with intracytoplasmic inclusions. Klaus Hummeler, with other colleagues from the Division of Experimental Pathology, The Children's Hospital of Philadelphia, School of Medicine, University of Pennsylvania and the Wistar Institute used anti-rabies virus conjugate labeled with ferritin and TEM to demonstrate the nature of the dense "matrix" of inclusion bodies and identified the strands and helical forms of viral nucleocapsid abundantly present [32]. Further clarification was made by Atanasiu and colleagues that more than the nucleocapsid replication occurred in the intracytoplasmic inclusions. The formation of the envelope and the assembly of the virus particle also occurred within the "matrix" of the inclusions [33]. The early events of rabies virus infection of BHK-21 were studied by Iwasaki and colleagues at the Wistar Institute in 1973 [34]. They discovered that within 5–30 min virus particles interacted with the host cell surfaces, and some were engulfed in phagocytes. It appeared that the viral envelope fused with the plasma membranes and the vacuolar membranes, first at the planar end of the virion. In this study, "no viral morphogenetic" events were observed after the adsorption and "uncoating" for 5 h. However, we now know that during this eclipse period, the RNP complex is responsible for transcription of genomic RNA into five polyadenylated mRNAs. These are translated into the structural proteins, and for viral replication by providing a template to synthesize complementary full-length (negative sense) genomic RNA. The G protein inserts into the host cell plasma membrane. As observed by EM, during assembly, there is coiling of RNP-M protein followed by binding with the G protein, as the completed virions bud from plasma membranes. Viral particle budding was noted after 6 h post-inoculation by Iwasaki and collaborators [34]. Budding of "rabies street viruses" usually occurs in the plasma membranes of the endoplasmic reticulum, though this was indiscriminate when fixed viruses were studied in cell cultures. Budding sites may vary between rabies street viruses, fixed viruses, and non-rabies rhabdoviruses, as noted by Shope and colleagues at Yale University and CDC and Matsumato and colleagues at the Federal Research Institute for Animal Virus Diseases [35, 36].

Before molecular methods of virus characterization, EM was used by Frederick Murphy, CDC, Atlanta, GA, in addition to serologic, physicochemical and antigenic properties, to classify Rhabdoviruses by their "morphologic-morphogenic" differences. Later, the 1989–1990 Rhabdovirus Study Group (William Wunner, Chairman) included EM studies to recommend the classification of 20 new (animal and plant) rhabdoviruses in addition to the 13 previously recognized rhabdoviruses. F Murphy and colleagues at CDC used EM for this purpose, while others at CDC (J Smith, W Bellini) in collaboration with Charles H. Calisher, Chief, CDC Reference Arbovirus Laboratory, Fort Collins, CO, and international partners (e.g., M. Lafon, N. Tordo, A. Jackson, and others) focused on serologic, antigenic or molecular properties of these rhabdoviruses. In addition, Murphy applied EM as well as DFA in pathogenesis studies to detect virions and antigens in tissues to monitor the clinical course of rabies. Similar studies were performed in skunks and foxes by Kenneth M. Charlton and colleagues in Canada using EM and peroxidase-labeled antibodies for rabies virus antigen detection, with particular interest in investigation of the spongiform lesions found in skunk rabies cases [37, 38]. The use of EM by investigators in the Americas enhanced knowledge of rabies virion structure, intracellular processes of infection and pathogenesis, and supported a basis for antigenic and molecular studies.

Antigen Detection Tests

Direct Fluorescent Antibody (DFA) Test

In 1958, Robert Goldwasser and Robert Kissling, at the CDC, Montgomery, AL facility, adapted "new" immunofluorescence techniques to the rapid detection of rabies virus antigens. Unlike the tests for detection of Negri bodies, the direct fluorescent antibody test utilized rabies virus-specific antibodies labeled with a fluorescent dye, fluorescein isothiocyanate (FITC), to detect rabies virus antigens, usually the abundant nucleoprotein. During replication of rabies virus in the cytoplasm, a single viral inclusion may contain 1000s of copies of RNP (the viral core proteins) that comprise the test target. Touch brain impressions or slip smears are air dried, fixed in cold acetone and anti-rabies virus FITC conjugate is then applied to the impression or smears. Following a 30-min incubation at 37 °C, if rabies virus antigens are present an antigen-antibody complex will be formed. After washing with phosphate buffered saline (PBS), cover glass mounted with glycerol mounting medium, and observed with a fluorescence microscope (equipped with filters to maximize excitation and emission of the fluorochrome), typical large and small rabies virus-specific inclusions, and antigenic dust will fluoresce with sparkling apple-green color [39]. Due to the sensitivity of the technique, antigens can be detected even in early stages of brain infection. Although the initial publication of Goldwasser and Kissling demonstrated the ability to detect rabies virus antigens in experimentally infected mice, in later publications they demonstrated detection of antigens in salivary glands of naturally infected animals in the field. Goldwasser, at the Israel Institute for Biological Research, Ness-Ziona, performed the DFA test on all samples submitted during a 12-month period [40] in comparison to Negri body detection and mouse inoculation. The DFA test demonstrated superior sensitivity of 100% compared to the Negri body test which was about 65% [40, 41]. Comparisons of Negri body detection and the DFA test in Canada and elsewhere in the 1960s confirmed the superiority of the DFA test [42]. This sensitive and specific procedure subsequently became the global gold standard test endorsed by the WHO and OIE [43–45]. In the USA, there are approximately 125 state public and veterinary diagnostic laboratories performing rabies diagnosis. To ensure the reliability of the DFA test, a National Standard Protocol for Post-mortem Diagnosis of Rabies (see https://www.cdc.gov/rabies/pdf/RabiesDFASPv2.pdf) was written in 2003. This standard included training recommendations, sample requirements (e.g., full-cross section of brain stem, aliquots of cerebellum), preparation of slides, reagents, methods to avoid cross-contamination, testing protocols, interpretation of results, repeat testing, and suggestions of when to send a sample to a reference laboratory. Provided fresh (unfixed) brain tissue containing a full brain stem cross-section is employed and testing is performed by competent individuals, the DFA test is very robust. In addition to brain, provided the samples are optimal, the DFA test is useful for rapid RABV antigen detection in antemortem nuchal (skin) biopsies (6 mm, full thickness from the hairline). The observation of typical RABV antigens in the highly innervated nerve endings associated with hair follicles is diagnostic. Although first described in 1983 by researchers Donald C. Blenden, and colleagues at Departments of Veterinary Microbiology and Medicine-Infectious Diseases, University of Missouri, Columbia, MO, in collaboration with J. "Fritz" Bell, Department of Microbiology, University of Montana, Missoula, MT, [46] as a possible test for early diagnosis in animals, the test is usually applied to human skin biopsies to provide one of the most reliable minimally invasiveness methods for human antemortem diagnosis [47]. The test can be completed in a few hours, which may be critical to management of the patient and potential contacts. Although the DFA test has demonstrated superior sensitivity and specificity for rabies diagnosis, the requirement for specialized equipment (fluorescence microscope, incubator, −20 °C freezer) as well as maintenance of a cold chain during transport and storage, makes the adaptation of the DFA test to field conditions difficult. As such, this limits implementation of the DFA test in some resource limited countries, due to the equipment cost as well as a less than ideal health network and transportation infrastructure to a central rabies laboratory.

Direct Rapid Immunohistochemistry Test (DRIT)

The direct rapid immunohistochemistry test (DRIT) is a highly sensitive and specific test, developed in 2006 by Michael Niezgoda and Charles E. Rupprecht at CDC, Atlanta, GA, as an alternative to the DFA test for decentralized rabies virus antigen detection in the field [48–50]. The DRIT is performed on brain touch impressions prepared the same as for the DFA test. Slides are fixed in 10% buffered formalin prior to processing in a multistep process involving: 1) incubation with anti-rabies virus biotinylated primary antibodies; 2) treatment with streptavidin-peroxidase enzyme which binds to biotin; 3) addition of enzyme substate; 4) followed by treatment with Gill's hematoxylin counterstain. Depending upon the chromogen used (e.g., AEC), observation of the slide under a standard light microscope reveals rabies virus antigens as bright red, large or small, oval or round inclusions, dust-like particles, or strands against a contrasting blue background [51].

Initially, the first DRIT used a cocktail of purified biotinylated anti-rabies virus nucleocapsid monoclonal antibodies (MAbs), generated from two hybridomas that originated from the Wistar Institute, to detect rabies virus antigens. Since then, other studies in laboratories from several countries, including China, India, Afghanistan, Iraq and Africa, have used this cocktail of monoclonal antibodies provided by the CDC/Wistar Institute to apply DRIT to their rabies surveillance studies [52–54]. However, since 2013 these antibodies were not readily available for CDC use, and new anti-rabies virus nucleocapsid MAbs were developed by CDC, biotinylated and optimized for use in the DRIT by Lillian Orciari and Pamela Yager. Coetzer and colleagues compared these biotinylated MAbs and a biotinylated polyclonal antibody (PAb) in both DRIT and the DFA test [55]. More recently, Ye Feng and collaborators at the Institute of Military Veterinary Medicine, Academy of Military Medical Sciences, China developed a biotinylated MAb conjugate against the RABV nucleoprotein for use in a DRIT [56]. As with the DFA test multiple DRIT conjugates were prepared around the world. For example, Daria L Manola and investigators at the Research Institute for Tropical Medicine (RITM) Manila, Philippines [57] as well as Bruno Stuart Castro together with Dr. Sandriana dos Ramos Silva and other researchers at the Pasteur Institute, São Paulo, Brazil [58], have independently developed polyclonal biotinylated anti-rabies antibodies for the DRIT. The DRIT has demonstrated sensitivity and specificity equal to that of the

DFA test in detecting rabies virus antigens. Moreover, the DRIT incurs a lower cost than the DFA test, mainly because it does not require a fluorescence microscope. Similar to the DFA test, primary antibodies for the DRIT may be either MAbs or PAbs.

In the USA and Canada, the DRIT is used primarily to test animals that have not exposed humans to rabies virus and to augment rabies surveillance within and bordering Oral Rabies Vaccination (ORV) zones. It's use has contributed to better monitoring and control of the eastern raccoon rabies virus variant [59], and facilitated enhanced wildlife surveillance in many jurisdictions across North America [60], thereby improving knowledge of disease prevalence beyond what would otherwise be possible through traditional surveillance regimes. This is especially important for remote northern communities in Canada and Alaska. Field use of this technique has also provided the ability to accurately diagnose rabies in remote tropical areas. The first such implementation was undertaken by Tiziana Lembo and her colleagues working in the Serengeti ecological region in Northwestern, Tanzania [48]. Other investigators, Salome Dürr and colleagues performed similar testing in Chad [52], and Carlos and Cecilia de Mattos performed the test in Amazonian regions of South America. Indeed, since rabies is a neglected and underreported disease in many developing countries, application of the DRIT could fill a significant void in rabies surveillance for regions that do not have laboratory infrastructure capable of undertaking the DFA test. To date the DRIT has been used on all affected continents, including islands of the Philippines, Taiwan, and Haiti, and is accepted by the World Health Organization (WHO) and World Organization for Animal Health (OIE) as an alternate test for rabies virus antigen detection [43, 51]. Advantages of the procedure include rapidity of the procedure (1 hour to completion); the inactivation of rabies virus in test brain impressions by fixation in formalin, (in contrast to acetone fixation in the DFA test which does not completely inactivate rabies virus); and the minimal equipment requirements (only standard light microscope). The sample requirements for DRIT are the same as those of DFA and other antigen-detection methods. However, the DRIT reliability is not altered by sample storage in 50% glycerol saline, a problem for the DFA test, allowing for transport and storage at ambient temperatures without compromising sample quality for antigen detection. The DRIT could also be used as a confirmatory method for the DFA test (and vice versa), if inconclusive results are obtained.

Serologic Tests

The serologic tests attempted for the detection of rabies virus antibodies over the decades are numerous. Humans and animals immunized against rabies may produce different types of antibodies: non-neutralizing that bind to specific rabies virus proteins, viral neutralizing, lytic antibodies (which cause rabies virus-infected cells to lyse in the presence of complement), etc. In efforts to find rapid tests for detection of rabies virus antibodies for immune status and rabies diagnosis, during the

1960s–1970s, a variety of serologic tests used for other agents were applied to rabies, but these were never adopted as a primary test for detection of such antibodies. Some of these may have been used successfully for specific research purposes and included: radioimmunoassay (RIA), complement fixation (CF), immune adherence hemagglutination, complement-mediated cell lysis, passive hemagglutination (PHA), counter electrophoresis, interference inhibition and hemagglutination inhibition (HAI) tests [61]. Of these tests, the PHA compared well to serum neutralization tests in mice, but non-specific hemagglutinins in serum of humans or animals prevented widespread use. Also, the HAI test was inferior when rabies virus antibody levels were low, and the CF test was also less sensitive. Alternatively, the RIA test was more sensitive than the serum neutralization test in mice but required use of radioactive-labeled virus. While the complement-mediated cell lysis test could indicate the presence of lytic antibodies, these might not represent neutralizing antibodies. Currently, the serologic tests for rabies virus antibodies include rapid rabies virus neutralization assays (which have replaced the 14-day serum neutralization test in mice). These are the rapid fluorescent foci inhibition test (RFFIT) and the fluorescent antibody virus neutralization test (FAVN). Others are the antibody-binding assay methods, such as the indirect fluorescent antibody test (IFA) and enzyme linked immunosorbent assay (ELISA) [43]. Each varies in sensitivity, specificity, type of antibody detected, and the viral antigens (protein) recognized. These methods are most robust when applied in parallel on sequential samples used to diagnose or monitor rabies in humans. Susan M. Moore and Cathleen A. Hanlon, then at the Rabies Laboratory, Kansas State University, provided an informative review of the current methods (neutralization tests and binding assays), usages, limitations, and needs for standardization and validation [62]. Current tests are described below in more detail.

Neutralization Tests

Neutralization tests (NT) are essential tools for the prediction of an adequate adaptive response to immunization against rabies virus by the recommendations of the Advisory Committee on Immunization Practices (ACIP), WHO and OIE [63–65]. The NT level of 0.5 IU/mL was first accepted as evidence of an appropriate immune response in the Eighth Report of the WHO Expert Committee in 1992 in response to the identification of Lyssaviruses species other than rabies virus in Europe in the 1980s. In addition to immune status by WHO or ACIP requirements, NTs performed on antemortem human serum and CSF may provide a specific diagnosis of rabies in humans, though in some cases these antibodies are present only in late clinical stages of the disease, or not at all. Monitoring of neutralizing antibody titers is a key component when human rabies cases are treated using intervention protocols, such as the "Milwaukee Protocol", and increased neutralizing antibody levels are an indicator, in some cases, of improvement or potential for viral clearance and survival [66]. The earliest of the neutralization tests was the serum neutralization test in mice

(SNM) described by Webster and Dawson in 1935. Neutralization of rabies virus relies on the presence of antibodies directed to the outer G protein antigens of rabies virus. These are functional assays measuring the performance of antibody dilutions after incubation with a known dose of challenge virus standard (i.e., CVS-11) to prevent the infection of cells in culture or historically, infection of mice. The SNM has been largely replaced by the *in vitro* RFFIT, inspired by Jean S. Smith while perusing a laboratory supply flyer advertising a "new" product, chamber slides. The RFFIT, described by Smith, Yager and Baer, CDC, Atlanta, GA in 1973 [67] and detailed in WHO publications in 1973 and 1996, measures the ability of rabies virus antibodies in serum or CSF samples to neutralize a known standard challenge virus dose (CVS-11). The test was originally performed in 8 well chamber slides, in which fivefold dilutions of serum were incubated for 90 min with CVS-11 (a standardized dose to provide between 32 and 100 fluorescent foci $_{50}$ (FF_{50}) in 20 h), and then BHK-21 cells were added. After incubation at 37 °C for 20 h the cells were fixed and subjected to the DFA test. Antibody titer was quantified based on the level of reduction of virus. An updated RFFIT procedure published in 1996 described the optional use of the more sensitive MNA cell line or DEAE-treated BHK-21 (clone 13) cells [68]. A modification of the RFFIT, the fluorescent antibody virus neutralization test (FAVN), developed by F. Cliquet, M. Aubert and L. Sagne, is performed in 96-well microtiter plates [69]. The RFFIT and FAVN are frequently used to determine the immunization status of vaccinated humans and animals, respectively. Deborah Briggs and colleagues at Kansas State University, Manhattan, KS in collaboration with Smith and colleagues at CDC, found no significant difference in sensitivity and specificity when the RFFIT and FAVN were compared [70]. When performed in qualified laboratories by qualified personnel, both test methods demonstrate acceptable and similar sensitivity and specificity in determining rabies virus neutralizing antibodies. These NT results are best compared when converted to international units per milliliter with criteria specifying the acceptable performance of an international reference serum standard, virus back titration and cell control in each run of the assay. Based on the RFFIT, in 1992, Favoretto and colleagues at the Pasteur Institute, São Paulo, Brazil developed a simplified cell culture virus neutralization test performed in 96-well microplates, called the Simplified Fluorescence Inhibition Microtest (SFIMT), with the advantage of easy and rapid reading [71]. Since 1995, the SFIMT has been used as a RABV serology test in Brazil for pre-exposure evaluation. Presently, more than 22,000 sera samples are analysed by the SFIMT each year.

Indirect Fluorescent Antibody Test (IFA)

The IFA tests are sensitive methods for detection of IgM and IgG antibodies that have bound to rabies virus antigens within infected cells. The procedure was described by Lois Leffingwell and colleagues at the Texas State Public Health Laboratory, Austin, Texas [72], and evaluated by J. "Ben" Thomas, R. "Keith" Sikes and researchers at the CDC, Atlanta, GA, in the 1960s [73]. Specific rabies viral (protein) antigen binding activity may be present in human serum and CSF and is the basis for the IFA inclusion in the battery of tests for antemortem diagnosis of rabies in humans, incorporated into such testing by Dane W. Sanderlin at CDC [47]. The IFA measures antibodies against internal RABV core proteins, such as RNP rather than rabies viral G surface (envelope) proteins. Since it is an indirect test, serum or CSF samples are first titrated and applied to acetone-fixed cell culture slides infected with RABV challenge virus strain (CVS-11) and incubated. Then secondary antibodies, either FITC-labeled anti-human IgG or IgM are added and incubated (other anti-species and specific isotypes of antibodies could be used as secondary antibody). The endpoint antibody titer is the last dilution demonstrating specific fluorescence (≥2+ intensity) when observed using a fluorescence microscope. The presence of rabies IgM and IgG in sera of persons with previous vaccination histories is rare which makes this test important in human diagnosis. The IFA titers are not comparable with levels of *in vivo* viral neutralizing antibody to the rabies virus G surface proteins. Of importance, the IFA test can be completed within a few hours, rather then 1 day for RFFIT (or 2 days for FAVN), and it may detect antigen binding IgM and IgG antibodies earlier in the clinical course of human rabies cases than the NT, thus enabling rapid antemortem diagnosis of rabies and informing subsequent patient management, infection control and prophylaxis for exposed individuals.

ELISA

The ELISA, like the IFA, are serologic tests for detection of antibodies in serum and CSF that bind to rabies virus specific proteins, G and N proteins being the most common targets. However, the target protein will be determined by the purpose of the assay, whether for immune status, diagnosis, or research. P. Atanasiu and colleagues at the Institut Pasteur, Paris, France, described an ELISA for detection of rabies virus antibodies [74]. Later M. Grassi and A. Wandeler, working in Switzerland, published a simple ELISA procedure using highly purified G protein as antigen [75]. Susan Realegeno and collaborators at CDC, developed an N-gene ELISA which could be used as an antemortem test for rabies diagnosis for detection of rabies anti-IgG and IgM, since these antibodies to the nucleoprotein appear earlier in the course of disease [76]. However, it appears the IFA test remains more sensitive for anti N protein antibodies. Methods can vary as to the type of the ELISA (indirect, indirect capture, or competitive), but all require a solid phase employing beads, a tube or microtiter plate. In most ELISAs a rabies virus antigen preparation, comprising purified viral proteins (N or G), whole virus or even vaccine, are coated directly onto the solid phase. Alternately, in a capture ELISA, antibodies (PAbs or MAbs) to specific rabies virus proteins are coated to the solid phase, then the target protein is added and incubated, resulting in an antibody-antigen complex bound to the solid phase. Using either of these methods, then serum or CSF is serially diluted (or screened at specific dilution) and added to the plate or tube containing the bound antigen and incubated. Rabies virus antibodies react with the target protein forming an antigen-antibody complex which is then allowed to bind to an enzyme (usually horseradish peroxidase or alkaline phosphatase) labeled anti-species IgG or IgM (or enzyme labeled streptococcus protein A) [76]. Following substrate addition the enzyme bound to the antigen-antibody complex catalyses a color change before the reaction is stopped by a pH change (addition of acid or base). After each reaction step, the excess unreacted components are removed by washing and test reactions can be read with a plate reader or spectrophotometer. Reagent control (blank) is usually subtracted from the reading and results are based on cut-off values for positive and negative samples obtained from a previous standardization curve. The advantages of ELISAs are that they can be automated for high thorough-put and used for large serosurveys. Most values obtained by ELISA testing are not equivalent to the neutralization assays which are functional tests and determine an individual's immune status by evaluating the ability of antibodies in serum to prevent infection (neutralization) of susceptible cell cultures by high titered challenge virus. However, a competitive ELISA procedure, published by L.D. Elmgren and Wandeler working at the Rabies Laboratory of the Animal Diseases Research Institute in Ottawa, Canada, in 1996, specifically measured serum neutralization capability [77]. This method involved mixing the serum under test with a neutralizing MAb conjugated to horseradish peroxidase and then allowing this mixture to compete for binding to a limited amount of immobilized viral antigen prepared from the Evelyn Rokitniki Abelseth (ERA) strain of rabies virus grown in BHK-21 C13 cells. After removal of non-bound antibody, enzyme substrate and chromogen were added, and colour intensity was measured after a 10-min interval. Sera containing rabies virus neutralizing antibodies compete with the MAb conjugate and reduce the intensity of the chromogenic reaction. This rapid test method facilitated high sample throughput and, in addition, the competitive format enabled testing of sera from a wide variety of species [77]. In studies evaluating field performance of two wildlife oral vaccines in Canada and the USA, direct comparison of this cELISA with viral neutralization assays found no significant difference in the proportion of antibody-positive animals identified by these methods, thereby establishing the cELISA's sensitivity and specificity and supporting its application to ORV programmes targeting both skunks and raccoons [78, 79].

Diagnosis by RT-PCR

The development of the automated polymerase chain reaction (PCR), by scientists working for a US-based biotechnology company in the 1980s [80], revolutionised the life sciences. Together with a pair of synthetic primers that bind to opposite strands of a DNA sequence, a thermostable DNA polymerase uses deoxynucleotide triphosphate (dNTP) substrates to synthesize new DNA. The process is driven by a two or three step temperature profile of 25–40 cycles, each cycle eliciting DNA denaturation, primer binding and DNA synthesis to generate thousands of copies of the DNA target. This technology can enable detection of RNA viruses by using a reverse transcription step to convert RNA to a cDNA copy prior to PCR (RT-PCR). The resulting PCR product or amplicon is relatively easily detected, either by traditional gel electrophoresis or with labelled probes, thus yielding a very sensitive, specific method. In the early 1990s many rabies laboratories developed RT-PCRs for both research and diagnostic use. Investigators at the Pasteur Institute in Paris described a RT-PCR targeting the N gene for potential diagnostic use and a second assay targeting the G-L region suitable for epidemiological application [81]. Smith and colleagues at the CDC also described an RT-PCR targeting the N gene for both rabies virus detection and epidemiological studies [82, 83] while Susan Nadin-Davis at the rabies laboratory in Ottawa, Canada, together with K. Charlton and colleagues, applied RT-PCR methodology to detect rabies virus in muscle tissue following long incubation periods in experimentally-infected skunks [84]. Versions of these methods proved especially useful for ante-mortem diagnosis of human rabies since they can be applied to saliva samples obtained non-invasively and nuchal skin biopsy material [85]. Such methods were transferred to many other countries [86, 87]. However, the caveat to the method's high sensitivity was the problem of carryover contamination of samples resulting in false positives. While RT-PCR was useful for secondary confirmatory testing it was not considered appropriate for routine diagnosis of animal samples. Developments in the technology to provide real-time amplicon detection (RT-qPCR) obviated the need for handling of post-PCR samples and significantly reduced the carryover contamination potential, thus facilitating recognition of such methods for use as primary rabies diagnostic tests by the OIE and WHO [43]. A protocol that detects all known rabies virus variants was developed in Canada [88] while the LN34 pan-lyssavirus assay was developed at the CDC [89, 90]. Both methods employ probes to detect the amplicon and ensure test specificity while the most recently developed European RT-qPCR relies on a DNA intercalating dye for pan-lyssavirus detection [91]. Studies to explore the replacement of the DFA test with RT-qPCR for routine post-mortem testing have been reported by the New York group [92, 93]. Efforts to develop a national standardized protocol for RT-qPCR are being explored by the Association of Public Health Laboratories.

Viral Typing

Antigenic Typing

Routine viral typing of laboratory-confirmed cases not only allows monitoring of the spread of indigenous viral variants but provides early warning of cross-border incursions of exogenous variants [94], information critical to timely implementation of control measures. With the development of hybridoma technology in the

1970s [95] production of MAbs that bound specific epitopes became possible. Investigators at the Wistar Institute were the first to be active in the generation of such tools for rabies virus characterisation [96]. The availability of panels of MAbs, each yielding a specific pattern of reactivity with different viral variants, provided a means by which the overall reactivity pattern of the panel identified a specific viral type. The WHO Collaborating Centers in the USA (Wistar Institute and CDC), Germany (Federal Institute for Animal Virus Diseases) and France (Pasteur Institute) initiated a special working group to evaluate MAbs useful for distinguishing rabies virus and other lyssaviruses. A consultation on MAbs in Diagnosis and Research, led to sharing of several MAbs (CDC 22, Wistar 8, Germany 2 and Pasteur 4) in 1983. The group, active until 1986, added other organizations, including the Veterinary Research Laboratory Causeway/Harare Zimbabwe, Central Veterinary Laboratory, Weybridge, UK, Centre National de la Recherche Scientifique (CNRS), Gif S/Yvette, France, and the Institute of Veterinary Virology, Berne, Switzerland. The utility of antigenic typing to discriminate between viruses circulating in different parts of the USA was demonstrated by Smith and colleagues at the CDC using MAbs to the N protein [97]. This approach, using a CDC panel of 20 MAbs, identified two distinct variants in skunks in the endemic North Central and South Central states, two unique variants in gray foxes in Arizona and Texas, a variant in Arctic and red foxes in Alaska which briefly circulated in red foxes in the northeast (NY, NH, VT and ME), a mongoose-associated variant in Puerto Rico, nine bat variants across the USA and identified the raccoon rabies outbreak in the southeastern states as a new epizootic [98]. Antigenic typing using this 20 MAb panel is still in use in the USA and remains useful for rapid identification of new emerging variants [99].

A collaboration between Anna-Maria Diaz, Instituto Panamericano de Protección de Alimentos y Zoonosis, Buenos Aires, Argentina, and Smith, CDC, explored the transfer of this technology for virus typing to other countries in Central, South America and the Caribbean. After review of the CDC data, and for convenience, Diaz, chose 8 MAbs to be used in the regional "Latin America Panel". The first study tested 288 samples from 17 countries, with identification of 11 antigenic variants in Central, South America and the Caribbean. Carlos and Cecilia de Mattos, while working at CDC, continued the mission started by Diaz and transferred the technology for both antigenic typing and genetic typing throughout the Americas. The reduced "Latin American Panel" has been valuable in identifying the primary variants responsible for many laboratory-confirmed rabies cases, particularly those associated with vampire bats and dogs, but has shown limitations in defining some of the rarer variants associated with insectivorous bats [100]. Since 1996, the Pasteur Institute, São Paulo, Brazil has undertaken the antigenic characterization of viral isolates from many animal species of various regions of Brazil, utilizing the eight MAb Latin American panel provided by CDC. Between 1989 and 2000, Favoretto and collaborators conducted a study on 330 isolates from humans and different animal species and demonstrated the presence of six antigenic variant (AgV) types as defined by the panel: AgV2 (dog), AgV3 (*Desmodus rotundus*), AgV4 (*Tadarida brasiliensis*), AgV5 (vampire bat from Venezuela), AgV6 (*Lasiurus cinereus*) and a type typical of laboratory strains that reacted to all MAbs. Six other profiles were not readily identified by this panel [101]. Further studies demonstrated that the viruses associated with different species of non-hematophagous bats, as well as those of other Brazilian rabies reservoirs in the northeastern region of the country, including *Cerdocyon thous* (crab-eating fox) and *Callithrix jacchus* (tufted marmoset), are not clearly defined by the CDC panel, although they have consistent antigenic profiles [102–104].

Antigenic typing of rabies viruses began in Canada during the early 1980s following a collaboration between Charlton of the Animal Diseases Research Institute, Ottawa, Canada, and Wiktor of the Wistar Institute. A panel of 43 anti-nucleoprotein MAbs identified eight antigenic groups across Canada, four in carnivores and another four in bats [105]. While some additional MAbs were acquired from Smith of the CDC, it was the arrival of Wandeler at the laboratory in 1989, together with several MAbs produced at the Swiss Rabies Centre of the Veterinary School in Berne, Switzerland, that led to a major initiative examining the utility of antigenic typing. Over many years this rabies laboratory, later designated as part of the Ottawa Laboratory Fallowfield (OLF) of the Canadian Food Inspection Agency (CFIA), compiled a collection of over 500 murine MAbs that were characterized against a large collection of lyssavirus isolates from Canada and other countries. Currently, a panel of 15 MAbs, that differentiates the ERA strain, used for many years in ORV programs in Ontario, from variants that circulate in wild carnivores and multiple bat species across eastern Canada, is routinely used to type all positive cases [106, 107]. A modified panel is used by the CFIA's Rabies laboratory in Lethbridge, Alberta, where diagnosis of cases due to the western skunk variant is common. The complete MAb collection has also proven useful in the development of panels capable of epidemiological surveillance of rabies cases of other countries [108, 109]. Although antigenic typing is relatively quick and inexpensive once established, it does have limitations in its discriminative capability. Moreover, generation of the MAb panels is time consuming and expensive so only a few laboratories are able to develop novel MAbs to suit local needs [102]. The Canadian group has used recombinant phosphoproteins as antigens for novel MAb generation. The modular organisation of this protein facilitates the generation of MAbs with a variety of properties some of which are highly discriminating and able to identify certain rabies virus variants [110, 111] or lyssavirus species [112], the latter of which may prove useful for epidemiological studies of rabies-related viruses in Africa.

Molecular Tools for Viral Variant Typing

The extensive target amplification realised by RT-PCR provides sufficient amplicon to enable direct genetic analysis of the detected virus. This capability revolutionized the field of molecular epidemiology and has been readily adopted by many laboratories across the Americas. Early studies of virus genetic diversity employed indirect methods of amplicon characterisation, such as restriction fragment length polymorphism (RFLP) that employed restriction endonucleases that cleaved dsDNA only at specific sequence motifs [83, 113]. However, over the last 30 years, technical improvements in Sanger sequencing enabled the generation of large amounts of lyssavirus sequence data, much of which has been deposited into public databases, such as GenBank, administered by the US National Center for Biotechnology Information. In addition, the development of freely available software tools to permit easy conversion of sequence data into phylogenetic trees, a cornerstone of molecular epidemiology, enables investigators to generate visually compelling illustrations of viral groupings and their interrelationships [114]. This wealth of information facilitated improvements to RT-PCR/RT-qPCR design to ensure targeting of sequences widely conserved amongst all circulating rabies viruses, an issue of concern in the Americas given the large number of distinct variants identified across the continent [115]. Indeed, the use of molecular genetic tools for both rabies diagnosis and viral typing have evolved in concert. More recently, development and application of highly parallel sequencing technologies, several of which were developed by US-based commercial companies, has facilitated complete genome sequencing of large viral collections, thereby improving our ability to identify small differences between viral isolates [116]. High resolution detection of viral sequence changes over time provides information on mechanisms of viral evolution and the geospatial factors that limit disease spread. The costs of such analyses presently restrict their application primarily to large national laboratories but the resulting knowledge of the mechanisms impacting viral spread can aid disease control through informed cost-effective prevention and control programs.

The rabies group at the CDC was among the first to explore the application of PCR technology for human diagnosis and viral typing of isolates both from the USA and around the world [82, 83, 117, 118]. These studies recognised the existence of two distinct lineages circulating in the Americas: an American indigenous lineage, composed of two mesocarnivore rabies viruses and multiple bat-associated viral variants, and a cosmopolitan lineage, harbored by multiple wildlife hosts, following introduction from the Old World [82, 119]. These studies illustrated the value of genetic typing in revealing the source of human infections [83] and identified the possibility of long incubation periods in some human cases [117]. A primary role of the CDC is to support state rabies laboratories through diagnostic assistance and viral typing of specific cases. Genetic typing of the viruses responsible for human rabies cases across North America has become routine with identification of both imported [120] and indigenously acquired cases, many of the latter being the result of exposure to the bat-associated viruses exclusive to the Americas [121, 122] while rapid identification of the importation of canine rabies from dog-rabies endemic countries [123] remains vital to stop its re-introduction into the USA. Genetic studies by CDC and others have also revealed ongoing emergence of new virus-host relationships with the identification of a host shift of a rabies viral variant from a big brown bat reservoir into the striped skunk in a localised area of Arizona [124], and another apparent host shift of the California skunk variant into a fox host [125]. Indeed, although once hotly debated, skunks represent a significant rabies reservoir in North America [126, 127] and are of special concern to jurisdictions involved in wildlife rabies control as a host that maintains several distinct rabies virus variants while also frequently becoming infected due to spill-over from other primary hosts.

The arrival of S. Nadin-Davis at the Canadian rabies laboratory of ADRI (later OLF) in Ottawa in 1988, together with Charlton's interest in the application of genetic methods of viral detection, facilitated ongoing studies in rabies molecular epidemiology. The long-standing association of the rabies laboratory at OLF with the provincial authority responsible for wildlife rabies control in Ontario led to the early use of RT-PCR methodology to explore the genetic diversity of the Ontario fox variant enzootic that circulated in parts of the province since the 1960s. The RFLP analysis of RT-PCR products identified four geographically segregated viral variants, Ontario types 1–4 [113], knowledge later used to demonstrate the involvement of two of these variants and a distinct northern variant in a rabies outbreak in central Ontario and western Quebec [128]. Such studies illustrated how molecular epidemiology could supplement and enhance surveillance data and identified the ongoing threat of fox rabies re-introduction into eastern Canada from arctic regions [129]. More recently, whole genome sequencing explored a host shift of the Ontario fox virus into the striped skunk in southwestern Ontario [130], knowledge of significance for wildlife rabies control. The raccoon rabies virus variant is of considerable concern due to its relatively recent emergence in the state of Florida, its rapid spread throughout the eastern seaboard of the US as far north as eastern Canada and its threat to public health due to the host's high population densities in urban areas. Investigators at Emory and Trent Universities, the rabies laboratory in Albany, NY and the Ottawa laboratory, often working in collaboration, have used the available molecular epidemiological tools to explore the emergence and spread of this variant. These studies have illuminated the variant's divergence into several lineages, the role of landscape features on disease spread and identified a long-distance incursion of raccoon rabies into the Canadian city of Hamilton, due to inadvertent anthropomorphic animal translocation [131–135].

As use of these molecular genetic tools became well established in the USA and Canada, many institutions in Latin American countries were keen to adopt them, often as an adjunct to or replacement for antigenic typing. Elimination of dog rabies, a problem resolved in the USA and Canada by the mid twentieth century [136], has been a primary focus throughout the region but as dog rabies cases have declined the importance of other rabies vector species including wild carnivores [137, 138] and multiple bat species [139] has been revealed. In Mexico, investigators at Instituto de Diagnostico y Referencia Epidemiologicos (InDRE), in collaboration with CDC, introduced the use of viral typing to discriminate the circulating wild carnivore and bat-associated viral variants [100, 140]. Subsequent studies by this group and researchers from the National Autonomous University of Mexico (UNAM) employed whole genome sequencing technology to investigate re-emergence of classical dog rabies in the country as well as the re-introduction of rabies into the dog population from a wildlife vector [141, 142]. Molecular genetic methods have also been transferred from CDC and OLF to institutions across Latin America [143–147] where bat rabies has assumed increasing importance as dog rabies prevalence has declined. Indeed, over the last 60 years the huge diversity of bat-associated rabies in the Americas has been recognized [107, 148]. The challenge of bat speciation was partially addressed in North America through the incorporation of genetic bar-coding, a species identification tool based on sequencing of the mitochondrial cytochrome oxidase subunit 1 gene developed at the University of Guelph in Canada [149]. Incorporation of genetic bar-coding into rabies molecular epidemiological studies helps to identify reservoir species [150–152] as well as the ecological and host factors impacting virus spread and evolution [153–155]. In South America, the Pasteur Institute in São Paulo, Brazil, has been a regional leader in the adoption of phylogenetic tools to characterise many rabies viruses, including dog rabies viruses and their inter-relationship with those of wild canids [156], and the viral variants associated with vampire [157] and non-hematophagous bats [158]. Multiple investigators at universities in São Paulo have also applied these tools for viral detection and characterisation in dogs [159] and bats [160–162]. Studies in Brazil and Colombia have revealed the potential role of frugivorous bats in maintaining viral variants closely related to those of vampire bats [137, 160, 163]. In addition, the white tufted marmoset, a non-human primate, has been recognised as a rabies reservoir [164] while other novel viral types have been identified in certain mammalian species [165, 166], raising questions as to the true host of these viruses. The Institutes of National Health in both Chile [145, 167, 168] and Argentina [169] have also employed genetic techniques to characterise rabies viruses associated with insectivorous bats while collaborative studies of vampire bat rabies have taken precedence in Peru [170] and Uruguay [171] due to the significant economic impact of this variant. Landscape features, especially elevation, and sex-biased dispersal have been linked to the spread of vampire bat rabies viruses [155, 161, 172]. Dog rabies remains a concern in Bolivia [173] and Paraguay [174] where re-emergence from a wildlife reservoir may have occurred. Isolated collaborative studies have explored the phylogeny of bat rabies in Guatemala [175], demonstrated multiple origins of vampire bat rabies in Trinidad [176] and revealed the independent emergence of the rabies cosmopolitan lineage in mongoose populations on several Caribbean islands [177–179].

The sequence information compiled from such studies can enable the design of rapid viral typing tools. For example, incorporation of variant-specific primers into a multiplex RT-PCR allows variant discrimination based on amplicon size only, a strategy first employed in Canada [180] and later used by investigators in Mexico [181], Brazil [182], South Africa [183] and Europe [184]. The molecular epidemiological expertise developed by American laboratories has been shared with investigators in many parts of the world [108, 109, 177, 178, 185–190], sometimes requiring novel means of sample transportation. For example, material from India was preserved and transported on Whatman FTA cards prior to recovery of RNA for analysis. These collaborations facilitate the amassing of extensive lyssavirus collections available for detailed study. In particular, the New World experience in recognizing the role of bat species as rabies reservoirs has stimulated increased bat surveillance for rabies and led to the identification of several novel bat-associated lyssaviruses world-wide [191]. Knowledge of the genetic diversity of lyssaviruses, much of it amassed by the CDC, has contributed significantly to current understanding of the taxonomy of this genus [192].

Diagnosis of Fixed Samples

Occasionally, brain tissue from a deceased patient or animal is fixed in 10% buffered formalin prior to the suspicion of a rabies diagnosis. Formalin-fixed CNS tissue samples cannot be tested by the routine antigen detection tests, such as the standard DFA or DRIT. The fixation process causes chemical cross-linking of proteins, which interferes with epitope recognition by the conjugated (fluorescent, biotin or enzyme) labeled anti-rabies antibodies. Methods to achieve a diagnosis using formalin fixed paraffin embedded (FFPE) tissues have been explored by investigators at both CDC, OLF and other research institutions, such as the University of Pennsylvania. For example, Makonnen Fekadu and colleagues at CDC reported on an immunohistochemical (IHC) test employing an avidin biotin peroxidase system that proved better than the standard immunofluorescence method for rabies virus antigen detection in formalin-fixed tissues [193]. Subsequently, highly sensitive and specific adaptations of the DFA test for formalin-fixed tissues (FFDFA), examined in a study of 800 samples (14 species), were published by Sylvia Whitfield and colleagues [194]. However, this method relied entirely on the availability of a highly reactive and specific commercial polyclonal anti-rabies FITC-labeled conjugate no longer available in the USA. This made the FFDFA test obsolete as the commercial Mab conjugates used in the DFA test recognise conformational epitopes and are unsuitable for use on formalin-fixed tissues. The IHC technique was further developed and used as a diagnostic tool in Canada [195] and by Amir Hamir at the University of Pennsylvania [196]. Although technically laborious and time consuming, a current IHC procedure employed at the CDC functions well [197]. Formalin-fixed tissue samples that have been processed, embedded in paraffin, and sectioned must undergo deparaffinization and rehydration steps, enzyme digestion (usually with proteinase K or pronase), treatment with hydrogen peroxide to remove endogenous peroxidases and incubation with a blocking agent (10% normal goat serum), extended incubations with anti-rabies virus antibodies, anti-species enzyme (biotin) labeled conjugate, streptavidin peroxidase and peroxidase substrate, followed briefly by a counterstain with Gill's hematoxylin, prior to viewing under a light microscope. In addition, IHC studies have been undertaken in South America. Fernandes and collaborators carried out an IHC study in human rabies cases resulting from an outbreak that occurred during 2004–2005, due to rabid vampire bat exposure in the Amazon region [198]. While such methods permit viral detection, the inability to use monoclonal antibodies for reliable viral typing meant that this type of sample was refractory to viral variant analysis.

Development of an alternative technique employed digoxigenin-labelled probes for detection of viral RNA by *in situ* hybridisation [199]. Further refinement of this technique employed a conserved N gene sequence for virus detection, while a divergent region of the P gene, identified by a study of lyssavirus diversity [200], proved useful for generating variant specific probes for viral typing of infected FFPE tissues [201]. While this procedure allowed epidemiological analysis of FFPE tissue samples, its technical challenges ensured that it was only used occasionally. More recently, RT-qPCR assays, which target short nucleotide sequences, have been shown suitable for viral RNA detection in FFPE samples [202]. Moreover, the resulting amplicon can be sequenced to aid in viral typing [203]. Such nucleotide-targeting methods currently appear ideal as a means of evaluating this tissue type.

Evolution of Rabies Vaccine Potency Testing

Although not identified as one of the laboratory methods used to routinely diagnose rabies in humans or animals, vaccine safety and potency is essential to maintenance of minimum standards for efficacious post-exposure and pre-exposure vaccine regimes. Safety tests are designed to ensure that contaminants, toxins, and residual infectious rabies virus which could be harmful to the recipient are not present in inactivated rabies vaccines. The WHO has published requirements for safety testing of rabies vaccines, including sterility testing for bacteria and fungi, innocuity testing for abnormal toxicity, as well as residual infectious virus [204]. After development of Pasteur's vaccine in 1885, production laboratories have attempted to test the potency (antigenicity) of biologics. Some of these evaluations and methods have been referenced in review of virus isolation procedures [3, 5, 8, 13]. Surprisingly, vaccine manufacturers did not always test products or individual lots after developing a standard protocol. It was not apparent until routine potency testing was performed that there could be deficiencies (including residual rabies virus) and lot to lot variations in potency.

The first rabies vaccine "potency testing" protocol was developed by Karl Habel, Chief, Laboratory of Biology of Viruses, National Institutes of Allergy and Infectious Diseases, NIH, Bethesda, MD. This protocol, for the potency testing of the Flury strain rabies vaccine, was published in 1953, known as the Habel Test [205, 206]. There were three general considerations which Habel believed were important to any vaccine potency test: (1) it should "reflect closely the ability of a vaccine to protect under natural conditions"; (2) "its practicality, ease in which the test can be carried out and the cost and time involved"; and (3) the "requirement for standardization of the test procedure so that there will be comparability of results from one vaccine to another in a single laboratory and between different laboratories." In addition, Habel mentioned that there were two important variables which can significantly affect the results of a potency test: the animal subjects (health, mouse strain used) and challenge virus used (CVS, Challenge Virus Standard). Historically, *in vivo* potency tests have the following basic components: an animal host highly susceptible to rabies (i.e., 4–6-week-old Swiss albino mice), an immunization protocol, virus challenge and determination of potency based on the 50% endpoint titer for the LD_{50} of protection. In the standard Habel Test, used primarily for human and veterinary inactivated vaccines, 60 mice are immunized intraperitoneally 6 times (e.g., Monday, Wednesday, and Friday) over two weeks with 0.25 ml of vaccine (1:10 or 0.5% if brain tissue vaccine). In the test, 30 mice are kept as a control group. On day 14 after initiation of the vaccine regimen, the challenge virus (CVS) is diluted in tenfold increments from 10^{-1} to 10^{-7}using inactivated 2% horse serum or rabbit serum diluent and kept cold (4 °C) during inoculations to maintain virus titer. Each of the challenge virus dilutions are inoculated into groups of 10 vaccinated mice intracerebrally (0.03 ml per mouse). The 30 unvaccinated mice are inoculated in groups of 10 with 10^{-5}, 10^{-6}, 10^{-7} viral dilutions to determine the LD_{50} using the Reed Muench method [207]. By subtracting the log of 50% endpoint in the vaccinated animals from log of the controls, the LD_{50} of protection can be determined. To meet potency standards by the Habel test, the difference should be 3 or Log 1000 determining that the vaccine protects against $1000LD_{50.}$

The "practical-mindedness" of Habel, led to the Modified Habel Test for use in laboratory facilities with limited resources (unable to perform the standard test) when it was important to just screen a vaccine for potency and use a non-quantitative assay. In the modified Habel test, the vaccination procedure is the same as the standard test but the number of mice is reduced to 20, and 15 mice are held for controls. On day 14, the challenge virus is diluted to contain $500LD_{50}$ per 0.30 ml and inoculated intracerebrally into the vaccinated mice. The control mice are divided into 3 groups of 5 and receive the viral dilutions 10^{-5}, 10^{-6}, 10^{-7}. For the modified Habel test to be valid the control titer of the challenge virus should be between 100 and 1000 LD_{50} and greater than 50% of the mice survive the challenge. If the modified Habel Test is used to screen vaccine potency, it was suggested that the standard test be performed periodically. The standard Habel Test was as described by Habel as an "immunity-breakdown test" in which all animals get the same vaccine dose, and the viral challenge dose varies.

In contrast to the Habel test, the NIH test is an "antigen-extinction test", where varying amounts of vaccine dilutions are used, and the challenge virus dose remains the same. The NIH test, also developed in 1953, is the most standardized of the *in vivo* potency tests. Besides the use of CVS the test includes a reference rabies vaccine which has been calibrated against the International Standard for Rabies Vaccine [208, 209]. A vaccine's potency is compared with the reference vaccine and results are reported in IU. Like the Habel test, the NIH test also uses many mice for the assay. In the NIH assay, 3 or more dilutions of vaccine under test are prepared in fivefold dilution increments, and 0.5 ml are inoculated per mouse (IP) into groups of 16 mice each, in two doses a week apart. Control mice (30–40) are set aside for titration of the challenge virus. All vaccinated mice are challenged 14 days after the initial vaccine dose. The mice are challenged IC with CVS containing between 5 and $50LD_{50}/0.03$ ml. Mice are observed for 14 days as done for the Habel tests. The 50% endpoint titers are calculated by the Reed and Muench Method. The endpoint is calculated as the (ED_{50}) effective dose. Dividing the ED_{50} of the reference vaccine by the test vaccine provides the antigenic value of the vaccine compared to the reference vaccine and should be 0.30 or higher. Other *in vivo* tests for vaccine potency that were developed for use with modified live vaccines, such as the Guinea-Pig Test for chicken-embryo vaccines by the Wistar Institute, and the New York Laboratories and Research (NYLAR) potency tests, were used less often over the decades [210, 211].

Although *in vivo* tests have primarily been used to determine vaccine potency, some *in vitro tests* for determination of vaccine potency have been developed, most likely in consideration of the principle of the 3R's for the "Replacement, Reduction and Refinement" of laboratory animal testing. Such *in vitro* potency tests will hopefully replace the use of animals in the above potency protocols. The earliest of the tests which could be performed either *in vitro* or *in vivo* was the Antigen Binding Test for Vaccine Potency, developed by Arko, Wiktor and Sikes as a collaboration between the CDC and Wistar Institutes in the 1970s. Reportedly, this antigen binding assay compared well with the NIH and Habel tests for consistency but it never displaced the NIH test for vaccine potency, even though it decreased the test completion time from 28 days to just 14 days [212–214].

Historically, several *in vitro* immunodiagnostic methods (e.g., antigen binding, single radial immunodiffusion, and ELISA tests) for potency testing (antigenicity) by detection of rabies virus-specific G-protein have been developed for this purpose [215–219]. The methods utilizing radial immunoassays are less informative, because pre-treatment of the samples can denature the G proteins, and only trimeric forms of the glycoprotein retain immunogenicity. A useful ELISA test for potency is an indirect capture ELISA developed at the Pasteur Institute which utilizes a monoclonal antibody for capture and detection of the intact trimeric glycoprotein within the vaccine and targeting the G protein (aa 330–338) antigenic site III, important for immunogenicity, pathogenicity and virulence [220]. As with the NIH test, the level of G protein detection by ELISA can be compared with that of the reference vaccine and report equivalent international units/ml (EIU/ml) for standardization of potency. Other ELISA tests for potency use different monoclonal antibodies for antigen capture and detection.

Tests Under Development

Future applications will increasingly employ point of care (POC) tests that avoid the requirement for transportation of samples to a central laboratory. In general, they are easily stored, rapid, inexpensive and simple to perform and interpret. Some of the more promising POC tests for rabies diagnosis employ lateral flow assays (LFAs) and isothermal nucleic acid amplification methods. Lateral flow assays, which are more formally known as rabies immunochromatographic diagnostic tests (RIDC), are most frequently used to detect rabies virus antigens. A brain suspension is added to an opening in a device cassette and is adsorbed into a sample pad. The sample flows through a conjugate pad containing anti-rabies virus antibodies labeled with gold nanoparticles. If rabies virus antigens are present, a nanoparticle-labeled antigen-antibody complex diffuses down a membrane strip and reacts with the test band containing a second anti-rabies virus antibody and a colored band appears. When the suspension flows to the control band, the anti-IgG will react with the anti-rabies virus antibodies and the control band will appear. For the test to be valid, the reagents would need to flow appropriately along the membrane strip, and the conjugate labeled with nanoparticles (either bound with antigens or unbound in case of a negative sample) react with the anti-IgG control band. While the concept of these assays is sound, in reality the tests have not performed as well as expected (i.e., lower sensitivity and specificity) when used according to manufacturers' recommendations [221]. Currently, CDC, with collaborators in the USA and international partners, is working on a method to improve the sensitivity of the LFAs for rabies virus antigen detection.

While PCR-based methods have dominated genetic methods of viral detection, efforts to develop alternatives not requiring temperature cycling have been explored. One technique, known as reverse transcription loop-mediated isothermal amplification (RT-LAMP), was applied to detect certain Brazilian rabies virus variants, but the assay as designed failed as a more broadly reactive diagnostic tool [222]. Several investigators have explored the use of reverse transcription recombinase polymerase amplification (RT-RPA) assays for detection of many African variants of rabies virus [223–225]. These assays involve a reverse transcription step to generate a cDNA copy of the RNA template, followed by binding of a recombinase/primer complex to its DNA target to generate a D loop structure from which DNA synthesis can be initiated using a DNA polymerase. This isothermal process is repeated multiple times with real-time detection of the product using a fluorescent probe. With some improvement in sensitivity, this simple, specific, cost-effective, and highly portable POC method promises to provide valuable diagnostic capability, especially in remote tropical regions lacking alternative testing options.

Improvements for detection of virus neutralizing antibodies are also underway by research on a high through-put RFFIT being performed at the CDC. The traditional procedure is a laborious semi-quantitative protocol performed by trained individuals and requiring a DFA test at completion of the incubation, manual reading of the test with a fluorescence microscope, and manual calculations of the titer and international units (IU/ml). A high through-put neutralization test (HTNT) was the concept of Sathesh Panayampalli, Team Lead, Immunodiagnostics and Proteomics Team, CDC while Lauren Greenberg and Jillybeth Burgado performed the research to develop this procedure [226]. The test utilizes a recombinant challenge virus, which is an ERA rabies virus variant expressing Green fluorescent Protein (GFP), with a nuclear localization signal (NLS) for efficient quantitation. The HTNT samples and controls are run in duplicate to obtain average readings. The basic steps of the test are performed in the same way, in volumes convenient for microtiter plates rather than chamber slides. A robotic or mechanical dilutor could potentially be utilized for large numbers of samples. Advantages of this technique are in the reading of the results at test completion. After fixation in paraformaldehyde, the plate is stained with a DAPI nuclear stain and read with an ArrayScan XTI HCS reader, which removes subjectivity because the reader obtains multiple photos of the entire surface of the well. Images can be stored for later review. The HTNT, blue and green channels are used to detect DAPI-stained nuclei (total cells) and ERA-NLS-hMGFP infected cells (green). The total number of DAPI-stained cells and GFP positive are counted. The total percent of the GFP cells are calculated using HCS Studio Software. Results are analysed to obtain the 50% endpoint titer of samples and reference serum for conversion to IU/ml.

Conclusions

The last two hundred years has seen remarkable advancements in laboratory techniques, which in turn enabled ongoing improvement in the ability to detect and characterize rabies virus. Historically, Fig. 5.1 illustrates the timelines for development of relevant diagnostic methods. Figure 5.2 depicts prominent investigators who made significant contributions to these advances. Figures 5.3 and 5.4 illustrate several test methods. Nevertheless, improvements to testing protocols are only useful when made widely available and applied in support of programs which seek to prevent, control and even selectively eliminate the disease. Accordingly, some reference laboratories in the Americas have also been extensively involved in evaluation of the safety and efficacy of vaccine candidates for use as oral rabies vaccines in wildlife, as detailed elsewhere in this volume. Another significant feature of rabies management in the Americas has involved the sharing of knowledge and technical capability throughout the continent. One such major initiative is the annual RITA meeting, initiated in 1989 by the late George Baer, with support from many colleagues, which has grown into the most prominent annual global rabies conference, attracting a world-wide audience, active in all aspects of the rabies field. Future efforts to further mitigate the burden of rabies will continue to depend on sensitive

Fig. 5.3 (continued) proteins in the rough endoplasmic reticulum (ribosomes appear as dark granules with associated RNA strands). Ribonucleoprotein RNP coils are contained in the "matrix" of the inclusions which can be observed if photo is enlarged. The assembly occurs when RNP-matrix protein bound to the cytoplasmic membranes containing the rabies virus glycoprotein inserts coils tightly and the budding of the complete virions from cytoplasmic membranes occurs as visualized here. (**f**) Detection of rabies virus antigen is demonstrated in a brain impression from an Arizona gray fox by the direct fluorescent antibody test (DFA) with 200× magnification using a fluorescence microscope. A duplicate brain impression slide was tested by DRIT with CDC DRIT Reagent 1 (see **g**). (**g**) Detection of rabies virus antigen is demonstrated in a brain impression from an Arizona gray fox by the direct rapid immunohistochemistry test (DRIT) with 200× magnification using a light microscope. Duplicate brain impression slide was tested by DFA with FDI anti-rabies conjugate (see **f**). (**h**). Photo of Jordona Kirby and Kevin Baker from the TN/KY USDA WS program performing DRIT in a lab converted from a horse trailer. This photo emphasizes the adaptability of the DRIT to field testing and mobile laboratory environments. (**i**) Detection of rabies virus antigen is demonstrated in the nerves of a hair follicle (cross-section) by the DFA test. (**j**) Detection of rabies virus antigen is demonstrated in the nerve endings of a hair follicle (sagittal section) by the DFA test. The arrow points out the specific (apple-green) fluorescent rabies virus antigen. (**k**) Detection of rabies virus antigen by immunohistochemistry (IHC) testing is demonstrated in formalin-fixed paraffin embedded tissue (FFPE) sections from the first human rabies case infected with the eastern raccoon rabies virus variant. A duplicate tissue section slide was stained by FFDFA (L). (**l**) Detection of rabies virus antigen by the formalin-fixed direct fluorescent antibody (FFDFA) test is demonstrated in formalin-fixed paraffin embedded tissue (FFPE) sections from the first human rabies case infected with the eastern raccoon rabies virus variant. (**m**) Detection of rabies virus antigen is demonstrated using a lateral flow device (LFD). With improved protocols the rapid test kits such as lateral flow assays (LFA) will be key to providing rabies diagnosis availability to resource limited areas. (**n**) Detection of rabies virus neutralizing antibodies by the serum neutralization test in mice. Jean Smith, developer of the *in vitro* RFFIT for rabies virus neutralizing antibodies, demonstrates that 1 cage of mice is required for every serum dilution of a chamber slide (See **o**). (**o**) Detection of rabies neutralizing antibodies can be performed by *in-vitro* RFFIT or FAVN or antigen binding assays for rabies specific IgM and IgG antibodies

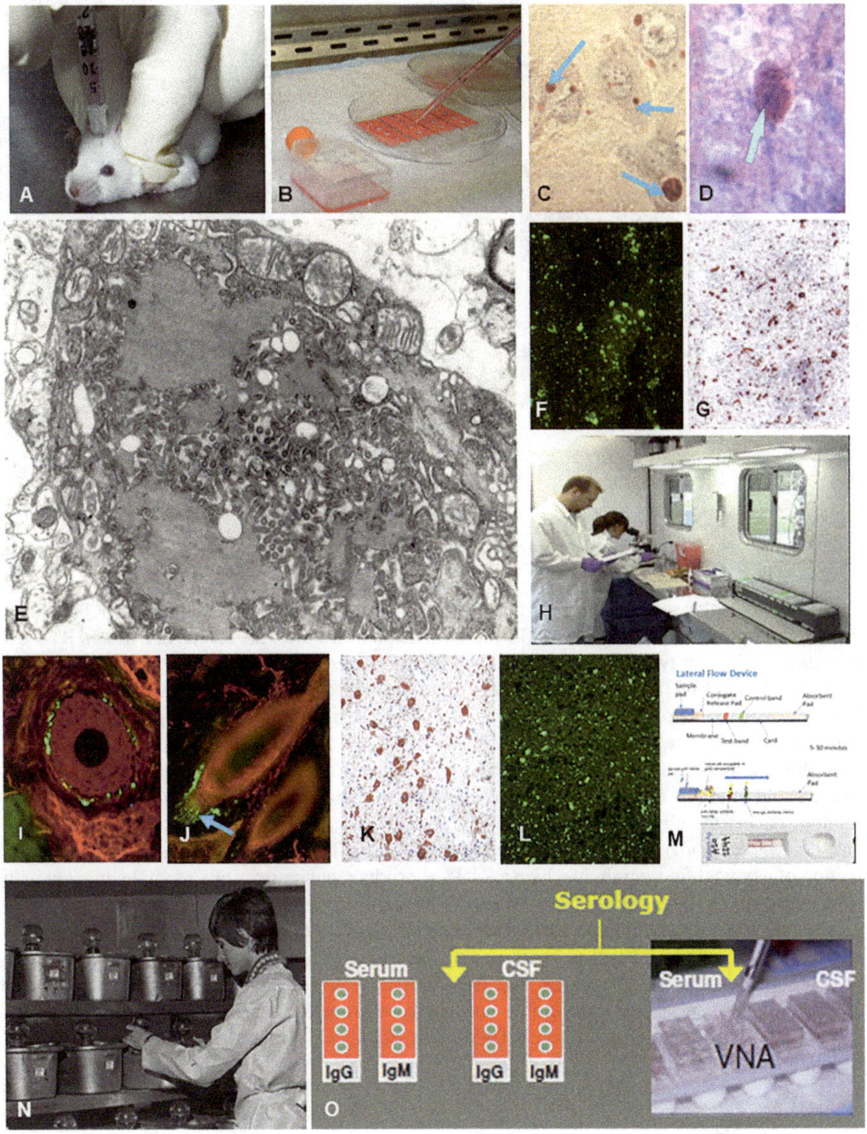

Fig. 5.3 (**a**) *In vivo* rabies virus isolation is demonstrated by intracerebral inoculation of 3-week-old Swiss albino mice as described by Webster and Dawson in 1935. (**b**) *In vitro* rabies virus isolation by inoculation of mouse neuroblastoma cells, is demonstrated using T25 flask and Teflon coated slides in a moist chamber; chamber slides and tissue culture plates may be used also. (**c**) Detection of Negri bodies from formalin-fixed processed paraffin embedded tissue sections is demonstrated. The arrows indicate examples of the abundant magenta colored Negri bodies with dark blue (basophilic granules) internal staining. (**d**) Detection of Negri bodies in brain impressions using the Sellers staining method is demonstrated by close up photo of a single Negri body with view of the classic internal basophilic granules. (**e**) Electron microscopy (thin section) is used to demonstrate the "morphogenesis of rabies" in an infected neuron. Careful examination of the photograph taken in Dr Frederick Murphy's Laboratory at CDC illustrates most steps in the rabies virus infection cycle. After adsorption, endocytosis occurs followed by unwrapping of virus inFig. 5.3 (continued) vacuoles, transcription to generate five mRNAs and full-length positive RNA strands as template for genome negative strand RNA replication and translation of mRNAs to

(continued)

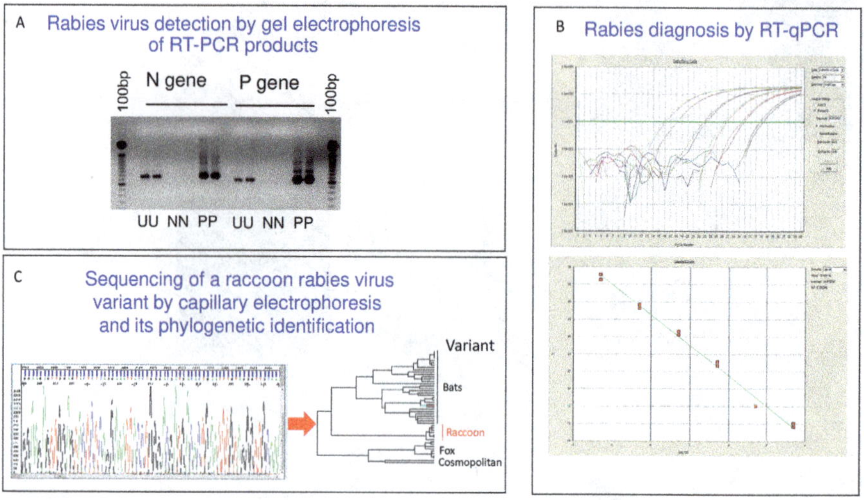

Fig. 5.4 Molecular methods of viral detection and characterization. (**a**) The use of RT-PCR for detection of rabies virus sequences is illustrated. The picture shows the results of a nested RT-PCR targeting portions of two viral genes, N and P. Products were electrophoresed through an agarose gel and detected by UV irradiation after ethidium bromide staining. U, unknown (which is positive); P, positive control; N, negative control. The 100 bp marker confirms the expected size of each amplicon. (**b**) The use of real-time RT-PCR (RT-qPCR) for rabies virus diagnosis. The top panel illustrates the amplification curves obtained over a 40-cycle assay employing a dual labelled probe and ten-fold dilutions of a positive sample; these curves can be used to generate a standard curve (bottom panel) thereby providing a quantitative analysis. Samples that fail to cross above the threshold (green horizontal line) are scored as negative. (**c**) This panel illustrates a screen shot of a section of a typical fluorogram in which a RT-PCR product, generated from a sample infected with a raccoon rabies virus variant, is sequenced by capillary electrophoresis. Unlike traditional sequencing gels using radioisotopes, this method employs four different fluorophores, each associated with one of the four bases to generate a distinct set of peaks, enabling the entire sequence to be determined in a single sample loading. The resulting sequence data can be combined with a set of reference sequences and analyzed by specialized software to produce a phylogenetic tree in which each viral variant type is clustered within a specific clade thereby allowing variant identification of the sample

and timely diagnosis of the disease in both humans and animals. Such methods based on the specific detection of rabies viral components will continue to play a major role. While laboratory-based methods have in the past provided the most accurate diagnostic tools, lower- and middle-income countries still encounter challenges due in part to a lack of basic infrastructure. Tests that are simple, inexpensive and that can be performed in the field will be increasingly important to more remote areas that constitute a significant rabies burden. Only when all countries of the world have strong rabies diagnostic capability will major advances in global elimination of this zoonosis be possible.

References

1. Zinke G (1804) Nue Anichten der Hundswuth, ihere Ursachen und Folgen, nebsteiner sichern Behandlungsart der von tollen Thieren gebissen Menschen. Gabler, Jena; 16:212
2. Galtier P (1879) Studies on rabies. Proc Acad Sci 89:444–446
3. Webster LT, Dawson JR (1935) Early diagnosis of rabies by mouse inoculation. Measurement of humoral immunity to rabies by mouse protection test. Proc Soc Exp Biol Med 32:570–573
4. Koprowski H (1996) The mouse inoculation test. In: Meslin FX, Kaplan MM, Koprowski H (eds) Laboratory techniques in rabies, 4th edn. World Health Organization, Geneva, pp 80–87
5. Fuenzalida E (1973) Suckling mouse brain -vaccine. In: Kaplan MM, Koprowski H (eds) Laboratory techniques in rabies. World Health Organization, Geneva, pp 216–220
6. Noguchi H (1913) Contribution to the cultivation of the parasite of rabies. J Exp Med 18(3):314–316
7. Webster L, Clow A (1937) Propagation of rabies virus in tissue culture. J Exp Med 66(1):125–131
8. Fox JP, Koprowski H, Conwell DP, Black J, Gelfand HM (1957) Study of antirabies immunization of man; observations with HEP Flury and other vaccines, with and without hyperimmune serum, in primary and recall immunizations. Bull World Health Organ 17(6):869–904
9. Koprowski H (1954) Biological modification of rabies virus as a result of its adaptation to chicks and developing chick embryos. Bull World Health Organ 10(5):709–724
10. Koprowski H, Black J, Nelsen DJ (1954) Studies on chick-embryo-adapted-rabies virus. VI. Further changes in pathogenic properties following prolonged cultivation in the developing chick embryo. J Immunol 72(1):94–106
11. Kissling RE (1958) Growth of rabies virus in non-nervous tissue culture. Proc Soc Exp Biol Med 98:223–225
12. Stoker M, MacPherson I (1964) Syrian hamster fibroblast cell line BHK21 and its derivatives. Nature 203:1355–1357
13. Kissling RE, Reese DR (1963) Anti-rabies vaccine of tissue culture origin. J Immunol 91:362–368
14. Wiktor TJ, Koprowski H (1965) Successful immunization of primates with rabies vaccine prepared in human diploid cell strain WI-38. Proc Soc Exp Biol Med 118:1069–1073
15. Larghi OP, Nebel AE, Lazaro L, Savy VL (1975) Sensitivity of bhk-21 cells supplemented with diethylaminoethyl-dextran for detection of street rabies virus in saliva samples. J Clin Microbiol 1(3):243–245
16. Rudd RJ, Trimarchi CV, Abelseth MK (1980) Tissue culture technique for routine isolation of street strain rabies virus. J Clin Microbiol 12(4):590–593
17. Iwasaki Y, Clark H (1977) Rabies virus infection in mouse neuroblastoma. Lab Investig 36(6):578–584
18. Smith AL, Tignor GH, Emmons RW, Woodie JD (1978) Isolation of field rabies virus strains in CER and murine neuroblastoma cell cultures. Intervirology 9(6):359–361
19. Rudd RJ, Trimarchi CV (1989) Development and evaluation of an in vitro virus isolation procedure as a replacement for the mouse inoculation test in rabies diagnosis. J Clin Microbiol 27(11):2522–2528
20. Negri A (1904) Contributo allo studio dell' eziologia della rabia. Bollettino della Societàmedico-chirurgica di Pavia 2:88–115
21. Williams AW (1905–1906) Negri bodies with special reference to diagnosis. Proc New Pathol Soc 5:155–162
22. Williams AW (1908) The diagnosis of rabies. Am J Public Hyg 18(1):10–15
23. Carini A (1911) Sobre a epizzotia de raiva observada no Estado de Santa Catarina: morcegos propagadores da moléstia? Chacaras e Quintaes III(6):11–15
24. Young C, Sellers T (1927) Laboratory: a new method for staining Negri bodies of rabies. Am J Public Health 17(10):1080–1081

25. Hummeler K, Atanasiu P (1973) Electron microscopy. In: Kaplan MM, Koprowski H (eds) Laboratory techniques in rabies, 3rd edn. World Health Organization, Geneva, pp 158–164
26. Knoll M, Ruska E (1932) Das elektronenmikroskop. Z Phys 78:318–339
27. Hottle GA, Morgan C, Peers JH, Wyckoff RW (1951) The electron microscopy of rabies inclusion (Negri) bodies. Proc Soc Exp Biol Med 77(4):721–723
28. Atanasiu P, Lépine P, Sisman J, Dauguet C, Wetten M (1963) Étude morphologique du virus rabique des rues en culture de tissu. Comptes Rendus de l'Academie des Sciences:3219–3221
29. Davies MC, Englert ME, Sharpless GR, Cabasso VJ (1963) The electron microscopy of rabies virus in cultures of chicken embryo tissues. Virology 21:642–651
30. Hummeler K, Koprowski H, Wiktor TJ (1967) Structure and development of rabies virus in tissue culture. J Virol 1(1):152–170
31. Matsumoto S (1962) Electron microscopy of nerve cells infected with street rabies virus. Virology 17:198–202
32. Hummeler K, Tomassini N, Sokol F, Kuwert E, Koprowski H (1968) Morphology of the nucleoprotein component of rabies virus. J Virol 2(10):1191–1199
33. Atanasiu P, Lépine P, Sisman J (1969) Réplication du virus rabique fixe et du virus des rues chez le souriceau nouveau-né. Comptes Rendus Hebdomadaires des Seances de l Academie des Sciences D: Sciences Naturelles 269(13):1226–1229
34. Iwasaki Y, Wiktor TJ, Koprowski H (1973) Early events of rabies virus replicaton in tissue cultures. An electron microscopic study. Lab Investig 28(2):142–148
35. Matsumoto S, Schneider LG, Kawai A, Yonezawa T (1974) Further studies on the replication of rabies and rabies-like viruses in organized cultures of mammalian neural tissues. J Virol 14(4):981–996
36. Shope RE, Murphy FA, Harrison AK, Causey OR, Kemp GE, Simpson DI et al (1970) Two African viruses serologically and morphologically related to rabies virus. J Virol 6(5):690–692
37. Charlton KM (1984) Rabies: spongiform lesions in the brain. Acta Neuropathol 63(3):198–202
38. Charlton KM, Casey GA, Webster WA, Bundza A (1987) Experimental rabies in skunks and foxes. Pathogenesis of the spongiform lesions. Lab Investig 57(6):634–645
39. Goldwasser RA, Kissling RE (1958) Fluorescent antibody staining of street and fixed rabies virus antigens. Proc Soc Exp Biol Med 98(2):219–223
40. Goldwasser RA, Kissling RE, Carski TR, Hosty TS (1959) Fluorescent antibody staining of rabies virus antigens in the salivary glands of rabid animals. Bull World Health Organ 20(4):579–588
41. Goldwasser RA, Kemron A, Nobel TA (1961) Fluorescent antibody in the diagnosis of rabies. Refuah Veterinarith Q Israel Vet Med Assoc 18(4):208–212
42. Beauregard M, Boulanger P, Webster WA (1965) The use of fluorescent antibody staining in the diagnosis of rabies. Can J Comp Med Vet Sci 29(6):141–147
43. World Health Organization (2018) WHO expert consultation on rabies: third report. World Health Organization, Geneva
44. World Health Organization (2018) The direct florescent antibody test (DFAT). In: Rupprecht CE, Fooks AR, Abela-Ridder B (eds) Laboratory techniques in rabies, 5th edn. World Health Organization, Geneva, pp 74–84
45. World Health Organization (1966) WHO expert committee on rabies: fifth report. World Health Organization, Geneva
46. Blenden DC, Bell JF, Tsao AT, Umoh JU (1983) Immunofluorescent examination of the skin of rabies-infected animals as a means of early detection of rabies virus antigen. J Clin Microbiol 18(3):631–636
47. Noah DL, Drenzek CL, Smith JS, Krebs JW, Orciari L, Shaddock J et al (1998) Epidemiology of human rabies in the United States, 1980 to 1996. Ann Intern Med 128(11):922–930
48. Lembo T, Niezgoda M, Velasco-Villa A, Cleaveland S, Ernest E, Rupprecht CE (2006) Evaluation of a direct, rapid immunohistochemical test for rabies diagnosis. Emerg Infect Dis 12(2):310–313

49. Rupprecht CE, Xiang Z, Servat A, Franka R, Kirby J, Ertl HCJ (2018) Additional progress in the development and application of a direct, rapid immunohistochemical test for rabies diagnosis. Vet Sci 5(2):e59
50. Niezgoda M, Rupprecht CE (2006) Standard operating procedure for the direct rapid immunohistochemistry test for the detection of rabies virus antigen. National Laboratory Training Network Course. US Department of Health and Human Services, Centers for Disease Control and Prevention, Atlanta, pp 1–16
51. World Health Organization (2018) The direct rapid immunohistochemistry test (DRIT) for the detection of lyssavirus antigens. In: Rupprecht CE, Fooks AR, Abela-Ridder B (eds) Laboratory techniques in rabies, 5th edn. World Health Organization, Geneva, pp 130–135
52. Dürr S, Naïssengar S, Mindekem R, Diguimbye C, Niezgoda M, Kuzmin I et al (2008) Rabies diagnosis for developing countries. PLoS Negl Trop Dis 2(3):e206
53. Madhusudana SN, Subha S, Thankappan U, Ashwin YB (2012) Evaluation of a direct rapid immunohistochemical test (dRIT) for rapid diagnosis of rabies in animals and humans. Virol Sin 27(5):299–302
54. Saturday GA, King R, Fuhrmann L (2009) Validation and operational application of a rapid method for rabies antigen detection. US Army Med Dep J 13(Jan–Mar):42–45
55. Coetzer A, Sabeta C, Markotter W, Rupprecht CE, Nel LH (2014) Comparison of biotinylated monoclonal and polyclonal antibodies in an evaluation of a direct rapid immunohistochemical test for the routine diagnosis of rabies in southern Africa. PLoS Negl Trop Dis 8(9):e3189
56. Feng Y, Wang M, Liu T, Zhang Y, Tu Z, Guo H (2018) Evaluation of monoclonal antibody-based direct, rapid immunohistochemical test for rabies diagnosis. J Virol Methods 256:12–16
57. Manalo DL, Gomez MRR, Jarilla BR, Ang MJC, Tuason LT, Demetria CS et al (2020) A preliminary evaluation of a locally produced biotinylated polyclonal anti-rabies antibody for direct rapid immunohistochemical test (DRIT) in the Philippines. Acta Trop 211:105610
58. Castro BS, Guedes F, Fernandes ER, Koike G, Katz ISS, Chaves LB et al (2020) Development of biotinylated polyclonal anti-ribonucleoprotein IgG for detection of rabies virus antigen by direct rapid immunohistochemical test. Biologicals 68:74–78
59. Kirby JD, Chipman RB, Nelson KM, Rupprecht CE, Blanton JD, Algeo TP et al (2017) Enhanced rabies surveillance to support effective oral rabies vaccination of Raccoons in the Eastern United States. Trop Med Infect Dis 2(3):34
60. Middel K, Fehlner-Gardiner C, Pulham N, Buchanan T (2017) Incorporating direct rapid immunohistochemical testing into large-scale wildlife rabies surveillance. Trop Med Infect Dis 2(3):21
61. World Health Organization (1973) WHO expert committee on rabies: sixth report. World Health Organization, Geneva
62. Moore SM, Hanlon CA (2010) Rabies-specific antibodies: measuring surrogates of protection against a fatal disease. PLoS Negl Trop Dis 4(3):e595
63. Manning SE, Rupprecht CE, Fishbein DB, Hanlon CA, Lumlertdacha B, Guerra M et al (2008) Human rabies prevention–United States, 2008: recommendations of the advisory committee on immunization practices. Morb Mortal Wkly Rep Recomm Rep 57(RR03):1–28
64. OIE (2013) Rabies. Manual of diagnostic tests and vaccines for terrestrial animals. Office international des épizooties, Paris, pp 1–26
65. World Health Organization (1992) WHO expert committee on rabies: eigth report. World Health Organization, Geneva
66. Ledesma LA, Lemos ERS, Horta MA (2020) Comparing clinical protocols for the treatment of human rabies: the Milwaukee protocol and the Brazilian protocol (Recife). Rev Soc Bras Med Trop 53:e20200352
67. Smith JS, Yager PA, Baer GM (1973) A rapid reproducible test for determining rabies neutralizing antibody. Bull World Health Organ 48(5):535–541
68. Smith JS, Yager PA, Baer GM (1996) A rapid fluorescent focus inhibition test (RFFIT) for determining rabies virus neutralizing antibody. In: Meslin FX, Kaplan MM, Koprowski H (eds) Laboratory techniques in rabies, 4th edn. World Health Organization, Geneva, pp 181–192

69. Cliquet F, Aubert M, Sagne L (1998) Development of a fluorescent antibody virus neutralisation test (FAVN test) for the quantitation of rabies-neutralising antibody. J Immunol Methods 212:79–87
70. Briggs DJ, Smith JS, Mueller FL, Schwenke J, Davis RD, Gordon CR et al (1998) A comparison of two serological methods for detecting the immune response after rabies vaccination in dogs and cats being exported to rabies-free areas. Biologicals 26:347–355
71. Favoretto SR, Carrieri ML, Tino MS, Zanetti CR, Pereira OA (1993) Simplified fluorescent inhibition microtest for the titration of rabies neutralizing antibodies. Rev Inst Med Trop Sao Paulo 35(2):171–175
72. Leffingwell L, Irons JV (1965) Rabies antibodies in human serums titrated by the indirect FA method. Public Health Rep 80(11):999–1004
73. Thomas JB, Sikes RK, Ricker AS (1963) Evaluation of indirect fluorescent antibody technique for detection of rabies antibody in human sera. J Immunol 91:721–723
74. Atanasiu P, Savy V, Perrin P (1977) Epreuve immunoenzymatique pour la detection rapide des anticorps anti-rabiques. Ann Microbiol 128A:489–498
75. Grassi M, Wandeler AI, Peterhans E (1989) Enzyme-linked immunosorbent assay for determination of antibodies to the envelope glycoprotein of rabies virus. J Clin Microbiol 27:899–902
76. Realegeno S, Niezgoda M, Yager PA, Kumar A, Hoque L, Orciari L et al (2018) An ELISA-based method for detection of rabies virus nucleoprotein-specific antibodies in human antemortem samples. PLoS One 13(11):e0207009
77. Elmgren LD, Wandeler AI (1996) Competitive ELISA for the detection of rabies virus-neutralizing antibodies. In: Meslin FX, Kaplan MM, Koprowski H (eds) Laboratory techniques in rabies, 4th edn. World Health Organization, Geneva, pp 200–208
78. Fehlner-Gardiner C, Rudd R, Donovan D, Slate D, Kempf L, Badcock J (2012) Comparing ONRAB® and RABORAL V-RG® oral rabies vaccine field performance in raccoons and striped skunks, New Brunswick, Canada, and Maine, USA. J Wildl Dis 48(1):157–167
79. Mainguy J, Fehlner-Gardiner C, Slate D, Rudd RJ (2013) Oral rabies vaccination in raccoons: comparison of ONRAB® and RABORAL V-RG® vaccine-bait field performance in Québec, Canada and Vermont, USA. J Wildl Dis 49(1):190–193
80. Saiki R, Gelfand D, Stoffel S, Scharf S, Higuchi R, Horn G et al (1988) Primer-directed enzymatic amplification of DNA with a thermostable DNA polymerase. Science 239(4839):487–491
81. Sacramento D, Bourhy H, Tordo N (1991) PCR technique as an alternative method for diagnosis and molecular epidemiology of rabies virus. Mol Cell Probes 5(3):229–240
82. Smith JS, Orciari LA, Yager PA, Seidel HD, Warner CK (1992) Epidemiologic and historical relationships among 87 rabies virus isolates as determined by limited sequence analysis. J Infect Dis 166:296–307
83. Smith JS, Fishbein DB, Rupprecht CE, Clark K (1991) Unexplained rabies in three immigrants in the United States. A virologic investigation. N Engl J Med 324(4):205–211
84. Charlton KM, Nadin-Davis S, Casey GA, Wandeler AI (1997) The long incubation period in rabies: delayed progression of infection in muscle at the site of exposure. Acta Neuropathol 94(1):73–77
85. Nadin-Davis SA (1998) Polymerase chain reaction protocols for rabies virus discrimination. J Virol Methods 75:1–8
86. Loza-Rubio E, Rojas-Anaya E, Banda-Ruíz VM, Nadin-Davis SA, Cortez-García B (2005) Detection of multiple strains of rabies virus RNA using primers designed to target Mexican vampire bat variants. Epidemiol Infect 133(5):927–934
87. Brito MG, Chamone TL, Silva FJ, Wada MY, Miranda AB, Castilho JG et al (2011) Antemortem diagnosis of human rabies in a veterinarian infected when handling a herbivore in Minas Gerais, Brazil. Rev Inst Med Trop Sao Paulo 53:39–44
88. Nadin-Davis SA, Sheen M, Wandeler AI (2009) Development of real-time reverse transcriptase polymerase chain reaction methods for human rabies diagnosis. J Med Virol 81:1484–1497

89. Gigante CM, Dettinger L, Powell JW, Seiders M, Condori REC, Griesser R et al (2018) Multi-site evaluation of the LN34 pan-lyssavirus real-time RT-PCR assay for post-mortem rabies diagnostics. PLoS One 13(5):e0197074
90. Wadhwa A, Wilkins K, Gao J, Condori Condori RE, Gigante CM, Zhao H et al (2017) A pan-lyssavirus Taqman Real-time RT-PCR assay for the detection of highly variable rabies virus and other lyssaviruses. PLoS Negl Trop Dis 11(1):e0005258
91. Marston DA, Jennings DL, MacLaren NC, Dorey-Robinson D, Fooks AR, Banyard AC et al (2019) Pan-lyssavirus Real time RT-PCR for rabies diagnosis. J Vis Exp 149:e59709
92. Appler K, Brunt S, Jarvis JA, Davis AD (2019) Clarifying indeterminate results on the rabies direct fluorescent antibody test using real-time reverse transcriptase polymerase chain reaction. Public Health Rep 134(1):57–62
93. Dupuis M, Brunt S, Appler K, Davis A, Rudd R (2015) Comparison of automated quantitative reverse transcription-PCR and direct fluorescent-antibody detection for routine rabies diagnosis in the United States. J Clin Microbiol 53(9):2983–2989
94. Wandeler AI, Salsberg E (1999) Raccoon rabies in eastern Ontario. Can Vet J 40:731
95. Köhler G, Milstein C (1975) Continuous cultures of fused cells secreting antibody of predefined specificity. Nature 256:495–497
96. Wiktor TJ, Koprowski H (1978) Monoclonal antibodies against rabies virus produced by somatic cell hybridization: detection of antigenic variants. Proc Natl Acad Sci USA 75(8):3938–3942
97. Smith JS, Reid-Sanden FL, Roumillat LF, Trimarchi C, Clark K, Baer GM et al (1986) Demonstration of antigenic variation among rabies virus isolates by using monoclonal antibodies to nucleocapsid proteins. J Clin Microbiol 24(4):573–580
98. Smith JS (1989) Rabies virus epitopic variation: use in ecologic studies. Adv Virus Res 36:215–253
99. Condori RE, Aragon A, Breckenridge M, Pesko K, Mower K, Ettestad P, et al. (2022) Divergent rabies virus variant of probable bat origin in 2 gray foxes, New Mexico, USA. Emerg Infect Dis 28(6):1137–1145
100. Velasco-Villa A, Orciari LA, Juarez-Islas V, Gomez-Sierra M, Padilla-Medina I, Flisser A et al (2006) Molecular diversity of rabies viruses associated with bats in Mexico and other countries of the Americas. J Clin Microbiol 44(5):1697–1710
101. Favoretto SR, Carrieri ML, Cunha ES, Aguiar EAC, Silva LHQ, Sodré MM et al (2002) Antigenic typing of brazilian rabies virus samples isolated from animals and humans, 1989–2000. Rev Inst Med Trop Sao Paulo 44:91–95
102. Chaves LB, Achkar SM, Rodrigues da Silva AC, Caporale GMM, Cruz PS, Batista AM et al (2015) Monoclonal antibodies for characterization of rabies virus isolated from non-hematophagous bats in Brazil. J Infect Dev Ctries 9(11):1238–1249
103. Carnieli PJ, Fahl WO, Castilho JG, Oliveira RN, Macedo CI, Durymanova E et al (2008) Characterization of rabies virus isolated from canids and identification of the main wild canid host in Northeastern Brazil. Virus Res 31:33–46
104. Favoretto SR, de Mattos CC, Morais NB, Alves Araújo FA, de Mattos CA (2001) Rabies in marmosets (Callithrix jacchus), Ceará, Brazil. Emerg Infect Dis 7:1062–1065
105. Webster WA, Casey GA, Charlton KM (1986) Major antigenic groups of rabies virus in Canada determined by anti-nucleocapsid monoclonal antibodies. Comp Immunol Microbiol Infect Dis 9(1):59–69
106. Fehlner-Gardiner C, Nadin-Davis S, Armstrong J, Muldoon F, Bachmann P, Wandeler A (2008) ERA vaccine-derived cases of rabies in wildlife and domestic animals in Ontario, Canada, 1989–2004. J Wildl Dis 44(1):71–85
107. Nadin-Davis SA, Huang W, Armstrong J, Casey GA, Bahloul C, Tordo N et al (2001) Antigenic and genetic divergence of rabies viruses from bat species indigenous to Canada. Virus Res 74:139–156
108. Bernardi F, Nadin-Davis SA, Wandeler AI, Armstrong J, Gomes AAB, Lima FS et al (2005) Antigenic and genetic characterization of rabies viruses isolated from domestic and wild animals of Brazil identifies the hoary fox as a rabies reservoir. J Gen Virol 86:3153–3162

109. Nadin-Davis SA, Simani S, Armstrong J, Fayaz A, Wandeler AI (2003) Molecular and antigenic characterization of rabies viruses from Iran identifies variants with distinct epidemiological origins. Epidemiol Infect 131:777–790
110. Nadin-Davis SA, Fehlner-Gardiner C, Sheen M, Wandeler AI (2010) Characterization of a panel of anti-phosphoprotein monoclonal antibodies generated against the raccoon strain of rabies virus. Virus Res 152(1):126–136
111. Nadin-Davis SA, Sheen M, Abdel-Malik M, Elmgren L, Armstrong J, Wandeler AI (2000) A panel of monoclonal antibodies targeting the rabies virus phosphoprotein identifies a highly variable epitope of value for sensitive strain discrimination. J Clin Microbiol 38(4):1397–1403
112. Nadin-Davis SA, Elmgren L, Sheen M, Sabeta C, Wandeler AI (2011) Generation and characterization of a panel of anti-phosphoprotein monoclonal antibodies directed against Mokola virus. Virus Res 160(1–2):238–245
113. Nadin-Davis SA, Casey GA, Wandeler A (1993) Identification of regional variants of the rabies virus within the Canadian province of Ontario. J Gen Virol 74:829–837
114. Nadin-Davis SA (2020) Molecular epidemiology. In: Fooks AR, Jackson AC (eds) Rabies: scientific basis of the disease and its management, 4th edn. Academic, London, pp 145–195
115. Diaz AM, Papo S, Rodriguez A, Smith JS (1994) Antigenic analysis of rabies-virus isolates from Latin America and the Caribbean. Zentralbl Veterinarmed B 41(3):153–160
116. Radford AD, Chapman D, Dixon L, Chantrey J, Darby AC, Hall N (2012) Application of next-generation sequencing technologies in virology. J Gen Virol 93:1853–1868
117. McColl K, Gould A, Selleck P, Hooper P, Westbury H, Smith J (1993) Polymerase chain reaction and other laboratory techniques in the diagnosis of long incubation rabies in Australia. Aust Vet J 70(3):84–89
118. Smith JS, Orciari LA, Yager PA (1995) Molecular epidemiology of rabies in the United States. Semin Virol 6:387–400
119. Smith JS, Seidel HD (1993) Rabies: a new look at an old disease. Prog Med Virol 40:82–106
120. Malerczyk C, DeTora L, Gniel D (2011) Imported human rabies cases in Europe, the United States, and Japan, 1990 to 2010. J Travel Med 18:402–407
121. De Serres G, Dallaire F, Cote M, Skowronski DM (2008) Bat rabies in the United States and Canada from 1950 through 2007: human cases with and without bat contact. Clin Infect Dis 46:1329–1337
122. Messenger SL, Smith JS, Rupprecht CE (2002) Emerging epidemiology of bat-associated cryptic cases of rabies in humans in the United States. Clin Infect Dis 35:738–747
123. Hercules Y, Bryant NJ, Wallace RM, Nelson R, Palumbo G, Williams JN et al (2018) Rabies in a dog imported from Egypt – Connecticut, 2017. Morb Mortal Wkly Rev 67(50):1388–1391
124. Kuzmin IV, Shi M, Orciari LA, Yager PA, Velasco-Villa A, Kuzmina NA et al (2012) Molecular inferences suggest multiple host shifts of rabies viruses from bats to mesocarnivores in Arizona during 2001–2009. PLoS Pathog 8(6):e1002786
125. Borucki MK, Chen-Harris H, Lao V, Vanier G, Wadford DA, Messenger S et al (2013) Ultradeep sequencing of intra-host rabies virus populations during cross-species transmission. PLoS Negl Trop Dis 7(11):e2555
126. Davis R, Nadin-Davis SA, Moore M, Hanlon C (2013) Genetic characterization and phylogenetic analysis of skunk-associated rabies viruses in North America with special emphasis on the central plains. Virus Res 174:27–36
127. Aranda M, Lopez-de Buen L (1999) Rabies in skunks from Mexico. J Wildl Dis 35(3):574–577
128. Nadin-Davis SA, Casey GA, Wandeler AI (1994) A molecular epidemiological study of rabies virus in Central Ontario and western Quebec. J Gen Virol 75(10):2575–2583
129. Nadin-Davis S, Muldoon F, Whitney H, Wandeler AI (2008) Origins of the rabies viruses associated with an outbreak in Newfoundland during 2002–2003. J Wildl Dis 44(1):86–98
130. Nadin-Davis SA, Fehlner-Gardiner C (2019) Origins of the arctic fox variant rabies viruses responsible for recent cases of the disease in southern Ontario. PLoS Negl Trop Dis 13(9):e0007699

131. Biek R, Henderson CI, Waller LA, Rupprecht CE, Real LA (2007) A high-resolution genetic signature of demographic and spatial expansion in epizootic rabies virus. Proc Natl Acad Sci USA 104:7993–7998
132. Szanto AG, Nadin-Davis SA, Rosatte RC, White BN (2011) Genetic tracking of the raccoon variant of rabies virus in eastern North America. Epidemics 3:76–87
133. Trewby H, Nadin-Davis SA, Real LA, Biek R (2017) Phylogeographic analysis of rabies virus incursions across US-Canada border. Emerg Infect Dis 23(9):1454–1461
134. Nadin-Davis S, Buchanan T, Nituch L, Fehlner-Gardiner C (2019) A long-distance translocation initiated an outbreak of raccoon rabies in Hamilton, Ontario, Canada. PLoS Negl Trop Dis 14(3):e0008113
135. Nadin-Davis SA, Fu Q, Trewby H, Biek R, Johnson RH, Real L (2018) Geography but not alternative host species explain the spread of raccoon rabies in Vermont. Epidemiol Infect 146:1977–1986
136. Velasco-Villa A, Escobar LE, Sanchez A, Shi M, Streicker DG, Gallardo-Romero NF et al (2017) Successful strategies implemented towards the elimination of canine rabies in the Western hemisphere. Antivir Res 143:1–12
137. Queiroz LH, Favoretto SR, Cunha EMS, Campos ACA, Lopes MC, de Carvalho C et al (2012) Rabies in Southeast Brazil: a change in the epidemiological pattern. Arch Virol 157(1):93–105
138. Favoretto SR, de Mattos CC, de Mattos CA, Campos ACA, Sacramento DRV, Durigon EL (2013) The emergence of wildlife species as a source of human rabies infection in Brazil. Epidemiol Infect 141(7):1552–1561
139. Escobar LE, Peterson AT, Favi M, Yung V, Medina-Vogel G (2015) Bat-borne rabies in Latin America. Rev Inst Med Trop Sao Paulo 57:63–72
140. Velasco-Villa A, Orciari LA, Souza V, Juárez-Islas V, Gomez-Sierre M, Castillo A et al (2005) Molecular epizootiology of rabies associated with terrestrial carnivores in Mexico. Virus Res 111:13–27
141. Garcés-Ayala F, Aréchiga-Ceballos N, Ortiz-Alcántara JM, González-Durán E, Pérez-Agüeros SI, Méndez-Tenorio A et al (2017) Molecular characterization of atypical antigenic variants of canine rabies virus reveals its reintroduction by wildlife vectors in southeastern Mexico. Arch Virol 162(12):3629–3637
142. Pérez-Agüeros S, Ortiz-Alcántara JM, Garcés-Ayala F, Mendieta-Condado E, González-Durán E, Aréchiga-Ceballos N et al (2018) Genome sequence of a rabies virus isolated from a dog in Chiapas, Mexico, 2013. Genome Announc 6(4):e01586–17
143. de Mattos CA, de Mattos CC, Smith JS, Miller ET, Papo S, Utrera A et al (1996) Genetic characterization of rabies field isolates from Venezuela. J Clin Microbiol 34(6):1553–1558
144. Warner CK, Zaki SR, Shieh WJ, Whitfield SG, Smith JS, Orciari LA et al (1999) Laboratory investigation of human deaths from vampire bat rabies in Peru. Am J Trop Med Hyg 60:502–507
145. de Mattos CA, Favi M, Yung V, Pavletic C, de Mattos CC (2000) Bat rabies in urban centers in Chile. J Wildl Dis 36(2):231–240
146. Bárcenas-Reyes I, Loza-Rubio E, Cantó Alarcón GJ, Luna-Cozar J, Enríquez-Vázquez A, Barrón-Rodríguez RJ et al (2017) Whole genome sequence phylogenetic analysis of four Mexican rabies viruses isolated from cattle. Res Vet Sci 113:21–24
147. Nadin-Davis SA, Loza-Rubio E (2006) The molecular epidemiology of rabies associated with chiropteran hosts in Mexico. Virus Res 117:215–226
148. Nadin-Davis SA, Feng Y, Mousse D, Wandeler AI, Aris-Brosou S (2010) Spatial and temporal dynamics of rabies virus variants in big brown bat populations across Canada: footprints of an emerging zoonosis. Mol Ecol 19:2120–2136
149. Hebert PDN, Cywinska A, Ball SL, deWaard JR (2003) Biological identifications through DNA barcodes. Proc R Soc Lond Ser B 270:313–322
150. Carnieli P Jr, Castilho JG, Oliveira RN, Brandão PE, Batista HBCR (2016) Identification of different species of mammalians involved in zoonoses as reservoirs or hosts by sequencing of the mitochondrial DNA cytochrome B gene. Annu Res Rev Biol 10(1):ARRB.25230

151. Streicker DG, Turmelle AS, Vonhof MJ, Kuzmin IV, McCracken GF, Rupprecht CE (2010) Host phylogeny constrains cross-species emergence and establishment of rabies virus in bats. Science 329:676–679
152. Nadin-Davis S, Alnabelseya N, Knowles MK (2017) The phylogeography of *Myotis* bat-associated rabies viruses across Canada. PLoS Negl Trop Dis 11(5):e0005541
153. Faria NR, Suchard MA, Rambaut A, Streicker DG, Lemey P (2013) Simultaneously reconstructing viral cross-species transmission history and identifying the underlying constraints. Philos Trans R Soc B 368:20120196
154. Streicker DG, Altizer SM, Velasco-Villa A, Rupprecht CE (2012) Variable evolutionary routes to host establishment across repeated rabies virus host shifts among bats. Proc Natl Acad Sci USA 109(48):19715–19720
155. Streicker DG, Winternitz JC, Satterfield DA, Condori-Condori RE, Broos A, Tello C et al (2016) Host-pathogen evolutionary signatures reveal dynamics and future invasions of vampire bat rabies. Proc Natl Acad Sci USA 113(39):10926–10931
156. Carnieli P Jr, Ruthner Batista HB, de Novaes Oliveira R, Castilho JG, Vieira LF (2013) Phylogeographic dispersion and diversification of rabies virus lineages associated with dogs and crab-eating foxes (*Cerdocyon thous*) in Brazil. Arch Virol 158(11):2307–2313
157. Barbosa TFS, Medeiros DBA, Travassos da Rosa ES, Casseb LMN, Medeiros R, Pereira AS et al (2008) Molecular epidemiology of rabies virus isolated from different sources during a bat-transmitted human outbreak occurring in Augusto Correa municipality, Brazilian Amazon. Virology 370(2):228–236
158. Oliveira RN, de Souza SP, Lobo RSV, Castilho JG, Macedo CI, Carnieli PJ et al (2010) Rabies virus in insectivorous bats: implications of the diversity of the nucleoprotein and glycoprotein genes for molecular epidemiology. Virology 405:352–360
159. Kobayashi Y, Suzuki Y, Itou T, Ito FH, Sakai T, Gojobori T (2011) Evolutionary history of dog rabies in Brazil. J Gen Virol 92(1):85–90
160. Allendorf SD, Cortez A, Heinemann MB, Harary CMA, Antunes JMAP, Peres MG et al (2012) Rabies virus distribution in tissues and molecular characterization of strains from naturally infected non-hematophagous bats. Virus Res 165(2):119–125
161. Kobayashi Y, Sato G, Mochizuki N, Hirano S, Itou T, Carvalho AAB et al (2008) Molecular and geographic analyses of vampire bat-transmitted cattle rabies in Central Brazil. BMC Vet Res 4(1):44
162. Kobayashi Y, Sato G, Kato M, Itou T, Cunha EMS, Silva MV et al (2007) Genetic diversity of bat rabies viruses in Brazil. Arch Virol 152:1995–2004
163. Calderón A, Guzmán C, Mattar S, Rodríguez V, Acosta A, Martínez C (2019) Frugivorous bats in the Colombian Caribbean region are reservoirs of the rabies virus. Ann Clin Microbiol Antimicrob 18(1):11
164. Kotait I, Oliveira RN, Carrieri ML, Castilho JG, Macedo CI, Pereira PMC et al (2019) Non-human primates as a reservoir for rabies virus in Brazil. Zoonoses Public Health 66(1):47–59
165. Aréchiga-Ceballos N, Velasco-Villa A, Shi M, Flores-Chávez S, Barrón B, Cuevas-Domínguez E et al (2010) New rabies virus variant found during an epizootic in white-nosed coatis from the Yucatan peninsula. Epidemiol Infect 138(11):1586–1589
166. Dell'Armelina Rocha PR, Velasco-Villa A, de Lima EM, Salomoni A, Fusaro A, da Conceição Souza E et al (2020) Unexpected rabies variant identified in kinkajou (*Potos flavus*), Mato Grosso, Brazil. Emerg Microbe Infect 9(1):851–854
167. Escobar LE, Peterson AT, Favi M, Yung V, Pons DJ, Medina-Vogel G (2013) Ecology and geography of transmission of two bat-borne rabies lineages in Chile. PLoS Negl Trop Dis 7(12):e2577
168. Yung V, Favi M, Fernandez J (2012) Typing of the rabies virus in Chile, 2002–2008. Epidemiol Infect 140(12):2157–2162
169. Pinero C, Dohmen FG, Beltran F, Martinez L, Novaro L, Russo S et al (2012) High diversity of rabies viruses asociated with insectivorous bats in Argentina: presence of several independent enzootics. PLoS Negl Trop Dis 6(5):e1635

170. Condori-Condori RE, Streicker DG, Cabezas-Sanchez C, Velasco-Villa A (2013) Enzootic and epizootic rabies associated with vampire bats, Peru. Emerg Infect Dis 19(9):1463–1469
171. Guarino H, Castilho JG, Souto J, Oliveira RDN, Carrieri ML, Kotait I (2013) Antigenic and genetic characterization of rabies virus isolates from Uruguay. Virus Res 173(2):415–420
172. Torres C, Lema C, Dohmen FG, Beltran F, Novaro L, Russo S et al (2014) Phylodynamics of vampire bat-transmitted rabies in Argentina. Mol Ecol 23:2340–2352
173. Favi M, Nina A, Yung V, Fernández J (2003) Characterization of rabies virus isolates in Bolivia. Virus Res 97(2):135–140
174. Amarilla ACF, Pompei JCA, Araujo DB, Vázquez FA, Galeano RR, Delgado LM et al (2018) Re-emergence of rabies virus maintained by canid populations in Paraguay. Zoonoses Public Health 65(1):222–226
175. Ellison JA, Gilbert AT, Recuenco S, Moran D, Alvarez DA, Kuzmina NA et al (2014) Bat rabies in Guatemala. PLoS Negl Trop Dis 8(7):e3070
176. Seetahal JFR, Velasco-Villa A, Allicock OM, Adesiyun AA, Bissessar J, Amour K et al (2013) Evolutionary history and phylogeography of rabies viruses associated with outbreaks in Trinidad. PLoS Negl Trop Dis 7(8):e2365
177. Nadin-Davis SA, Torres G, de Los Angeles Ribas M, Guzman M, de la Paz RC, Morales M et al (2006) A molecular epidemiological study of rabies in Cuba. Epidemiol Infect 134:1313–1324
178. Nadin-Davis SA, Velez J, Malaga C, Wandeler AI (2008) A molecular epidemiological study of rabies in Puerto Rico. Virus Res 131:8–15
179. Zieger U, Marston DA, Sharma R, Chikweto A, Tiwari K, Sayyid M et al (2014) The Phylogeography of rabies in Grenada, West Indies, and implications for control. PLoS Negl Trop Dis 8(10):e3251
180. Nadin-Davis SA, Huang W, Wandeler AI (1996) The design of strain-specific polymerase chain reactions for discrimination of the raccoon rabies virus strain from indigenous rabies viruses of Ontario. J Virol Methods 57:141–156
181. Bastida-González F, Ramírez-Hernández DG, Chavira-Suárez E, Lara-Padilla E, Zárate-Segura P (2016) Development of primer pairs from molecular typing of rabies virus variants present in Mexico. Biomed Res Int 2016:4659470
182. Sato G, Tanabe H, Shoji Y, Itou T, Ito FH, Sato T et al (2005) Rapid discrimination of rabies viruses isolated from various host species in Brazil by multiplex reverse transcription-polymerase chain reaction. J Clin Virol 33(4):267–273
183. Nel LH, Bingham J, Jacobs JA, Jaftha JB (1998) A nucleotide-specific polymerase chain reaction assay to differentiate rabies virus biotypes in South Africa. Onderstepoort J Vet Res 65(4):297–303
184. Picard-Meyer E, Bruyère V, Barrat J, Tissot E, Barrat MJ, Cliquet F (2004) Development of a hemi-nested RT-PCR method for the specific determination of European bat lyssavirus 1: comparison with other rabies diagnostic methods. Vaccine 22(15):1921–1929
185. Nadin-Davis SA, Turner G, Paul JPV, Madhusudana N, Wandeler AI (2007) Emergence of arctic-like rabies lineage in India. Emerg Infect Dis 13(1):111–116
186. David D, Yakobson B, Smith JS, Stram Y (2000) Molecular epidemiology of rabies virus isolates from Israel and other middle- and near-eastern countries. J Clin Microbiol 38(2):755–762
187. Nanayakkara S, Smith JS, Rupprecht CE (2003) Rabies in Sri Lanka: splendid isolation. Emerg Infect Dis 9:368–371
188. Smith JS, Yager PA, Orciari LA (1993) Rabies in wild and domestic carnivores of Africa: epidemiological and historical associations determined by limited sequence analysis. Onderstepoort J Vet Res 60(4):307–314
189. Kuzmin IV, Wu X, Tordo N, Rupprecht CE (2008) Complete genomes of Aravan, Khujand, Irkut and west Caucasian bat viruses, with special attention to the polymerase gene and non-coding regions. Virus Res 136(1–2):81–90
190. Hooper PT, Lunt RA, Gould AR, Samaratunga H, Hyatt AD, Gleeson LJ et al (1997) A new lyssavirus – the first endemic rabies-related virus recognized in Australia. Bulletin du Institut Pasteur 95:209–218

191. Banyard AC, Davis A, Gilbert AT, Markotter W (2020) Bat rabies. In: Fooks AR, Jackson AC (eds) Rabies: scientific basis of the disease and its management, 4th edn. Academic, London, pp 231–276
192. Kuzmin IV, Hughes GJ, Botvinkin AD, Orciari LA, Rupprecht CE (2005) Phylogenetic relationships of Irkut and west Caucasian bat viruses within the *lyssavirus* genus and suggested quantitative criteria based on the N gene sequence for lyssavirus genotype definition. Virus Res 111:28–43
193. Fekadu M, Greer PW, Chandler FW, Sanderlin DW (1988) Use of the avidin-biotin peroxidase system to detect rabies antigen in formalin-fixed paraffin-embedded tissues. J Virol Methods 19(2):91–96
194. Whitfield SG, Fekadu M, Shaddock JH, Niezgoda M, Warner CK, Messenger SL et al (2001) A comparative study of the fluorescent antibody test for rabies diagnosis in fresh and formalin-fixed brain tissue specimens. J Virol Methods 95(1–2):145–151
195. Balachandran A, Charlton K (1994) Experimental rabies infection of non-nervous tissues in skunks (*Mephitis mephitis*) and foxes (*Vulpes vulpes*). Vet Pathol 31(1):93–102
196. Hamir AN, Moser G, Fu ZF, Dietzschold B, Rupprecht CE (1995) Immunohistochemical test for rabies: identification of a diagnostically superior monoclonal antibody. Vet Rec 136:295–296
197. Niezgoda M, Satheshkumar PS (2021) Immunohistochemistry test for the lyssavirus antigen detection from formalin-fixed tissues. J Vis Exp 176:60138
198. Fernandes ER, de Andrade HF, Lancellotti CLP, Quaresma JAS, Demachki S, da Costa Vasconcelos PF et al (2011) In situ apoptosis of adaptive immune cells and the cellular escape of rabies virus in CNS from patients with human rabies transmitted by Desmodus rotundus. Virus Res 156(1):121–126
199. Jackson AC (1992) Detection of rabies virus mRNA in mouse brain by using in situ hybridization with digoxigenin-labelled RNA probes. Mol Cell Probes 6(2):131–136
200. Nadin-Davis SA, Abdel-Malik M, Armstrong J, Wandeler AI (2002) Lyssavirus P gene characterisation provides insights into the phylogeny of the genus and identifies structural similarities and diversity within the encoded phosphoprotein. Virology 298:286–305
201. Nadin-Davis SA, Sheen M, Wandeler AI (2003) Use of discriminatory probes for strain typing of formalin-fixed, rabies virus-infected tissues by in situ hybridization. J Clin Microbiol 41(9):4343–4352
202. Warner CK, Whitfield SG, Fekadu M, Ho H (1997) Procedures for reproducible detection of rabies virus antigen mRNA and genome in situ in formalin-fixed tissues. J Virol Methods 67(1):5–12
203. Condori RE, Niezgoda M, Lopez G, Matos CA, Mateo ED, Gigante C et al (2020) Using the LN34 pan-lyssavirus Real-time RT-PCR assay for rabies diagnosis and rapid genetic typing from formalin-fixed human brain tissue. Viruses 12(1):120
204. WHO Expert Committee on Biological Standardization [Internet]. World Health Organization. 2008 [cited October 1, 2018]. Available from: http://apps.who.int/iris/bitstream/handle/10665/70593/WHO_BS_08.2087_eng.pdf
205. Habel K (ed) (1953) Potency test of flury vaccine. Expert Committee on Rabies, Second Session, 1953 14–19 September. World Health Organization, Rome
206. Habel K (1966) Laboratory techniques in rabies: Habel test for potency. Monograph Series No. 23, 2nd edn., World Health Organization, Geneva,140–143
207. Reed LJ, Muench HA (1938) A simplified method of estimating fifty percent endpoint. Am J Hyg 27:493–497
208. Seligmann EB (1966) Laboratory techniques in rabies: the NIH test for potency. Monograph Series No. 23, 2nd edn. World Health Organization, Geneva, 145–151
209. World Health Organization (2018) The NIH test for potency detection of vaccines. In: Rupprecht CE, Fooks AR, Abela-Ridder B (eds) Laboratory techniques in rabies, 5th edn. World Health Organization, Geneva, pp 180–186
210. Dean DJ (1966) Laboratory Techniques in Rabies: the NYLAR text for measuring potency of antirabies vaccine. Monograph Series No. 23, 2nd edn., World Health Organization 157

211. Koprowski H (1973) Laboratory techniques in rabies: guineapig potency test for chicken-embryo vaccine, Monograph series No. 23, 3rd edn. World Health Organization, Geneva, 287–291
212. Arko RJ, Wiktor TJ, Sikes RK (1973) Laboratory techniques in rabies: the antibody binding test for vaccine potency. Monograph Series No. 23, 3rd edn., World Health Organization, Geneva, 292–294
213. Fitzgerald E, Green O, Seligmann E (1974) Rabies vaccine potency testing: a comparison between the antibody-binding test and the NIH test. Symp Ser Immunobiol Stand 21:300–307
214. Oei HL, Krauss H (1975) Wirksamkeitsprüfung von Tollwutimpfstoffen aus inaktivierten Viren mit dem Antikörperbindungstest im Vergleich zum NIH- und Habeltest [Potency testing of inactivated rabies vaccines with the antibody binding test in comparison with the NIH and Habel test]. Zentralblatt für Bakteriologie Orig A 231(1–3):1–14
215. Atanasiu P, Perrin P, Delagneau J (1980) Use of an enzyme immunoassay with a protein A for rabies antigen and antibody determination. Dev Biol Stand 46:207–215
216. Lafon M, Perrin P, Versmisse P, Sureau P (1985) Use of monoclonal antibody for quantitation of rabies vaccine glycoprotein by enzyme immunoassay. J Biol Stand 13(4):295–301
217. Libeau G, Lafon M, Rollin P (1984) Etude de la spécificité d'anticorps monoclonaux obtenus avec la souche de virus de rage Pasteur PV. Revue d'élévage et de médicine vétérinaire des pays tropicaux 37:383–394
218. Perrin P, Morgeaux S, Sureau P (1990) In vitro rabies vaccine potency appraisal by ELISA: advantages of the immunocapture method with a neutralizing anti-glycoprotein monoclonal antibody. Biologicals 18:321–330
219. Wang Z, Sun Y, Wu X, Carroll DS, Lv W, You L et al (2018) Development of a relative potency test using ELISA for human rabies vaccines. Biologicals 55:59–62
220. Jallet C, Tordo N (2020) In vitro ELISA test to evaluate rabies vaccine potency. J Vis Exp 159. https://doi.org/10.3791/59641
221. Klein A, Fahrion A, Finke S, Eyngor M, Novak S, Yakobson B et al (2020) Further evidence of inadequate quality in lateral flow devices commercially offered for the diagnosis of rabies. Trop Med Infect Dis 5(1):13
222. Saitou Y, Kobayashi Y, Hirano S, Mochizuki N, Itou T, Ito FH et al (2010) A method for simultaneous detection and identification of Brazilian dog- and vampire bat-related rabies virus by reverse transcription loop-mediated isothermal amplification assay. J Virol Methods 168:13–17
223. Coertse J, Weyer J, Nel LH, Markotter W (2019) Reverse transcription recombinase polymerase amplification assay for rapid detection of canine associated rabies virus in Africa. PLoS One 14(7):e0219292
224. Faye M, Abd El Wahed A, Faye O, Kissenkötter J, Hoffmann B, Sall AA et al (2021) A recombinase polymerase amplification assay for rapid detection of rabies virus. Sci Rep 11(1):3131
225. Schlottau K, Freuling CM, Müller T, Beer M, Hoffmann B (2017) Development of molecular confirmation tools for swift and easy rabies diagnostics. Virol J 14(1):184
226. Burgado J, Greenberg L, Niezgoda M, Kumar A, Olson V, Wu X et al (2018) A high throughput neutralization test based on GFP expression by recombinant rabies virus. PLoS Negl Trop Dis 12(12):e0007011

Chapter 6
Historical Progression from Nerve Tissue-Based Rabies Vaccines to Recombinant Biologics for Humans and Companion Animals

Hildegund C. J. Ertl

Abstract Annually, vaccines against infectious agents prevent countless cases of human misery and death. One of the first vaccines that was developed and tested in humans was against rabies, an infectious disease which causes a fatal encephalitis in unvaccinated, exposed humans. The initial vaccine, first used successfully during 1885 in a child by Pasteur and colleagues in France, was derived from nerve tissue of experimentally infected rabbits. Shortly afterwards, such products were produced globally, including in the New World, throughout the early twentieth century. Although they protected humans who had been exposed to a rabid animal, most commonly a dog, these biologics were not only poorly immunogenic (requiring multiple, painful injections), but also caused serious neurological adverse events in a sizable fraction of recipients. Early attempts at improvement included the use of different animal species for viral propagation, and various physical and chemical alterations, to help inactivate rabies virus and lessen the risk of serious adverse events, including vaccine-associated rabies. Upon the development of techniques to maintain and expand embryonal cells in culture, modern methods for rabies vaccine propagation were developed. Cell culture-derived, inactivated rabies vaccines were more immunogenic and importantly, they were safe and replaced the need for nerve tissue-derived rabies vaccines. By the mid-twentieth century, rabies immune globulin (RIG) was added to the postexposure prophylaxis (PEP) protocol for humans after severe exposure to rabies virus, as a means of providing passive immunity, in the period before the active induction of virus neutralizing antibodies. Concomitant with the development of human cell culture-derived rabies vaccines, preexposure (PrEP) vaccination programs expanded for companion animals, most notably dogs. These biologics reduced or even eliminated canine rabies. Nevertheless, human rabies vaccines and RIG are costly and continue to be underutilized. Furthermore,

H. C. J. Ertl (✉)
The Wistar Institute, Philadelphia, PA, USA
e-mail: ertl@wistar.org

C. E. Rupprecht (ed.), *History of Rabies in the Americas: From the Pre-Columbian to the Present, Volume I*, Fascinating Life Sciences,
https://doi.org/10.1007/978-3-031-25052-1_6

rapid PEP is essential to prevent disease, which is often not possible in remote areas with limited access to health care. The use of currently available vaccines for more wide-spread PrEP rabies vaccination, which would reduce the human death toll due to rabies, is not considered cost-effective, necessitating the development of less expensive vaccines that induce protective immune responses after a single dose. Several next generation vaccines that may allow for greater dose-sparing and less costly vaccine regimens are in development.

Keywords Antibodies · Antigens · Biologics · Prophylaxis · Recombinant · Vaccine

Introduction

Each year, countless number of humans die of rabies, a nearly always fatal encephalitic infection [1]. Rabies is caused by infection with different species of lyssaviruses, negative-stranded, bullet-shaped RNA viruses in the Rhabdovirus family, within the order Mononegavirales. Lyssaviruses encode five structural proteins in the form of a glycoprotein, which is expressed on the viral surface, a nucleoprotein, a matrix protein, a phosphoprotein, and an RNA-dependent polymerase. Lyssaviruses are grouped into different phylogroups, most of which can be isolated from bats [2]. Rabies in humans is most commonly caused by rabies virus, a phylogroup 1 virus, which also circulates in many mammalian species, such as dogs, foxes, jackals, raccoons, skunks, and others [3]. Human rabies is most frequently the result of viral transmission through dogs and nearly half of the reported deaths tragically affect children below the age of 15 [4, 5]. Other lyssaviruses can cause infections and fatal disease in humans, but this is rare compared to rabies virus and its burden in Africa and Eurasia. Throughout the Americas, only rabies virus is present. Accordingly, all currently available vaccines and other biologics, such as rabies immune globulin (RIG) that is used together with rabies vaccines during postexposure prophylaxis (PEP), are based on rabies virus. Vaccines against rabies virus can cross-protect against lyssaviruses of phylogroup 1 (such as the Australian bat lyssavirus and European bat lyssaviruses), but not against the genetically more distant lyssaviruses of other phylogroups found in Africa and Eurasia [6, 7].

Protection against rabies virus is well defined. Neutralizing antibodies directed against the viral glycoprotein appear within 7–14 days after vaccination. Antibody titers of 0.5 international units are defined arbitrarily as adequate against rabies virus. Prevention againt infections with other phylogroup 1 lyssaviruses may require higher levels of neutralizing antibodies, but this remains to be investigated in more depth [8–10]. Vaccines against rabies virus are given in preexposure prophylaxis (PrEP) to domestic animals or to humans at occupational risk of exposure, such as veterinarians and staff, rabies diagnosticians, cavers, certain travelers, etc., but most commonly they are administered to persons after exposure to a rabid animal. Rabies is one of the few diseases that can reliably be prevented by PEP, which in case of a bite, or viral exposure to mucosal membranes or to damaged skin has to be combined with RIG [11]. The reasons are simple. First, rabies has a lengthy incubation period ranging in general from 20–60 days [12]. Shorter incubation times of 5–7 days have been described in patients with severe bite wounds to the face or other locations close to the brain. Longer incubation periods of several years are also possible, although it remains unknown where the virus remains during this period and what prevents invasion of the brain and onset of symptomatology. Secondly, rabies virus is transmitted via an infected animal, most commonly through the bite by an infected dog. Such events are quite memorable, unlike transmission of respiratory viruses, which can be acquired by inhaling the breath of an infected person. Thus, PEP is given routinely to humans exposed to a potentially infected animal. However, PEP is not practiced routinely in unvaccinated animals. In most developed countries where rabies is common, laws require prophylactic rabies vaccination for dogs and cats. In case of an exposure, vaccinated animals should receive a booster, whereas most unvaccinated pets are quarantined or euthanized.

Rabies is an ancient disease that has terrorized humans for millennia, although where and how remains debatable. One of the first potential descriptions of rabies may date back to about 2300 BCE, where in the Mosaic Eshnunna Code of Babylon owners had to pay a fine if their 'mad' dog killed a person [13]. Historically, many treatments were attempted, such as holding rabies patients under water, ingesting different concoctions, burning the wound or using so-called 'madstones' (typically hairball concretions found in the stomach of ruminants, frequently used in traditional folk medicine to draw out various venoms or poisons). Even today, some people use home-made herbal teas instead of vaccines. None of these treatments worked. The fate of humans exposed to a rabid animal changed dramatically during 1885 when Louis Pasteur tested successfully the first rabies vaccine in a child, Joseph Meister, who had been bitten by a rabid dog. This first generation vaccine, that in some variations was used for more than a century, was derived from infected nerve tissue, initially using infected rabbits. Such products were poorly immunogenic, had to be given multiple times and caused serious adverse events in a sizable fraction of vaccine recipients. This original rabies vaccine underwent a remarkable evolution and current vaccines are much more immunogenic and safer. Nevertheless, human rabies vaccines are quite costly and thus underutilized. Efforts are underway to develop additional effective, safe and inexpensive rabies vaccines based on recombinant technology. Such vaccines may be particularly suited for wide-spread PrEP of children in highly endemic areas, which is not cost effective with currently available products.

In this chapter, I provide a history of key milestones in the historical progression from the original nerve tissue-based rabies vaccines to modern biologics for humans and domestic animals, and from RIG to monoclonal antibodies, as well as highlight some of the potential future recombinant vaccines, focusing especially upon contributions from multiple investigators throughout the Americas.

Human Rabies Vaccines

Nervous Tissue-Derived Rabies Vaccines

Rabies virus, like all lyssaviruses, is a neurotropic RNA virus that replicates in the central nervous system. The initial rabies vaccines were derived from spinal cords or brains of infected animals. They contained myelin basic protein, which in a sizable fraction of vaccine recipients (i.e., ~1 in 300 to 7000 depending on the vaccine prototype), triggered serious adverse events in the form of meningoencephalomyelitis, mononeuritis, lower limb paralysis, Landry type ascending paralysis, or Guillain-Barre syndrome [11, 14]. In addition to being unsafe, nerve tissue-derived rabies vaccines were poorly immunogenic and had to be given daily for 14–21 days. This was the first and only vaccine type available in the Americas during the late 19th and early 20th centuries. With eventual development of safer and effective alternatives, the World Health Organization recommended that the use of nerve tissue-derived rabies vaccines should be discontinued [11]. Most countries complied, but they were still being used in some resource-poor countries, such as Ethiopia and parts of Latin America [15].

The First Rabies Vaccine

The first rabies vaccine was developed essentially by Pierre Roux, a French physician who from 1878–1883 served as an assistant to Louis Pasteur (historically given credit for his team's work). The vaccine was derived from the spinal cord of rabies virus-infected rabbits (which replaced the use of dogs, based on the alternative animal model as used by Pierre-Victor Galtier, a French veterinarian). Pasteur serially passaged rabies virus many times in rabbits, until the incubation time became fixed to about 7 days [16]. Such laboratory-derived variants are referred to as 'fixed' rabies virus strains as opposed to 'street' (or wild-type) rabies virus isolates, which are viruses obtained from naturally infected animals. To inactivate the virus, infected spinal cords were suspended for ~12 days in a so-called Roux bottle, which had a hole at the top with a stopper from which the cord dangled and another opening at the lower side. Inactivation was linked to the duration of the drying process over potash. Spinal cords that were dried for shorter periods of time retained higher levels of virulent rabies virus [17]. These desiccated spinal cords were minced, suspended in a solution, and used as a vaccine [1, 18]. The vaccine was tested in dogs, which in 2 daily intervals received several doses of vaccine that had undergone successively less drying. Using this regimen, Roux and Pasteur showed that the vaccine protected dogs against a lethal rabies virus. During 1885, before conducting extensive animal postexposure trials, vaccination began on the first patient, a Parisian named Girard, with one dose of rabies vaccine. The hospital where Girard was being treated had concerns about the vaccine, so no further doses were given. Girard recovered, but it is unlikely that he had rabies. A second patient was 11-year-old Julie-Antoinette Poughon, who was already exhibiting symptoms of rabies, when she received two doses of vaccine. She died one day later [19]. On July sixth of 1885, a request for help came to Pasteur's laboratory: a mother and her 9-year old son, Joseph Meister. Joseph had been bitten by a dog that was considered rabid. The child had multiple bite wounds, received more than 2 days earlier. After much hesitation and consultations with a physician (Jacques-Joseph Grancher), but against the advice of Roux, Pasteur agreed. Meister received 13 injections of the spinal cord-derived rabies vaccine, and survived (he became a caretaker at the newly formed Pasteur Institute until 1940) [20, 21]. Another exposed boy, 15 year-old Jean-Baptiste Jupille, was the second patient treated successfully, several months later. Thereafter, responding to the news, a steady stream of victims of dog or wolf bites arrived at the laboratory (from throughout France, Russia, and the celebrated 'New Jersey boys' from the US) to receive Pasteur's vaccine [22]. Rather than have patients travel to Paris, institutions began forming throughout the world, to emulate the production and use of nervous tissue-derived rabies vaccines locally [23]. For example, as reported by the New York Times on May 17, 1886 "...among the arrivals from Europe by the steamship Etruria yesterday was Dr. Valentine Mott, who spent a week or more with M. Pasteur. Dr. Mott brought with him a memento of his visit to the Paris physician in the shape of a rabbit which was inoculated with rabies virus on the day he left Paris. He was too tired yesterday to talk at any length of the result of his visit to Pasteur, but intimated that he was in every way equipped to perform the work in America which Pasteur is doing in Europe...". On July 5, 1886, Mott supervised vaccination of the first patient in the US and the incorporation of the 'American Pasteur Institute' in New York City [23]. During the early twentieth century, the use of the Pastuerian method for human rabies prevention became widespread throughout the Americas. This included the US where besides production at the National Hygienic Facility and more than a dozen commercial firms (that mailed kits of nerve tissue preserved in glycerin to physicians), multiple state public health laboratories manufactured rabies vaccine (e.g., California, Georgia, Michigan, Mississippi, North Carolina, South Carolina, etc.), located in part to the main distribution of canine rabies at that time, primarily concentrated to the eastern part of the country.

Fermi Vaccine

Modifications of the original protocol began quickly as institutions refined their own techniques, which continued during the twentieth century. For example, during 1908, Fermi changed the production protocol of Pasteur's rabies vaccine by including phenol to the infected brains from a variety of animals, such as horses, sheep, goats, rabbits and monkeys, instead of only using the Roux bottle for drying [24]. Inactivation tended to be incomplete and the vaccine occasionally caused rabies. However, use of such biologics for human prophylaxis continued into the twenty-first century.

Semple Vaccine

During 1911, Lieutenant-Colonel Sir David Semple, an officer in the British army who was stationed in India, improved on the Pasteur vaccine by using brain tissue first from rabies virus-infected rabbits and then from infected sheep or goats [25]. He initially inactivated the virus with carbolic acid and then switched to phenol (similar to the ideas of Fermi), using a protocol that was believed to completely inactivate rabies virus. This action was included to minimize the possibility that the vaccine could inadvertently cause rabies. It also allowed for improved shipment of the vaccine and for its standardization. For later nerve tissue-derived vaccines, rabies virus was inactivated by treatment with beta-propionolactone (BPL), an organic compound of the lactone family, that through modifications of nucleosides and some amino acids inactivates viruses. Compared to other chemical additives, BPL was better than phenol at maintaining the structural integrity of proteins. As such, the BPL-derived vaccines were considered a variation of the Semple vaccine in concept.

Suckling Mouse Brain (SMB) Vaccine

Myelinogenesis of the central nervous system starts after birth. To reduce the content of myelin basic protein (MBP) in rabies vaccines, Fuenzalida developed a UV-light inactivated rabies vaccine in Chile that was prepared from the brains of infected suckling mice, which was used widely in Latin America [26]. Nevertheless, depending on production procedures, the Fuenzalida SMB vaccine could still contain contaminants, including MBP, that occasionally triggered adverse reactions [27].

Embryonated Egg Vaccines (EEV)

Inactivated purified rabies virus grown in embryonated avian eggs were developed in the mid- twentieth century. The EEV started to replace the nerve tissue-derived vaccines [28]. They had a superior safety profile and in most vaccine recipients caused only minor local reactions, or mild and transient systemic symptoms.

Duck Embryo Vaccine (DEV)

During 1957, a rabies vaccine was developed from embryonated duck eggs (DEV). The vaccine was inactivated by BPL and purified by ultracentrifugation. Ethyl(2-mercaptobenzoato-(−2)-O,S mercurate(1-)sodium (thiomersal) was added as a stabilizer. The vaccine was lyophilized, which facilitated storage and shipment. This vaccine had a better safety record than nerve tissue-based vaccine. The initial vaccines contained small amounts of MBP and occasionally caused allergic encephalomyelitis. This was addressed by improved purification methods resulting in a vaccine that was well tolerated [29]. Initial reports claimed that the vaccine was as immunogenic as the Semple vaccine [25], which was refuted by others [30, 31]. Overall, the original DEV had low immunogenicity, requiring 14 initial immunizations followed by 3 booster injections [32]. A later improved, purified DEV version was safer than the original biologic from the 1950s, was devoid of MBP, and was more immunogenic, requiring the same number of doses for PrEP and PEP as other cell culture vaccines [33].

Cell Culture-Derived Vaccines

During the late nineteenth century William Roux, a German zoologist, pioneered the use of cell culture. He was the first to show that living cells could be maintained for a few days *in vitro* [34]. During the first decade of the twentieth century, the American Ross Granville Harrison was able to extend frog embryonic cells *in vitro* [35]. Montrose Burrows and Alexis Carrel further refined this technique and established cell cultures from embryonic and adult tissues from numerous species [36]. During 1951, Henrietta Lacks, an African American resident of Virginia, was diagnosed with cervical cancer at John Hopkins, the only hospital in the area willing to treat black patients. A biopsy of her cancer was taken by her physician and given without her permission (a common practice at the time) to George Otto Gey, a cancer researcher at John Hopkins. He used the biopsy material to establish a cell line, called HeLa, that continuously grew *in vitro* and is still widely used today [37]. During 1949, John Franklin Enders, together with Thomas Huckle Weller and Frederick Chapman Robbins, used HeLa cells to propagate poliovirus thus starting a new area in vaccine development that allowed for the global pharmaceutical production of tissue culture-derived vaccines [38].

Human Diploid Cell Rabies Vaccine (HDCV)

During 1962, a Swedish woman obtained an abortion in the fourth month of her pregnancy. The fetus was dissected at the Karolinska Institute and the lungs were flown to Philadelphia, where Leonard Hayflick at the Wistar Institute used them to establish the WI-38 cell line. Heyflick identified the so-called Hayflick limit with WI-38 cells, which stated that a normal cell can only replicate a finite number of times [39]. Hilary Koprowski, Director of the Wistar Institute, reported in 1964 that his group had been able to adapt several fixed strains of rabies virus, including the High Egg Passage (HEP), Pitman Moore (PM), and Challenge Virus Standard

(CVS), to grow in WI-38 cells. The HEP rabies virus strain, upon growth in WI-38, became apathogenic and caused no disease upon intracerebral inoculation into non-human primates. The other two rabies virus strains were inactivated prior to their testing in animals. All three vaccines were found to be highly immunogenic [40]. During 1966, J.P. Jacobs developed the Medical Research Council cell strain 5 (MRC-5) from lung tissue of an aborted fetus [41]. This cell line was also shown to support the growth of fixed strains of rabies virus. The first HDCV, based upon the PM strain, was licensed in Europe during 1976 and in the Americas during 1980. The cell culture-derived rabies vaccines were more potent compared to nerve tissue-derived rabies vaccines and also outperformed DEV in clinical trials. Adequate antibody titers against rabies virus were achieved after 3–4 doses [42]. The vaccine, given when required with RIG, protected rabies virus-exposed humans [41] and over time replaced the far more reactogenic nerve tissue-derived rabies vaccines, including the SMB vaccines.

Rabies Vaccines Grown in Primary Animal Cells or Continuous Cell Lines

After HDCV, rabies virus was subsequently adapted to other cell lines, such as embryonated chick egg cells (e.g., PCECV) [43], Vero cells (e.g., PVRV) [44], or primary baby hamster kidney cells [45]. The safety and potency profiles of rabies virus vaccines propagated in different cell lines were largely comparable, which led to their wide scale application throughout the Americas.

Rabies Immune Globulin

Historically, PEP with a rabies vaccine alone was not always successful in preventing death in individuals who developed rabies after severe bites, which prompted the development of additional products and protocols. During 1908, Fermi pioneered serum transfer as a method to combat rabies virus by antibodies [46], which during 1890 had been discovered as therapeutic serum components in animals that had been immunized against diphtheria [46, 47]. At the same time, Marie reported that serum transfer alone did not effectively protect animals from rabies virus [48]. During 1929, Stuart and Krikorian presented a study that measured antibodies that developed in rabbits after vaccination with different rabies vaccines [49]. At that time, it was still debated if indeed resistance against rabies virus was caused by rabicidal serum factors [50]. Additional studies conducted during 1935 by Hoyt [51], 1942 by Yen [52], 1951 by Koprowski [53], and 1954 by Habel [54] showed in several animal models that rabies immune sera could transfer protection. Additional studies again conducted in animals explored combinations of vaccines with RIG transfer [55]. The final proof that RIG combined with rabies vaccination benefited humans exposed to a rabid animal came during 1954. A confirmed rabid wolf bit 29 people in an Iranian village and 18 of them suffered severe bite wounds to their heads or necks. In the latter, PEP was initiated immediately, five individuals received rabies vaccine only, and the others were treated with serum and vaccine. Sixty percent of those that were only vaccinated succumbed to rabies while only 7% (1 out of 13) died in the group that had also received serum [56, 57]. During 1957, WHO incorporated the use of serum into PEP for severe exposures [58]. Initially it was given systemically as it was assumed that protection was provided by circulating antibodies. During 1963, Dean et al. reported on the importance of injecting antibodies locally into the rabies virus-infected wound [59]. This procedure was eventually incorporated into WHO recommendations. During 1971, Cabasso et al. defined the doses for human RIG at 20 mg/kg [60], while an equine product (ERIG) with a shorter half-life, was recommended to be given at 40 mg/kg. Remaining sera that could not be injected into the wound due to space restraints were given intramuscularly (IM). During 2016, Bharti showed that protection could be achieved without systemic injection of RIG [61]. During 2017, WHO changed their recommendation, so that up to 20 or 40 IU/kg of RIG (depending on its homologous or heterologous source) was injected into the wound. In case the full dose could not be injected, no leftover was applied to a distant site [62].

Monoclonal antibodies, originally developed at the Wistar Institute during the 1970s, were considered as a potential replacement for RIG (which is habitually in short supply), and were later shown to be effective in animals [63]. To date, several different monoclonal antibody preparations have been tested in humans [63]. A phase I trial conducted with a single monoclonal antibody in humans showed that the approach was safe [64]. A phase II/III trial in exposed individuals showed non-inferiority to human RIG [65]. This antibody was licensed in India, produced by the Serum Institute of India and sold under the name Rabishield. Zydus Cadia in India initially conducted phase I/II and the phase III trials with a mixture of two murine monoclonal antibodies, called RabiMabs that were also licensed [66]. North China Pharmaceutical company completed a phase I/II study with one monoclonal antibody [67]. Ideally, use of a cocktail targeting different viral epitopes would be preferable to broaden the coverage against naturally occurring street rabies viruses (as well as phylogroup I lyssaviruses) and minimize the risk of viral escape. As of yet, no monoclonal antibodies have been licensed in the Americas for human PEP.

Licensed Human Rabies Vaccines

Currently licensed human rabies vaccines are suitable for PrEP and PEP. Most rabies vaccines do not contain an adjuvant. Some were supplemented with an adjuvant such as aluminium phosphate used in the Rabies Vaccine, Adsorbed which was produced originally by the Biologics Products Program, Michigan Department of Public Health, but no longer available. Two other modern rabies vaccines were licensed in the US. One is IMOVAX™ from Sanofi Pasteur, which is based on the

PM-1503-3 M rabies virus strain grown on MRC5 cells. The other is RabAvert™ from Bavarian Nordic, using the fixed rabies virus strain Flury low egg passage (LEP) grown in primary cultures of chicken fibroblasts. This vaccine contains polygeline and potassium glutamate as stabilizers. In both vaccines, rabies virus is inactivated by BPL. Vaccines are supplied in a freeze-dried form, which should be used immediately after reconstitution. IMOVAX™ and RabAvert™ are also used abroad, such as in the UK where they are called Rabies Vaccine BP™ and Rabipur™, respectively.

India, one of the countries where dog-transmitted rabies is very common, has several licensed rabies vaccines. For example, these include: RAVIVAX™ (Serum Institute of India) based on the PM strain grown on human diploid cells; Abhayrab™ (Indian Immunologicals, Ltd) based on the Pasteur virus 2061 strain grown on Vero cells; INDIRAB™ (BHARAT Biotech International, Ltd.), using a PM strain virus grown on Vero cells; and VaxiRab™ (Zydus Cadila), with a purified DEV vaccine based on the PM strain [68]. Additional rabies vaccines are produced in other countries, such as China.

In addition to routine quality controls that need to be conducted for all vaccines, rabies vaccines have to be tested for effective inactivation as any residual live virus could cause a fatal infection [69]. In addition, rabies vaccines are subjected to a potency test, which is typically conducted in mice [70]. Alternative *in vitro* approaches are being explored [71]. Although most rabies vaccines are thoroughly tested before they are released for use in humans, this is not always a given. A routine audit of Changchun Changsheng Biotechnology, the second largest manufacturer of rabies vaccines in China, by the China Food and Drug Administration conducted in 2018 revealed that the company had falsified documentation for ~113,000 doses of their rabies vaccine. Unfortunately, this is not the only example of the release of flawed rabies vaccines [72]. In contrast, due in part to strict regulatory oversight in the Americas, no recent instances of counterfeit vaccines have been reported.

Regimens for Human Rabies Vaccines

The regimen for PrEP calls for 2–3 doses, while several regimens were initially recommended for PEP including the 5-dose Essen regimen in which vaccine is given IM on days 0, 3, 7, 14, and 30 or the 4-dose Zagreb regimen based on 2 IM doses given on day 0 and additional doses on days 7 and 21 [73]. Since then, WHO has approved the use of shortened IM regimens and the use of intradermal (ID) vaccination which requires 1/5th to 1/10th of the IM dose and thus allows for substantial cost-savings in regions where rabies virus exposure is common [74, 75]. The use of ID immunization was shown in several studies to achieve titers of neutralizing antibodies that are similar to those induced by IM vaccination [11, 43, 76]. Nevertheless, due to logistical reasons, licensed rabies vaccines are currently approved for IM administration and accordingly, each vial contains one IM dose.

Typically, ID rabies vaccination remains an off-label use, which has potential legal consequences. Although ID rabies vaccination is widely used in resource-poor countries, it is not being applied in developed countries nor have manufacturers adjusted their rabies vaccine fill volume to ID doses. Given the large numbers of bitten people in Africa and Asia, where specialized rabies bite clinics predominant, the reduction of canine rabies in the Americas has minimized much of the rationale for ID vaccination in this region.

Rabies PrEP as Part of Routine Childhood Immunization?

Except for individuals at high risk, most humans are not vaccinated prophylactically but rather after exposure to a rabid animal. Nearly half of the world's children get bitten by a potentially rabid dog by the time they reach adulthood in rabies endemic areas, where they are at high risk to contract the disease, which could be prevented by PrEP [77]. Inclusion of PrEP into childhood immunization programs, given together with other vaccines, effectively induces rabies virus-neutralizing antibodies and would likely save many lives [78–80]. Routine rabies immunizations of children residing in highly endemic areas are currently not being pursued actively, because they are not deemed cost-effective [81].

Two notable exceptions for considerations of childhood PrEP are Peru and the Philippines [82, 83]. Periodic rabies outbreaks due to vampire bats are common in the Peruvian Amazon. These regions are remote, which makes timely rabies PEP challenging, and consequently humans, mainly children, die of rabies. Following a vampire bat rabies outbreak during 2011, Peru initiated a mass rabies vaccination program, which successfully stopped further human deaths related to vampire rabies. Today, PrEP is given to Peruvian children in highly endemic, remote areas, as it should in other affected parts of Amazonia. In contrast today in the Americas, dog-transmitted rabies is common throughout Africa and Asia, such as in the Philippines. During 2007, the Government recommended that free routine PrEP should be provided to school children living in high-risk areas. Due to vaccine shortages, the program was stopped temporarily during 2011. More wide-spread inclusion of rabies vaccines into childhood immunization programs will likely require less costly vaccines that achieve neutralizing antibody titers after a single dose.

Potential Next Generation Rabies Vaccines

Although rabies vaccines have been available for well over a century and those that are currently being used are highly effective and have very few side effects, rabies continues to claim tens of thousands of human lives each year. The actual number of human deaths is unknown. Several factors contribute to the unremittingly high death toll of rabies, including lack of public education, which leads to a lack of PEP (or use of ineffective therapies), lack of rapid access to health care, lack of RIG (which is rarely used in resource-poor countries), and most notably, the lack of inexpensive rabies vaccination. The rabies vaccine regimen composed of vaccine and RIG is expensive. Furthermore, the vaccine must be given several times. This adds costs for patient travel and loss of wages. Less expensive rabies vaccines that achieve immunity after a single dose would likely increase both PrEP of children in highly endemic areas and PEP after a suspected exposure.

A plethora of rabies vaccines, ranging from peptides, proteins from yeast, bacteria, mammalian cells, or even plants, simple genetic materials such mRNA or DNA plasmids, to more complex recombinant viruses, have undergone pre-clinical testing in mice or other mammals [84]. Some of these vaccines, such as protein vaccines or genetically modified rabies virus vaccines, could be used both for PrEP and PEP. So-called genetic vaccines that administer the genetic material of the rabies virus glycoprotein, which produces the vaccine antigen with a delay, are only suited for PrEP, as rapid induction of immune responses is of the essence upon PEP. In this section, I focus on next generation rabies vaccines that are close to, or have undergone, clinical testing, including continued contributions from investigators in the Americas.

Protein Vaccines

Developing an effective vaccine based on the rabies virus glycoprotein is hampered by its complex three dimensional trimeric structure and by antibody recognition of primarily conformation-dependent epitopes [85, 86]. Although numerous expression systems have attempted to develop an immunogenic rabies virus glycoprotein preparation, only one based on a recombinant baculovirus-derived protein, developed by CPL Biological, India, has undergone clinical testing [87]. The glycoprotein was modified to allow for its assembly into nanoparticles. This vaccine has been licensed, requiring a three dose regimen, similar to other conventional inactivated rabies vaccines.

PIKA Rabies Vaccine

Most human rabies vaccines do not contain adjuvants which serve to boost immune responses by activating innate sensors of pathogen-associated molecular patterns. The PIKA vaccine is a traditional inactivated rabies vaccine based on the CTN-1 stain of a Chinese isolate of rabies virus grown in Vero cells, formulated with a chemical analogue of a double-stranded RNA, that activates the Toll-like receptor-3 pathway [88]. In a phase II clinical trial, the PIKA vaccine given in an accelerated regimen with 2 doses on days 0 and 3, followed by a single dose on day 7, showed non-inferiority to Rabipur™, given in the classical regimen of 1 dose each on days 0, 3, 7 and 14 [89, 90].

mRNA Rabies Vaccine

An mRNA vaccine expressing the rabies virus glycoprotein was tested in healthy adults [91]. The vaccine was given three times IM or ID either by needle or by a needle-free device. Several individuals experienced grade three adverse events. The most immunogenic form of application (i.e., ID delivery by the needle-free device), achieved neutralizing antibody titers in only 71% of vaccine recipients. More worrisome, a booster immunization given with a needle-free device to responders 1 year later resulted in adequate titers in only 54% of the human volunteers. Clearly, based upon these preliminary data, this type of vaccine needs improvement.

Canarypox Rabies Vaccine

A canarypox vector expressing the rabies virus glycoprotein (ALVAC-RG) was tested at several concentrations in comparison to HDCV in healthy human adults [92]. Vaccinees received two vaccine doses, and those injected with the higher dose of ALVAC-RG or HDCV were boosted 6 months later with the same or the heterologous vaccine. The ALVAC-RG was well tolerated. At the higher doses the vaccine achieved rabies virus-neutralizing antibodies in all recipients by day 35 after vaccination, compared to elicitation by HDCV within 28 days. Titers were higher in the HDCV group than in the high dose ALVAC-RG group. A booster immunization increased antibody titers in all groups. No follow-up trials were conducted.

Adenoviral Vector Vaccines

Adenoviral vectors are exceptionally immunogenic. Adenoviruses are a large family of double-stranded DNA viruses with an approximately 32 kb genome. They are species-specific but different serotypes infect mammals as well as birds, reptiles, and amphibians. There are 57 serotypes of adenovirus, which can infect humans. They are grouped into families A-G. Adenoviruses can be rendered replication-defective by depletion of the E1 domain, which encodes polypeptides that are essential for the transcription of the other viral genes. In replication-defective adenovirus vectors that are used for gene transfer, the transgene expression cassette is placed into the deleted E1 domain. Other modifications may include deletions of E2 or replacement of E4 with that of the most commonly human serotype 5, which was used to generate the cell line that transcomplements the deleted genes during virus propagation [93]. The E4 replacement was shown to increase virus yields [94].

Most humans become infected with multiple serotypes during their lifetime and develop neutralizing serotype-specific antibodies, which can blunt the immunogenicity of a vaccine based on the same serotype by preventing the virus from infecting cells, which is a pre-requisite for the production of the transgene product. Instead of using common human serotypes such as serotype 5 or 26, to which humans commonly have neutralizing antibodies, several groups are exploring adenoviruses from chimpanzees, which are phylogenetically related to human serotypes [95–97]. Replication-defective chimpanzee adenovirus vectors expressing the rabies virus glycoprotein have been tested in mice, dogs, and non-human primates (NHP) [98–100]. After a single dose, rabies virus-neutralizing antibodies were sustained, with complete protection against a subsequent challenge with a street rabies virus that in a NHP trial was given more than a year later to animals that had received a single modest dose of 10^9 virus particles (vp) of the vaccine. The challenge resulted in a robust anamnestic response, indicating that vaccination had stimulated a long-term memory B cell response. The chimpanzee adenovirus vector failed to provide complete protection to NHP when used in PEP. The initial experiments were performed with a chimpanzee adenovirus vector of serotype 68, which belongs to the family E of adenoviruses [101]. A phase one clinical trial with this vaccine, which was further modified to replace part of the viral E4 domain with that of human 5 adenovirus, was conducted by Oxford University in collaboration with the Wistar Institute. An economic analysis indicated that the vaccine would likely cost no more than $ 2–3 per dose [102]. Such competitive pricing would allow a cost-effective way to expand PrEP to children in remote rabies endemic areas. Another advantage of this vaccine is that it can be rendered stable at ambient temperatures, thus avoiding the need for costly cold chains [103].

More recently, these studies were partially repeated with a family C chimpanzee adenovirus vector termed CH-Ad155-RG [104]. The glycoprotein insert for this vector represents a medoid sequence selected from 2,060 glycoprotein sequences from lyssaviruses of phylogroups 1 and 2. Accordingly, the vaccine induces antibodies that neutralize these viruses. Immunogenicity of CH-Af155-RG in NHP, which were injected with a fairly high dose of 5 x 10^{10} vp, was non-inferior to that of Rabipur™. Challenge studies were only conducted in mice, but due to lack of complete infection of control animals, these investigations were not entirely conclusive. While the authors claim that CH-Ad55-RG is suitable for PEP, this statement that is not supported by actual data.

Rabies Management in Companion Animals

Canine rabies was described for centuries. While bat lyssaviruses may have evolved initially, canine rabies seems to have originated in Eurasia, but spread globally, including wildlife involvement [105, 106]. Canine rabies was likely introduced to the Americas by European colonizers [107]. Some of the first reports of canine rabies came from Central America during 1703, followed by an outbreak during 1741 in Barbados [108, 109]. In the US, canine rabies was described in the eastern region of the country [46]. Initial first reports came during 1753 from Virginia, 1762 from North Carolina, 1768 from Boston, and 1779 from Philadelphia. Canine rabies virus then spread to the South Atlantic States, and by 1909, most US States reported cases of dog rabies [46]. Throughout the Americas, canine rabies virus adapted to wildlife species such as foxes, coyotes, and skunks [110]. During 1946, the US established the Public Health Service (USPHS), which by today has several divisions, including the Centers for Disease Control and Prevention (CDC). Originally the US Bureau of Animal Industry, and later the USPHS, started to collect epidemiological data on human, companion, and farm animal rabies. By the mid-twentieth century, records included foxes, skunks, and a variety of other species. Although early canine rabies vaccination began during the 1920s, programmatic attention improved post-World War II with the better quality and immunogenicity of veterinary vaccines. Between 1946 and 1965, canine rabies declined from 8,384 to 412 cases annually, due to such increased immunization, management of stray dogs, and public education [111]. By 1970, following laws that mandated vaccination of dogs in most US states, canine rabies ceased to be widely enzootic in the US, and was considered eliminated by the early twenty-first century. Thereafter, all rabies cases in dogs were caused by spill-over infections from wildlife.

During 1983, South America and the Caribbean pledged to eliminate canine rabies by aggressive mass vaccinations of dogs, which resulted in a 91% and 93% reduction in human and canine rabies cases, respectively, between 1982 and 2003 [112]. During 2018, 28 of 35 Latin American countries reported zero human deaths due to canine rabies virus [113]. Based upon this regional example, Asian and African countries are following suit, although their dog rabies immunization programs are hampered by the large numbers of community-owned or free-ranging dogs. The WHO has developed a strategic plan to eliminate human deaths mediated via canine rabies through mass dog vaccination programs by 2030 [114].

Companion Animal Rabies Vaccines

Clearly the most efficient way to reduce the human death toll of rabies is to eliminate canine rabies, which causes the vast majority of human cases. The initial rabies vaccine that was developed by Roux and Pasteur was tested in dogs and was shown to protect them against a rabies virus challenge. As a rule, rabies vaccines that protect humans also protect dogs and other companion animals, such as cats, or farm animals, such as cattle or horses. Multiple products are available today throughout the Americas. For example, in the US, 10 rabies vaccines are approved for use in dogs, cats, and other animals [115]. They include RABVAC1™ and RABVAC3™ from Boehringer Ingelheim, Defensor 1™ and Defensor 3™ from Zoetis, Nobivac 1™ and Nobivac 3™, from Merck Animal Health and IMRAB 1™, IMRAB 1

TF™, IMRAB 3™, and IMRAB 3 TF™ from Merial, Inc. Two other rabies vaccines are available for horses (Equirab™ from Merck Animal Health), cattle, horses, and sheep (IMRAB, Large Animals™ from Merial, Ins.). All of these vaccines contain inactivated virus from fixed rabies virus strains. Some contain gentamicin and most contain an adjuvant (that has been linked to cause soft-tissue sarcoma in cats). A non-adjuvanted rabies vaccine is available for cats. This vaccine, PureVax Feline™ from Merial, Inc., is based on a recombinant canarypox vector that expresses the rabies virus glycoprotein.

Rabies Vaccine Regimens for Dogs and Cats

Pet dogs and cats are vaccinated prophylactically to rabies virus. Neonates born to rabies virus-immune dams may be protected for a short period of time by maternally transferred antibodies [116]. Dogs and cats receive a single dose of a rabies vaccine once they are about 2 to 3 months old. Depending on the vaccine, animals need to be boosted in 1–3 yearly intervals [117].

PEP for Domestic Animals

Unlike with humans, PEP is not practiced routinely for companion or farm animals. Instead, unvaccinated animals that are exposed to rabies virus usually are euthanized or are quarantined for several months. Nevertheless, studies have shown that depending on the vaccine and protocol, PEP can protect animals after they were experimentally challenged with rabies virus. One study showed solid protection of dogs that were vaccinated and administered a cocktail of rabies virus-neutralizing monoclonal antibodies 6 hours after a challenge given to the masseter muscle [7]. Another study showed that cell culture-derived vaccine could protect cattle that had been infected with rabies virus [118]. According to current guidelines, vaccinated animals that are exposed to rabies virus should be revaccinated immediately and placed under observation by the owner for 45 days [119]. Recommendations for PEP in exposed naïve animals vary by jurisdiction within North America, including case by case indications for quarantine plus multiple vaccinations over time, such as in Texas [120].

Summary

Rabies is the deadliest viral disease of all – killing more than 99% of humans who develop a symptomatic infection. This malady is also one of the first diseases for which an efficacious vaccine was developed, more than 130 years ago by Pasteur.

While ground-breaking in rabies prevention (and replicated throughout the world), this original nerve-tissue derived vaccine had serious side effects. As such, this first-generation product was modified gradually for improved safety. Investigators throughout the Americas were instrumental in building upon this legacy. By the mid-twentieth century, application of tissue culture techniques improved vaccine immunogenicity and reduced the need for production in animals. In addition, use of RIG was added during PEP, after severe human exposures to a potentially rabid animal. Modern rabies vaccines are not only very pure, potent, and efficacious, with virtually no vaccine failures if they are given correctly and combined with wound cleaning and RIG, but they are also very safe. However, such vaccines are costly and require multiple doses. Moreover, RIG is quite expensive and habitually in limited supply. Both biologics are underutilized, resulting in untold tens of thousands of human deaths and tens of millions of viral exposures each year. Next generation biologics, in the form of novel vaccines and replacements for RIG, are in clinical development. Hopefully, with improved dog rabies management, use of these tools will substantially reduce the global burden of this neglected zoonosis in a One Health context.

References

1. Hampson K, Coudeville L, Lembo T, Sambo M, Kieffer A, Attlan M et al (2015) Estimating the global burden of endemic canine rabies. PLoS Negl Trop Dis 9(4):e0003709
2. Hayman DTS, Fooks AR, Marston DA, Garcia-R JC (2016) The global phylogeography of lyssaviruses – challenging the "out of Africa" hypothesis. PLoS Negl Trop Dis 10(12):e0005266
3. Kuhn JH, Adkins S, Alioto D, Alkhovsky SV, Amarasinghe GK, Anthony SJ et al (2020) 2020 taxonomic update for phylum Negarnaviricota (Riboviria: Orthornavirae), including the large orders Bunyavirales and Mononegavirales. Arch Virol 165(12):3023–3072
4. Fooks AR, Banyard AC, Horton DL, Johnson N, McElhinney LM, Jackson AC (2014) Current status of rabies and prospects for elimination. Lancet 384(9951):1389–1399
5. Department of Health, Philippines. National Rabies Prevention and Control Program Manual of Operations (2012) [Internet]. 2012. Available from: https://www.doh.gov.ph/sites/default/files/publications/FINALMOP6.4.13WORDRADMay30.pdf
6. Horton DL, Banyard AC, Marston DA, Wise E, Selden D, Nunez A et al (2014) Antigenic and genetic characterization of a divergent African virus, Ikoma lyssavirus. J Gen Virol 95(5):1025–1032
7. Hanlon CA, Niezgoda M, Rupprecht CE (2002) Postexposure prophylaxis for prevention of rabies in dogs. Am J Vet Res 63(8):1096–1100
8. Xiang ZQ, Knowles BB, McCarrick JW, Ertl HC (1995) Immune effector mechanisms required for protection to rabies virus. Virology 214(2):398–404
9. Moore SM, Hanlon CA (2010) Rabies-specific antibodies: measuring surrogates of protection against a fatal disease. PLoS Negl Trop Dis 4(3):e595
10. Malerczyk C, Freuling C, Gniel D, Giesen A, Selhorst T, Müller T (2014) Cross-neutralization of antibodies induced by vaccination with Purified Chick Embryo Cell Vaccine (PCECV) against different lyssavirus species. Hum Vaccin Immunother 10(10):2799–2804
11. World Health Organization (2010) Rabies vaccines: WHO position paper—Recommendations. Vaccine 28(44):7140–7142
12. Bleck TP, Rupprecht CE (2002) Rhabdoviruses. In: Richman DD, Whitley RJ, Hayden FG (eds) Clinical virology, 2nd edn. ASM Press, Ann Arbor, pp 857–873

13. Tarantola A (2017) Four thousand years of concepts relating to rabies in animals and humans, its prevention and its cure. Trop Med Infect Dis 2(2):5
14. Murthy JMK (1998) MRI in acute disseminated encephalomyelitis following Semple antirabies vaccine. Neuroradiology 40(7):420–423
15. Coetzer A, Kidane AH, Bekele M, Hundera AD, Pieracci EG, Shiferaw ML et al (2016) The SARE tool for rabies control: current experience in Ethiopia. Antivir Res 135:74–80
16. Roux EPP (1883) Thèse de doctorat: Médecine: Des Nouvelles Acquisitions sur la Rage, Paris, p 55
17. Pasteur L (1885) Méthode pour prévenir la rage après morsure. C R Acad Sci 101:765–774
18. Pasteur's Method (1896) Pioneer mail Indian wkly news. 26–7
19. Geison GL (2016) Private science of Louis Pasteur. Princeton University Pres, Princeton
20. Dubail A (1985) Joseph Meister, le premier être humain sauvé de la rage. Annu Société Hist Val Villé 10:93–148
21. Steige M de. Joseph Meister [Internet]. Available from: http://steige.fr/le-village/joseph-meister-1876-1940/
22. Rotivel Y, Goudal M, Perrin P, Tordo N (2002) Une histoire de la vaccination contre la rage. Virologie 6(2):89–104
23. Hoenig LJ, Jackson AC, Dickinson GM (2018) The early use of Pasteur's rabies vaccine in the United States. Vaccine 36(30):4578–4581
24. Fermi C (1908) Wirkung der Antiwutimpstoffe und sera je nach der Tierspecies aus welcher sie estammen und welcher sie verabreicht werden. ZHyg InfektKr 58:223
25. Semple SD (1911) The preparation of a safe and efficient antirabic vaccine. Superintendent Govt Print, Kolkata
26. Fuenzalida E, Palacios R, Borgono JM (1964) Antirabies antibody response in man to vaccine made from infected suckling-mouse brains. Bull World Health Organ 30:431–436
27. Nogueira YL (1998) Adverse effect versus quality control of the Fuenzalida-Palacios antirabies vaccine. Rev Inst Med Trop São Paulo 40(5):295–299
28. Dean DJ, Evans WM, Thompson WR (1964) Studies on the low egg passage flury strain of modified live rabies virus produced in embryonating chicken eggs and tissue culture. Am J Vet Res 25:756–763
29. Glück R, Matthieu JM, Wegmann A, Méan F (1986) Absence of myelin basic protein in an improved purified duck embryo rabies vaccine. Neurochem Pathol 4(2):69–75
30. Greenberg M (1960) Vaccination against rabies with duck-embryo and semple vaccines. J Am Med Assoc 173(4):333–337
31. Veeraraghavan N, Subrahmanyan TP (1963) The value of duck-embryo vaccine and high-egg-passage flury vaccine in experimental rabies infection in Guinea-pigs. Bull World Health Organ 29:323–330
32. Keller H, Glück R, Wegmann A, Wandeler AI (1984) Immunogenicity of a new, highly purified, highly concentrated duck-embryo rabies vaccine. Schweiz Med Wochenschr 114(19):648–653
33. Mahendra BJ, Madhusudana SN, Ashwathnarayana DH, Sampath G, Datta SS, Sudarshan MK et al (2007) A comparative study on the immunogenicity, safety and tolerance of purified duck embryo vaccine (PDEV) manufactured in India (Vaxirab) and Switzerland (Lyssavac-N): a randomized simulated post-exposure study in healthy volunteers. Vaccine 25(50):8405–8409
34. Rodriguez-Hernandez CO, Torres-Garcia SE, Olvera-Sandoval C, Ramirez-Castillo FY, Muro AL, Avelar-Gonzalez FJ et al (2014) Cell culture: history, development and prospects. Int J Curr Res Acad Rev 12:188–200
35. Harrison RG, Greenman MJ, Mall FP, Jackson CM (1907) Observations of the living developing nerve fiber. Anat Rec 1(5):116–128
36. Malinin TI (1996) Remembering Alexis Carrel and Charles A. Lindbergh. Tex Heart Inst J 23(1):28–35
37. Gey GO, Coffman WD, Kubicek MT (1952) Tissue culture studies of the proliferative capacity of cervical carcinoma and normal epithelium. Cancer Res 12:264–265

38. Enders JF, Weller TH, Robbins FC (1949) Cultivation of the lansing strain of poliomyelitis virus in cultures of various human embryonic tissues. Science 109(2822):85–87
39. Hayflick L, Moorhead PS (1961) The serial cultivation of human diploid cell strains. Exp Cell Res 25(3):585–621
40. Wiktor TJ, Fernandes MV, Koprowski H (1964) Cultivation of rabies virus in human diploid cell strain wi-38. J Immunol 93:353–366
41. Kuwert EK, Marcus I, Werner J, Scheiermann N, Höher PG, Thraenhart O et al (1976) Post-exposure use of human diploid cell culture rabies vaccine. Dev Biol Stand 37:273–286
42. Plotkin SA, Wiktor TJ, Koprowski H, Rosanoff EI, Tint H (1976) Immunization schedules for the new human diploid cell vaccine against rabies. Am J Epidemiol 103(1):75–80
43. Kamoltham T, Thinyounyong W, Phongchamnaphai P, Phraisuwan P, Khawplod P, Banzhoff A et al (2007) Pre-exposure rabies vaccination using purified chick embryo cell rabies vaccine intradermally is immunogenic and safe. J Pediatr 151(2):173–177
44. Briggs DJ, Banzhoff A, Nicolay U, Sirikwin S, Dumavibhat B, Tongswas S et al (2000) Antibody response of patients after postexposure rabies vaccination with small intradermal doses of purified chick embryo cell vaccine or purified Vero cell rabies vaccine. Bull World Health Organ 78(5):693–698
45. Kallel H, Jouini A, Majoul S, Rourou S (2002) Evaluation of various serum and animal protein free media for the production of a veterinary rabies vaccine in BHK-21 cells. J Biotechnol 95(3):195–204
46. Steele JH, Fernandez PJ (1991) History of rabies and global aspects. In: Baer GM (ed) The natural history of rabies, 2nd edn. CRC Press Inc, Boca Raton, pp 1–26
47. Behring E (1890) Ueber das Zustandekommen der Diphtherie-Immunitat und der Tetanus-Immunitat bei thieren. Deutsche medizinsche Wochenschrift 16:1113–1114
48. Marie A (1908) Recherches sur le serum antirabique. Ann Inst Pasteur 22:271
49. Stuart G, Krikorian KS (1929) Studies in Anti-rabies Immunisation. J Hyg 29(1):1–34
50. Biglieri R, Villigas C (1926) Immunite locale de la rage. C R Soc Biol 95:1176
51. Hoyt A, Gurley MK (1937) Experimental street-virus rabies in white mice. Studies on passive immunization. Proc Soc Exp Biol Med 37:454–455
52. Yen CH (1942) Protective value of antiviral serum in experimental rabies infection. Proc Soc Exp Biol Med 49:533–537
53. Koprowski H (1951) Recent developments in the prophylaxis of rabies. Am J Public Health Nations Health 41(12):1483–1489
54. Habel K (1954) Antiserum in the prophylaxis of rabies. Bull World Health Organ 10(5):781–788
55. Habel K (1966) Vaccination and serum prophylaxis of rabies in man. In: International symposium on rabies. S. Karger, Basel, pp 293–305
56. Baltazard M, Bahmanyar M (1955) Field trials with rabies vaccine on persons bitten by rabid wolves. Bull World Health Organ 13(5):747–772
57. Habel K, Koprowski H (1955) Laboratory data supporting the clinical trial of anti-rabies serum in persons bitten by a rabid wolf. Bull World Health Organ 13(5):773–779
58. Executive Board 19 (1957) Expert committee on rabies: third report. World Health Organization. (Third meeting, 16 January 1957)
59. Dean DJ, Baer GM, Thompson WR (1963) Studies on the local treatment of rabies-infected wounds. Bull World Health Organ 28(4):477–486
60. Cabasso VJ, Loofbourow JC, Roby RE, Anuskiewicz W (1971) Rabies immune globulin of human origin: preparation and dosage determination in non-exposed volunteer subjects. Bull World Health Organ 45(3):303–315
61. Bharti OK, Madhusudana SN, Gaunta PL, Belludi AY (2016) Local infiltration of rabies immunoglobulins without systemic intramuscular administration: an alternative cost effective approach for passive immunization against rabies. Hum Vaccin Immunother 12(3):837–842
62. WHO (2018) WHO expert consultation on rabies. Third report. World Health Organ Tech Rep Ser 982:1–139

63. Müller T, Dietzschold B, Ertl H, Fooks AR, Freuling C, Fehlner-Gardiner C et al (2009) Development of a mouse monoclonal antibody cocktail for post-exposure rabies prophylaxis in humans. PLoS Negl Trop Dis [Internet] 3(11) Available from: https://www.ncbi.nlm.nih.gov/pmc/articles/PMC2765635/
64. Sparrow E, Torvaldsen S, Newall AT, Wood JG, Sheikh M, Kieny MP et al (2019) Recent advances in the development of monoclonal antibodies for rabies post exposure prophylaxis: a review of the current status of the clinical development pipeline. Vaccine 37:A132–A139
65. Gogtay NJ, Munshi R, Ashwath Narayana DH, Mahendra BJ, Kshirsagar V, Gunale B et al (2018) Comparison of a novel human rabies monoclonal antibody to human rabies immunoglobulin for Postexposure prophylaxis: a phase 2/3, randomized, single-blind, noninferiority. Controlled Study Clin Infect Dis 66(3):387–395
66. Desai M (2016) CTRI/2016/08/007137 Immunogenicity and safety study of rabies G protein vaccine administered as a simulated post-exposure immunization in healthy volunteers [Internet]. Available from: http://www.ctri.nic.in/Clinicaltrials/pdf_generate.php?trialid=13580&EncHid=&modid=&compid=%27,%2713580det%27
67. Molecular Targeting Technologies, Inc. Development pipeline [Internet]. Available from: <http://www.mtarget.com/mm5/pdfs/pipeline/Asset%20Rabies%20mAb.pdf>
68. Bharati K, Vrati S (2012) Viral vaccines in India: an overview. Proc Natl Acad Sci Sect B Biol Sci 82(1):181–198
69. Mitchell JR, Everest RE, Anderson GR (1971 Oct) Sensitive procedure for detecting residual viable virus in inactivated rabies vaccine. Appl Microbiol 22(4):600–603
70. WHO (2007) WHO expert committee on biological standardization. World Health Organ Tech Rep Ser 941:1–340
71. Poston R, Hill R, Allen C, Casey W, Gatewood D, Levis R et al (2019) Achieving scientific and regulatory success in implementing non-animal approaches to human and veterinary rabies vaccine testing: a NICEATM and IABS workshop report. Biologicals 60:8–14
72. Taylor E, Banyard AC, Bourhy H, Cliquet F, Ertl H, Fehlner-Gardiner C et al (2019) Avoiding preventable deaths: the scourge of counterfeit rabies vaccines. Vaccine 37(17):2285–2287
73. Shi N, Zhang Y, Zheng H, Zhu Z, Wang D, Li S et al (2017) Immunogenicity, safety and antibody persistence of a purified vero cell cultured rabies vaccine (Speeda) administered by the Zagreb regimen or Essen regimen in post-exposure subjects. Hum Vaccin Immunother 13(6):1–8
74. Kerdpanich P, Chanthavanich P, De Los Reyes MR, Lim J, Yu D, Ma Ama C et al (2018) Shortening intradermal rabies post-exposure prophylaxis regimens to 1 week: results from a phase III clinical trial in children, adolescents and adults. PLOS Negl Trop Dis 12(6):e0006340
75. Permpalung N, Wongrakpanich S, Korpaisarn S, Tanratana P, Angsanakul J (2013) Trend of human rabies prophylaxis in developing countries: toward optimal rabies immunization. Vaccine 31(38):4079–4083
76. Madhusudana SN, Sanjay TV, Mahendra BJ, Sudarshan MK, Narayana DHA, Giri A et al (2006) Comparison of safety and immunogenicity of purified chick embryo cell rabies vaccine (PCECV) and purified vero cell rabies vaccine (PVRV) using the Thai red cross intradermal regimen at a dose of 0.1 ML. Hum Vaccin 2(5):200–204
77. Chen HH, Neumeier AT, Davies BW, Durairaj VD (2013) Analysis of pediatric facial dog bites. Craniomaxillofac Trauma Reconstr 6(4):225–231
78. Lang J, Hoa DQ, Gioi NV, Tho LT, Vien NC, Rouyrre N et al (1999) Immunogenicity and safety of low-dose intradermal rabies vaccination given during an expanded Programme on immunization session in Viet Nam: results of a comparative randomized trial. Trans R Soc Trop Med Hyg 93(2):208–213
79. Lang J, Feroldi E, Vien NC (2009) Pre-exposure purified vero cell rabies vaccine and concomitant routine childhood vaccinations: 5-year post-vaccination follow-up study of an infant cohort in Vietnam. J Trop Pediatr 55(1):26–31

80. Pengsaa K, Limkittikul K, Sabchareon A, Ariyasriwatana C, Chanthavanich P, Attanath P et al (2009) A three-year clinical study on immunogenicity, safety, and booster response of purified chick embryo cell rabies vaccine administered intramuscularly or intradermally to 12- to 18-month-old Thai children, concomitantly with Japanese encephalitis vaccine. Pediatr Infect Dis J 28(4):335–337
81. Chulasugandha P, Khawplod P, Havanond P, Wilde H (2006) Cost comparison of rabies pre-exposure vaccination with post-exposure treatment in Thai children. Vaccine 24(9):1478–1482
82. Kessels JA, Recuenco S, Navarro-Vela AM, Deray R, Vigilato M, Ertl H et al (2017) Pre-exposure rabies prophylaxis: a systematic review. Bull World Health Organ 95(3):210–219C
83. Estrategia nacional de zoonosis-MINSA (2014) Resultados plan de vacunación antirrábica de pre-exposición en comunidades en riesgo de rabia de la Region Amazonas: Perú 2011–2014. Ministerio de Salud del Perú, Lima, p 2015
84. Ertl HCJ (2019) New rabies vaccines for use in humans. Vaccine 7(2):54
85. Walker PJ, Kongsuwan K (1999) Deduced structural model for animal rhabdovirus glycoproteins. J Gen Virol 80(5):1211–1220
86. Buthelezi SG, Dirr HW, Chakauya E, Chikwamba R, Martens L, Tsekoa TL et al (2016) The lyssavirus glycoprotein: a key to cross-immunity. Virology 498:250–256
87. Zhao L, Seth A, Wibowo N, Zhao C-X, Mitter N, Yu C et al (2014) Nanoparticle vaccines. Vaccine 32(3):327–337
88. Zhang Y, Zhang S, Li W, Hu Y, Zhao J, Liu F et al (2016) A novel rabies vaccine based-on toll-like receptor 3 (TLR3) agonist PIKA adjuvant exhibiting excellent safety and efficacy in animal studies. Virology 489:165–172
89. Wijaya L, Tham CYL, Chan YFZ, Wong AWL, Li LT, Wang L-F et al (2017) An accelerated rabies vaccine schedule based on toll-like receptor 3 (TLR3) agonist PIKA adjuvant augments rabies virus specific antibody and T cell response in healthy adult volunteers. Vaccine 35(8):1175–1183
90. Kalimuddin S, Wijaya L, Chan YFZ, Wong AWL, Oh HML, Wang L-F et al (2017) A phase II randomized study to determine the safety and immunogenicity of the novel PIKA rabies vaccine containing the PIKA adjuvant using an accelerated regimen. Vaccine 35(51):7127–7132
91. Alberer M, Gnad-Vogt U, Hong HS, Mehr KT, Backert L, Finak G et al (2017) Safety and immunogenicity of a mRNA rabies vaccine in healthy adults: an open-label, non-randomised, prospective, first-in-human phase 1 clinical trial. Lancet 390(10101):1511–1520
92. Fries LF, Tartaglia J, Taylor J, Kauffman EK, Meignier B, Paoletti E et al (1996) Human safety and immunogenicity of a canarypox-rabies glycoprotein recombinant vaccine: an alternative poxvirus vector system. Vaccine 14(5):428–434
93. Tatsis N, Ertl HCJ (2004) Adenoviruses as vaccine vectors. Mol Ther 10(4):616–629
94. Dicks MDJ, Spencer AJ, Edwards NJ, Wadell G, Bojang K, Gilbert SC et al (2012) A novel chimpanzee adenovirus vector with low human seroprevalence: improved systems for vector derivation and comparative immunogenicity. PLoS ONE 7(7):e40385
95. Xiang ZQ, Li Y, Cun A, Yang W, Ellenberg S, Switzer WM et al (2006) Chimpanzee adenovirus antibodies in humans, sub-Saharan Africa. Emerg Infect Dis 12(10):1596–1599
96. Chen H, Xiang ZQ, Li Y, Kurupati RK, Jia B, Bian A et al (2010) Adenovirus-based vaccines: comparison of vectors from three species of adenoviridae. J Virol 84(20):10522–10532
97. Rux JJ, Kuser PR, Burnett RM (2003) Structural and phylogenetic analysis of adenovirus hexons by use of high-resolution x-ray crystallographic, molecular modeling, and sequence-based methods. J Virol 77(17):9553–9566
98. Xiang Z, Gao G, Reyes-Sandoval A, Cohen CJ, Li Y, Bergelson JM et al (2002) Novel, chimpanzee serotype 68-based adenoviral vaccine carrier for induction of antibodies to a transgene product. J Virol 76(6):2667–2675
99. Wang X, Fang Z, Xiong J, Yang K, Chi Y, Tang X et al (2019) A chimpanzee adenoviral vector-based rabies vaccine protects beagle dogs from lethal rabies virus challenge. Virology 536:32–38

100. Xiang ZQ, Greenberg L, Ertl HC, Rupprecht CE (2014) Protection of non-human primates against rabies with an adenovirus recombinant vaccine. Virology 450–451:243–249
101. Farina SF, Gao GP, Xiang ZQ, Rux JJ, Burnett RM, Alvira MR et al (2001) Replication-defective vector based on a chimpanzee adenovirus. J Virol 75(23):11603–11613
102. Vellinga J, Smith JP, Lipiec A, Majhen D, Lemckert A, van Ooij M et al (2014) Challenges in manufacturing adenoviral vectors for global vaccine product deployment. Hum Gene Ther 25(4):318–327
103. Wang C, Dulal P, Zhou X, Xiang Z, Goharriz H, Banyard A et al (2018) A simian-adenovirus-vectored rabies vaccine suitable for thermostabilisation and clinical development for low-cost single-dose pre-exposure prophylaxis. PLoS Negl Trop Dis 12(10):e0006870
104. Napolitano F, Merone R, Abbate A, Ammendola V, Horncastle E, Lanzaro F et al (2020) A next generation vaccine against human rabies based on a single dose of a chimpanzee adenovirus vector serotype C. PLoS Negl Trop Dis 14(7):e0008459
105. Nel LH, Rupprecht CE (2007) Emergence of lyssaviruses in the old world: the case of Africa. Curr Top Microbiol Immunol 315:161–193
106. Bourhy H, Kissi B, Audry L, Smreczak M, Sadkowska-Todys M, Kulonen K et al (1999) Ecology and evolution of rabies virus in Europe this paper is dedicated to the memory of the late Katariina Kulonen. J Gen Virol 80(10):2545–2557
107. Smith JS, Orciari LA, Yager PA, Seidel HD, Warner CK (1992) Epidemiologic and historical relationships among 87 rabies virus isolates as determined by limited sequence analysis. J Infect Dis 166(2):296–307
108. Malaga AA (1962) Epidemiological factors governing the control of rabies. Bol Oficina Sanit Panam 53:105–114
109. Fleming G (1872) Rabies and hydrophobia; their history, nature, causes, symptoms, and prevention. Chapman and Hall, London
110. Smith JS, Seidel HD (1993) Rabies: a new look at an old disease. Prog Med Virol 40:82–106
111. Pieracci EG, Pearson CM, Wallace RM, Blanton JD, Whitehouse ER, Ma X et al (2019) Vital signs: trends in human rabies deaths and exposures – United States, 1938–2018. MMWR Morb Mortal Wkly Rep 68(23):524–528
112. Vigilato MAN, Clavijo A, Knobl T, Silva HMT, Cosivi O, Schneider MC et al (2013) Progress towards eliminating canine rabies: policies and perspectives from Latin America and the Caribbean. Philos Trans R S B: Biol Sci 368(1623):20120143
113. PAHO. Rabies persists in only four countries of Latin America and the Caribbean [Internet]. Available from: https://www.paho.org/hq/index.php?option=com_content&view=article&id=14665:rabies-persists-in-only-four-countries-of-latin-america-and-the-caribbean&Itemid=135&lang=en
114. WHO Rabies Modelling Consortium (2020) Zero human deaths from dog-mediated rabies by 2030: perspectives from quantitative and mathematical modelling. Gates Open Res 3:1564
115. USDA United States Department of Agriculture (USDA) (2016) Approved animal rabies vaccines [Internet]. USDA. Available from: http://www.floridahealth.gov/diseases-and-conditions/rabies/_documents/13-rabies-vaccines-licensed-and-marketed-in-the-united-states-2016.pdf
116. Wang Y, Xiang Z, Pasquini S, Ertl HC (1998) Effect of passive immunization or maternally transferred immunity on the antibody response to a genetic vaccine to rabies virus. J Virol 72(3):1790–1796
117. Ford RB, Larson LJ, McClure KD, Schultz RD, Welborn LV (2017) AAHA Canine vaccination guidelines [Internet]. Available from: https://www.aaha.org/globalassets/02-guidelines/canine-vaccination/vaccination_recommendation_for_general_practice_table.pdf
118. Basheer AM, Ramakrishna J, Manickam R (1997) Evaluation of post-exposure vaccination against rabies in cattle. New Microbiol 20(3):289–294
119. CDC (2020) Caring for a client's animal that may have been exposed to rabies [Internet]. CDC. Available from: https://www.cdc.gov/rabies/specific_groups/veterinarians/potential_exposure.html
120. Wilson PJ, Rohde RE, Oertli EH, Willoughby RE (2019) Rabies: clinical considerations and exposure evaluations. Elsevier, Inc, St. Louis, pp 1–146

Chapter 7
Wildlife Rabies Management in the New World: Prevention, Control and Elimination in Mesocarnivores

Richard B. Chipman, Amy T. Gilbert, and Dennis Slate

Abstract Wildlife disease management is a growing and challenging One Health paradigm for scientists and program managers working to resolve human-wildlife conflicts. Managing wildlife diseases is complex and accomplished primarily through the manipulation of populations at a landscape scale. Rabies lyssavirus (RABV) perpetuates in dogs and wildlife, including mesocarnivores and bats, with the greatest diversity of viral variants (e.g., independent epizootic or enzootic foci) associated with wildlife throughout the Americas. From Argentina to the Canadian Arctic, prevention and control of rabies in carnivores has historically focused on managing dog-mediated human rabies through public health messaging, implementation of local animal control strategies to reduce free-ranging dog populations and mass rabies vaccination campaigns and local vaccination clinics. Multiple host shifts of canine RABV variants to a diverse array of wildlife has occurred throughout the Americas, creating a suite of unique management challenges. Successful management of RABV at the animal source across broad and complex landscapes requires knowledge of how reservoir populations use these landscapes and the development of coordinated management strategies that effectively target mesocarnivores across a variety of habitats and population densities. Historically, attempts at disease control in furbearers used population reduction. By the mid-twentieth century, alternative techniques focused upon oral rabies vaccination (ORV) of free-ranging animals. In the Americas, management of wildlife RABV through ORV of mesocarnivore populations has not occurred outside of

R. B. Chipman (✉)
US Department of Agriculture, Animal and Plant Health Inspection Service, Wildlife Services, National Rabies Management Program, Concord, NH, USA
e-mail: Richard.B.Chipman@usda.gov

A. T. Gilbert
US Department of Agriculture, Animal and Plant Health Inspection Service, Wildlife Services, National Wildlife Research Center, Fort Collins, CO, USA

D. Slate
Chippewa Bay Wildlife Art and Science LLC, Chippewa Bay, NY, USA

C. E. Rupprecht (ed.), *History of Rabies in the Americas: From the Pre-Columbian to the Present, Volume I*, Fascinating Life Sciences,
https://doi.org/10.1007/978-3-031-25052-1_7

Canada and the United States. A limited ORV field trial is being planned targeting mongoose populations in Puerto Rico. For three decades, integrated ORV programs have focused on coyotes, red and gray foxes, raccoons, and striped skunks to manage RABV variants in North America. Key components of successful wildlife rabies management include communication and coordination, enhanced RABV surveillance, ORV, program monitoring, contingency actions, and applied research. Science-based adaptive management and the development of new tools and technology aligned with sustained public support are critical to continued progress in controlling and eliminating RABV variants in mesocarnivore populations at local, regional, and national scales.

Keywords Adaptive management · Control · Mesocarnivore · Monitoring · Oral rabies vaccination · Enhanced rabies surveillance · Population reduction · Rabies virus · Wildlife management

Introduction

Understanding the source and spread of rabies lyssavirus (RABV) in wildlife populations in the Americas, and subsequent attempts to manage the disease in wild mesocarnivores, is an evolving story of One Health collaboration, landscape level oral rabies vaccination (ORV), knowledge gaps, applied research, innovation, risk taking, adaptive management, and funding challenges. Local and regional management of wildlife rabies is an important and growing focus for a community of scientists and managers working to resolve human-wildlife conflicts to protect human and animal health, food supplies and to mitigate the economic impacts of living with sylvatic rabies in the Americas [1–3]. Successful management of wildlife RABV across a diversity of landscapes requires a detailed understanding of how reservoir and nontarget populations use such landscapes and the development of management strategies to scale up to target populations that occur across a variety of regions and habitats [4].

Historical documentation of rabies and its control is scant in the New World [5]. The first written accounts did not occur until the early eighteenth century in Mexico [6–8]. Whether RABV existed among New World wildlife prior to European colonization and introduction of domestic dogs (*Canis lupus familiaris*) and livestock remains debatable [9–11]. However, both folklore and anecdotal records from Central America, the Canadian Arctic, and the Pacific Northwest in the U.S. report a disease clinically consistent with rabies, described by indigenous people and settlers [12–15]. Moreover, phylogenetic studies of some mesocarnivore RABV demonstrate a New World origin from bats [16]. The introduction and subsequent spread of RABV in multiple mesocarnivores throughout the New World was likely in part a direct consequence of the translocation of infected dogs from the Old World during colonization [7, 10]. Such animal translocation poses a significant risk of disease introduction into naïve areas and jeopardizes decades of successful rabies prevention and control efforts [17–19].

Globally, most human rabies deaths and related economic impacts result from viral transmission via dog bites [20, 21]. Canine rabies has been eliminated in all developed countries and is increasingly under control in less developed countries [22]. Rabies prevention and control in carnivores has historically focused on managing domestic dogs and dog-mediated human rabies throughout the Americas. Initially during the 1920s and resuming post World War II, when more potent rabies vaccine became available, a concerted effort to manage canine rabies began, particularly in North America [23]. Public health and veterinary educational campaigns increased awareness of the threat of rabies and encouraged vaccination for companion animals and livestock, which later included messaging that encouraged responsible pet ownership. Animal control regulations and enforcement, including quarantine and muzzling reduced risks, while capture and removal of free ranging dogs reduced populations and the risk of RABV exposure at the local level [24]. Government-funded mass vaccination campaigns and local reduced-cost and free vaccination clinics were essential for canine rabies management successes in the Americas [21, 25, 26]. Integrated rabies management programs led to the elimination of canine rabies in Canada during the 1950s, in the U.S. during the 1970s (and again in 2004), and in Mexico, canine mediated human rabies was eliminated in 2019, with validation by the World Health Organization (WHO) [27–30]. Reintroduction of canine rabies remains a threat that is partially mitigated by regulation and regulatory enforcement reducing the likelihood of importation of rabid dogs from canine rabies enzootic countries [31, 32]. Where canine rabies has been controlled or eliminated, the focus may transition to include rabies management efforts targeting wildlife populations to protect human and animal health [3].

Unlike rabies intervention strategies for humans and dogs, improved tools and technology for rabies management in wildlife only became available during the latter part of the twentieth century and early twenty-first century. In North America, population reduction was the primary method for managing rabies in carnivores during much of the eighteenth – twentieth centuries, under the premise that reduced population densities could eliminate RABV transmission [33, 34]. Shooting, bounties, trapping, gassing of dens, and toxicants proved ineffective as a stand-alone population reduction strategy in eliminating RABV transmission at broad landscape scales [34, 35].

Population reduction is credited with very limited, more localized rabies control success in a few cases involving red foxes (*Vulpes vulpes*) and striped skunks (*Mephitis mephitis*) in Canada and the U.S. [36–38]. Contemporary and integrated disease management strategies may remove individual animals or small groups of mesocarnivores when public health and safety and livestock are at risk of exposure, but population reduction is no longer considered a primary strategy to manage wildlife rabies [39, 40]. A management strategy focused on live-capture and release of wildlife for mass parenteral vaccination was determined to be expensive and largely impractical at a landscape scale [41]. During the 1970s, the concept of ORV led to a paradigm shift in management of rabies in wild mesocarnivores [42]. Although the concept as a potential method to control rabies in free-ranging wildlife populations originated in the U.S., ORV has a longer applied history in response to rabies epizootics in Europe and Canada [43–45].

Several recent reviews detail the history and success of rabies management in free-ranging wild carnivores in the Americas [34, 41, 44, 46, 47]. During the 1980s and 1990s, Canada and the U.S. began field testing and incorporating ORV into integrated wildlife rabies management targeting coyote (*Canis latrans*), red fox, gray fox (*Urocyon cinereoargenteus*), raccoon (*Procyon lotor*), and striped skunk populations. The use of vaccines in wildlife disease management, while not without controversy, became a sound example where pragmatic conservationism met fit-for-purpose rabies management to help protect human and animal health. Despite documented management successes, only two (Canada and U.S.) of 35 countries in the Americas have attempted to manage rabies in wildlife populations with ORV at the landscape scale [34, 48]. Management of wildlife rabies is currently a low priority for most countries in the Americas faced with competing public health priorities, including ongoing challenges to first prevent and eliminate human deaths due to canine rabies and mitigate the impacts of vampire bat rabies [3, 25].

Sustained public support and funding are essential for effective wildlife rabies management programs. A coordinated One Health approach that balances the art and science of wildlife management and melds traditional and novel strategies and methods enhances the chance for success. Key program components include communication and coordination, enhanced rabies surveillance (ERS), bait distribution logistics, program monitoring, contingency actions, applied research, and adaptive management. In Canada and the U.S., project managers plan and oversee large scale control programs across a variety of developed and rural landscapes in three Canadian provinces and across 17 U.S. states. These programs progressed through application of refined management strategies based on a coordinated approach including cross-border collaboration focused on ERS, strategic planning, leveraging of resources and information, and cooperative research under the umbrella of the Canadian National Rabies Plan, the U.S. National Plan for Wildlife Rabies Management, and the North American Rabies Management Plan (NARMP) [24, 34]. In essence, wildlife rabies management entails refinements to historical natural resource concepts, where management is broadly defined as the act of manipulating the structure and dynamics of wildlife populations to achieve specific human objectives by means of the wildlife resource [49, 50]. Contemporary wildlife rabies management in the Americas spans several decades and efforts by a relatively small group of passionate and dedicated scientists, public health officials, veterinarians and wildlife biologists collaborating to advance awareness about a neglected disease and foster sustained public support for large scale coordinated programs and applied research [34, 44, 51, 52].

Challenges of Managing Diseases in Wildlife Populations

Human-wildlife conflicts are often driven by changing human values and an imbalance between basic needs of society to produce food, provide shelter and sustain economic growth and a human-altered landscape that may increase human-wildlife interactions and impact disease spread in wildlife populations. The basic philosophy and regulatory framework for wildlife management varies among countries and may be impacted by cultural norms that affect government response to managing diseases. For example, in Canada and the U.S., the North American Model for Wildlife Conservation is based around seven core principles that promote conservation and science-based management where government, as trustee, protects and manages wildlife for the benefit of the resource and the public [53]. This model calls for adapting and adopting management strategies to protect agriculture, property, natural resources and human health and safety, including management of zoonotic and non-zoonotic wildlife diseases. Like other wildlife management programs, the prevention and control of wildlife diseases is complex, multifaceted and at times contentious, requiring program managers to consider relevant alternatives in formulating the best strategies to meet objectives that are primarily accomplished through the manipulation of populations at a landscape scale [48, 54]. Management focuses on preventing introduction or reintroduction of diseases, controlling, or limiting their spread, and may include elimination as a goal. Surveillance-based or 'no action' alternatives may be appropriate if strategies and tools are not available to effectively manage disease in certain species – e.g., rabies in Arctic foxes (*Vulpes lagopus*) or avian influenza in waterfowl and other wild bird populations [55, 56]. Shifting from foundational research on disease ecology to landscape level disease management requires applied research to develop and refine management strategies to overcome multiple logistical and operational challenges, uncertainties, and calculated risks. Rabies management strategies may target reservoir and vector wildlife populations with vaccination or may involve exclusion strategies to reduce local contacts among infected and susceptible animals [57]. Disease prevention and control in wildlife populations at a landscape scale is still a relatively new concept, with supporters and detractors regardless of disease focus [58, 59]. Those against disease management claim it is not "natural" and that removing a mortality factor would interfere with population dynamics that exacerbates human-wildlife conflicts or could limit natural selection for disease resistance. However, novel hosts, diseases and parasites have been introduced in rural and urban environments in various combinations, often facilitated by intentional or accidental wildlife translocation. Advocates for disease management believe wildlife managers should focus on priority goals of protecting human and animal health, while recognizing that mesocarnivore populations are common to abundant in human-altered habitats. Disease management strategies are more likely to succeed using a One Health approach that emphasizes cooperation and coordination that builds consensus among experts in human, animal, and environmental health into disease prediction, planning and project implementation [60].

Coordinated programs are necessary to prevent introduction and spread of zoonotic diseases in wildlife, and where practical to progressively work toward their elimination in wildlife reservoirs to protect human and animal health. Development of effective rabies control programs can be impeded by a lack of adequate ecological information for target populations, difficulty in estimating mesocarnivore population abundance across habitats and landscapes, and challenges in understanding population dynamics in the presence and absence of other diseases (e.g., canine distemper). Traditional veterinary approaches to surveillance and control are rarely adequate for disease management at its source in wildlife populations. Parenteral vaccination approaches applied to domestic animals or livestock seldom translate into practical applications for free-ranging wildlife populations. Conversely, oral vaccine delivery applications for free-ranging wildlife have received increasing attention for targeting inaccessible and free-ranging domestic dogs, as part of a comprehensive strategy towards the global elimination of human deaths mediated by canine rabies [61–64]. Planning and subsequent implementation of disease management programs often face strong head winds, considering conflicting wildlife management priorities in an environment of increasing costs and shrinking budgets for projects that fundamentally require multi-year investments to meet long-term management goals [65, 66]. Other hurdles may include jurisdictional and legal issues involving operational programs across geopolitical landscapes. Wildlife disease control programs also require maintaining and enhancing coalitions while pushing to increase the pace of innovation and, in the case of rabies management, the application of new technology and refinement of vaccine-baits and delivery systems. Successful wildlife disease management programs for rabies prevention, control, and elimination in the Americas require comprehensive public engagement, intersectoral communication, coordination and collaboration, practical and contingency planning, data transparency, ERS, adaptive management, timely diagnostics, program monitoring, and applied research [34].

Concepts and Strategies for Wildlife Rabies Management

From the earliest human migrations into the Americas and the subsequent development of agrarian-based communities, an ever-growing footprint across the landscape has modified terrestrial and aquatic ecosystems [67, 68]. Introduction of livestock and other domestic animals via the Columbian exchange starting early in the sixteenth century [69], as well as introduction of exotic carnivores (e.g., small Indian mongoose, *Herpestes javanicus*), impacted extant ecosystems and continue to increase the likelihood of disease transmission at the human-domestic animal-wildlife interface [70–72]. More recently, a growing body of evidence points to the impact of climate change on wildlife habitats and movement patterns, including range expansions, leading to increased vulnerability and reduced resiliency in some species to established and emerging diseases [73–75]. Red and Arctic foxes are the primary reservoirs for Arctic fox RABV variant and are responsible for spillover infection to domestic dogs, yet the red fox geographic range continues to expand into new areas of the Canadian Arctic and Alaska in the U.S., where previously only Arctic foxes occurred [74].

Management practices from the sixteenth–nineteenth centuries in the Americas to address rabies in wildlife are largely undocumented, but likely relied on removal of sick-acting animals to mitigate threats to the health of people and domestic animals. More recent development and refinement of techniques and integrated strategies to manage wildlife rabies during the twentieth and early twenty-first centuries resulted in significant advances in a One Heath collaboration to reduce risks to human and domestic animal populations. Canada and the U.S. have shifted from a strategy of surveillance for monitoring wildlife rabies to active management in which ORV represents the central method applied to mesocarnivore populations in an integrated approach at local and regional scales [76]. In the Neotropics, key priorities continue to focus on eliminating dog-mediated rabies and controlling vampire bat rabies to protect public health and livestock, although RABV circulation has been documented in multiple wildlife species in several countries [3].

The novel approach to rabies management in fox and other mesocarnivore populations through ORV, as conceived by the late George Baer in the 1960s and later field tested in Europe during the 1970s, created a unique opportunity to eliminate RABV variants locally and reduce further spread of the disease to naive wildlife populations [43, 77, 78]. Despite documented success in managing rabies in red fox populations in Europe as early as 1978, it was not until the 1980s and the early 1990s that Canada and the U.S., respectively, initiated proof of concept ORV field testing [77, 79].

Rabies prevention and control requires extensive technical and bureaucratic collaboration and a sustained interface among diverse disciplines across multiple agencies and organizations [47, 48, 80]. Interdepartmental rabies committees, such as in New York and Ontario, go back to at least the 1960s and serve as the precursor to the now more formalized One Health concept of collaboration across disciplines for addressing zoonotic diseases [51]. During the build-up to the inaugural distribution of oral rabies vaccines for landscape level management in North America, the planning process often involved frequent and sometimes contentious meetings where researchers, wildlife biologists, veterinarians and public health professionals held spirited debates. The red and gray fox, raccoon, and coyote were the primary management focus under early consideration for ORV by a handful of visionaries and creative thinkers from academia, government, and industry who could see the possibilities for prevention, control, and elimination of RABV variants and the benefits to North American citizens and beyond [15, 81]. These individuals understood the personal and professional stakes and perseverance required to develop and adopt science-based ORV strategies for managing rabies in wildlife populations across diverse landscapes in North America (Fig. 7.1). They also understood the art and science of wildlife management and could see the opportunities for blending the aspirational with the practical to complement conventional public health disease prevention and control.

Fig. 7.1 Early (deceased) visionaries in the development and application of oral rabies vaccine to manage wildlife rabies in North America, including (clockwise from top left): George Baer; John "Jack" Debbie; Keith Clark; Samuel B. "Sam" Linhart; and Charles "Charlie" MacInnes

Despite the growing momentum for broader application of ORV during the 1980s and 1990s, apprehension around the use of vaccines in wildlife populations remained, especially among wildlife biologists regarding management priorities, costs, and the potential impact on traditional furbearer management [51, 79]. Concerns expressed publicly at meetings and conferences included the fear of mutations in the vaccine virus vector, as well as the practice of "wildlife medicine" at the population level and its impact on traditional wildlife conservation. Questions were raised about the effects of rabies management on carnivore populations and the likelihood of artificially increased populations resulting in greater levels of human-wildlife conflict, impacts for rare or endangered species, and disruption of natural selective forces that could enhance disease resistance in wildlife populations [79]. Practical considerations were expressed over the degree of the rabies threat to humans from wildlife, given the ease of access to timely administration of post-exposure prophylaxis (PEP) in North America, compared to other public and animal health priorities. Perhaps the most common concern was that rabies management at the wildlife source through ORV was too expensive, considering the time it would take to achieve long term goals. Understanding the economic impact of rabies in the Americas and the benefits and costs of rabies prevention and control targeting reservoir species with ORV became essential for decision-makers to consider (i.e., whether sustained, long-term investment was warranted to reduce social and financial costs of living with rabies) [82–85]. Counter arguments were made that wildlife populations were already artificially increased in human-altered habitats and that certain viruses, such as raccoon RABV, represented a new mortality factor as it spread rapidly through the mid-Atlantic and northeastern U.S. raccoon populations following a translocation of infected animals during the 1970s [86, 87]. An additional compelling point was that an inevitably fatal disease such as rabies should not be considered a viable wildlife management "tool" to reduce abundant raccoon populations which pose a risk to humans and domestic animals. Ultimately, it was acknowledged that because tools were available to safely and effectively manage a neglected deadly disease impacting people, pets and livestock, there existed a responsibility to develop and implement control interventions by ORV in specific areas and wildlife species. The expansive use of ORV in North America ultimately became a reality based on 12 key factors: (1) scientific evidence of vaccine efficacy and field effectiveness; (2) human and animal safety; (3) comprehensive planning and clear management objectives; (4) cooperator and public support; (5) environmental compliance; (6) timely and sustained vaccine-bait availability; (7) overcoming multiple logistical and operational challenges; (8) valid metrics for measuring success, such as ERS; (9) preparedness for contingency actions; (10) applied species ecology and vaccine-bait research; (11) positive economic forecasting; and (12) regulatory approval.

The complexity of managing RABV in diverse mesocarnivore reservoir species in North America cannot be overstated. The Arctic, gray, and red foxes, coyote, striped skunk, raccoon, and small Indian mongoose illustrate the challenges of varying host size, foraging behavior, ability to manipulate and consume vaccine-baits, population abundance and ecological diversity [88–90]. Seasonal and environmental variables impact vaccine distribution strategies, adding complexity to landscape scale ORV [88]. Establishing species-specific management strategies and objectives through a focused risk assessment is essential prior to implementing a wildlife rabies management program [91], where risk and uncertainty can be mitigated, in part, by following an adaptive management strategy [34, 92]. Key ecological, biological, logistical, and environmental hurdles across species, RABV variants, and habitats require applied research and comprehensive strategic planning to provide the necessary and flexible framework to achieve management objectives [41, 48]. Conceptually, adaptive management allows managers to implement rabies prevention and control programs despite data gaps that can be systematically addressed through research and monitoring that accompany management. Modeling also fills a valuable niche that can predict a range of management outcomes, help identify knowledge gaps and generate hypotheses for future applied research [93–101].

Rabies prevention strategies may include site-specific habitat management and recommendations to alter human behaviors that discourage contact between people and domestic animals and wildlife. Examples include discouraging supplemental feeding of wildlife (e.g., reducing bird feeding and feeding of companion animals inside) and exclusion techniques that prevent wildlife from entering houses and other structures. Individual animals that pose an immediate risk to human health and safety should be removed and tested for rabies as recommended by public health and veterinary authorities [39]. Integration of local population reduction into a broader management framework such as the "Point Infection Control" (PIC) was developed in Ontario, Canada as a response strategy in the event raccoon rabies spread into the Province from New York. The goal was to achieve focal population reduction of animals that may be incubating rabies, followed by trap-vaccinate-release (TVR) and ORV around foci where raccoon population density reduction had been achieved. Euthanized raccoons served to enhance rabies surveillance prior to TVR and ORV around foci [24, 77, 84, 102]. Modifications of PIC continue to be applied in Canada as a contingency response. However, a singular focus on widespread reduction of wildlife populations at a landscape scale has proven ineffective in controlling or eliminating RABV and may potentially enhance animal contacts, movements and RABV spread [34, 46, 76, 103]. During rabies contingency actions, a trap, euthanize and test (TET) approach may be appropriate for focused ERS to better understand and define the rabies epizootic focus and inform detailed response planning [104]. However, the implementation of a TET strategy to increase surveillance during an emergency response must be weighed against the likelihood of increased movement and mixing of naïve individuals and those incubating rabies, with the potential to exacerbate the spread of localized epizootics.

Primary strategies for RABV management in wildlife populations continue to rely on ORV by aerial distribution at prescribed bait densities and patterns in specified habitats to meet management objectives of preventing spread or achieving regional elimination [98, 105, 106]. Implementation of TVR for target species like raccoons in tandem with ORV may enhance the reduction of RABV transmission if adequate in implementation scale, scope, and duration [24]. Integration of ORV and TVR have proven effective in mitigating local rabies epizootics and breaches of ORV zones as a component of contingency actions in both Canada and the U.S. [106]. Improving rabies management strategies through collaborative field trials with intensive pre- and post-project monitoring has been critical to advancing effective rabies management [107–110]. Tracking the spatial-temporal distribution of RABV variants in near real-time through exposure-based public health surveillance in tandem with ERS has played a key role in understanding rabies dynamics in wildlife populations and the degree of control program success [111].

Management Focus and Rabies Epidemiology in Wild Carnivores

All mammals are susceptible to RABV infection, yet the major wildlife reservoirs are found primarily in the Orders Chiroptera and Carnivora in the Americas [57, 112]. The first mention of rabies in wildlife in the Americas occurred with red foxes in Virginia during 1753 and 1762 [14]. The first major epizootic reported in North America involving wild mesocarnivores occurred in 1768 in Boston, Massachusetts where red foxes and free roaming dogs attacked pigs and other domestic animals

[6]. Rabies moved west with the pioneers in the U.S. and was reported in skunks by Mountain Men on the Great Plains during the 1830s and had spread to California by the 1850s [113]. By the end of the Civil War in 1865, wildlife rabies was widespread throughout the U.S. [15, 114–116]. Contemporary carnivore reservoirs may encompass one or all potential host populations within an ecosystem, where RABV perpetuates independently within specific host species, most often at the regional population scale in discrete geographic areas [57, 117–119]. Diverse mesocarnivores, many with overlapping geographic ranges, have been documented as reservoirs in the Americas and play a role in the long-term maintenance of RABV in the region [16, 74, 118, 120]. Cross species transmission 'spillover' infections of RABV are well-documented in association with established wildlife foci [121, 122], yet the likelihood for success in establishing onward transmission in the novel recipient may scale inversely with relatedness to the donor involved in transmission [123].

Enzootic (or eliminated) RABV variants in mesocarnivore reservoirs have been reported from the Canadian Arctic to the tropics of Brazil, including: (1) coyote variant ; (2) mongoose variant ; (3) raccoon variant ; (4) north central (U.S.) skunk variant; (5) south central skunk variant ; (6) California skunk variant; (7) Baja California sur skunk variant; (8) North central (Mexico) skunk variant; (9) Arizona gray fox variant; (10) Texas gray fox variant; (11) Arctic fox variant; (12) crab-eating fox variant; (13) hoary fox variant; (14) coati variant; (15) marmoset variant. Mesocarnivore RABV variants are established in wildlife populations across a variety of rural and developed landscapes, host ecological strategies, and population structures that contribute to the complexity of managing rabies in wildlife populations. Although small fractions of carnivore species are recognized RABV reservoirs, these animals share common traits including flexible diets, larger litter sizes, earlier sexual maturity (reproduce more rapidly), and relatively short life spans [118].

In North America, there are at least nine mesocarnivore species that serve as RABV reservoirs including the invasive small Indian mongoose introduced throughout the Caribbean region [24, 34, 124, 125]. Native reservoir species include red foxes, coyotes, gray foxes, Arctic foxes, raccoons, striped skunks, eastern spotted skunks (*Spilogale putorius*), and the white-nosed coati (*Nasua narica*) [126]. In contrast, in South America only a limited number of mesocarnivores are described as reservoirs including the crab-eating fox (*Cerdocyon thous*) and hoary fox (*Lycalopex vetulus*). Rabies in gray foxes infected with canine RABV variant has been reported from Columbia [127], whereas the crab-eating and hoary fox species are reported to be reservoirs for the cosmopolitan dog RABV in Brazil [3, 128–133]. However, control efforts have not been reported targeting foxes in these regions of South America [34]. The common marmoset (*Callithrix jacchus*), a unique primate reservoir of RABV, has also increased in importance in Brazil [134], but without reports of control.

Beginning in the 1960s in Canada and the U.S, rabid wildlife became reported more frequently than rabid domestic animals [15, 28]. This change was a direct result of aggressive measures beginning in the early part of the twentieth century to address periodic canine rabies outbreaks in Canada and enzootic canine rabies in the U.S. Prevention, control, and elimination of canine rabies in Canada and the

U.S. occurred as a cumulative effect of local dog rabies control programs, including vaccination with safe, potent, and efficacious vaccines, leash and muzzle laws, quarantine and stray animal removal [24, 135].

Wildlife RABV was reported in the Arctic during 1945–1947 in an extensive epizootic among Alaskan sled dogs and wild canids [136]. Rabies was also confirmed among Arctic foxes and gray wolves (*Canis lupus*) in the Northwest Territories, Canada in the mid-1900s [34, 137]. During the late 1940s, Arctic fox RABV spread south and became widespread in Canada, involving all provinces by the 1950s. An apparent host shift of RABV from Arctic to red foxes occurred in multiple locations throughout Canada where both species coexist but was most prominent in southern Ontario and western Quebec, where Arctic fox RABV eventually spread into the northeastern U.S. [24, 27]. Arctic fox RABV variant was enzootic in Ontario and Western Quebec for over 40 years until wildlife rabies control was implemented during the 1990s, involving a joint Ontario-Quebec control program [27, 52, 138–140]. Ontario had no reported cases of Arctic fox RABV from May 2012 until December 2015, when new cases were detected as part of ERS activities associated with management of the reemergence of raccoon RABV variant [120, 141]. New cases in red foxes and skunks appear to be related genetically to the previous Arctic fox RABV variant circulating in the province [120, 142]. Red foxes were an important RABV reservoir in the eastern U.S. during the mid to late twentieth century, while striped skunks remained the primary reservoir in the Great Plains for the north central skunk RABV variant and in the central part of the U.S. for the south-central skunk RABV variant [34, 113, 143]. Skunks are also an important reservoir of a unique RABV variant in California and Baja California sur Mexico. Other RABV variants have been reported from multiple skunk species in Mexico [144, 145], but none of these skunk RABV foci have been subject to intensive management.

Raccoon RABV was first confirmed in Florida during 1947 and, before 1977, was primarily restricted to Florida and a few neighboring states, including Georgia, eastern Alabama, and southern South Carolina [86, 146, 147]. During the late 1970s, translocations of infected raccoons shipped to West Virginia and western Virginia for hunting purposes led to the unintentional and rapid spread of raccoon RABV variant across the eastern U.S. and eventually north into the Canadian provinces of Ontario, Quebec, and New Brunswick [148–150]. Raccoon RABV remains enzootic in at least 18 U.S. states (Fig. 7.2). Since 1997, raccoon RABV has been the primary focus of coordinated and intensive management using ORV and other control strategies to prevent westward spread in the U.S. [30, 34, 48, 88, 106]. Despite demonstrated success in controlling spread and local elimination of raccoon RABV variant in some areas, multiple incursions have been detected in several provinces along the shared border with Canada [151]. Cases of raccoon RABV were detected in Ontario in 1999, in New Brunswick in 2000, and in Quebec in 2006. These incursions were controlled and eliminated through selective application of PIC (Ontario) and its modifications and aggressive ORV and TVR programs in 2005, 2002 and 2009, respectively [51, 52, 152, 153]. Quebec has remained raccoon rabies-free since 2009, except for a single case in 2015 (Belanger et al. 2020). New Brunswick

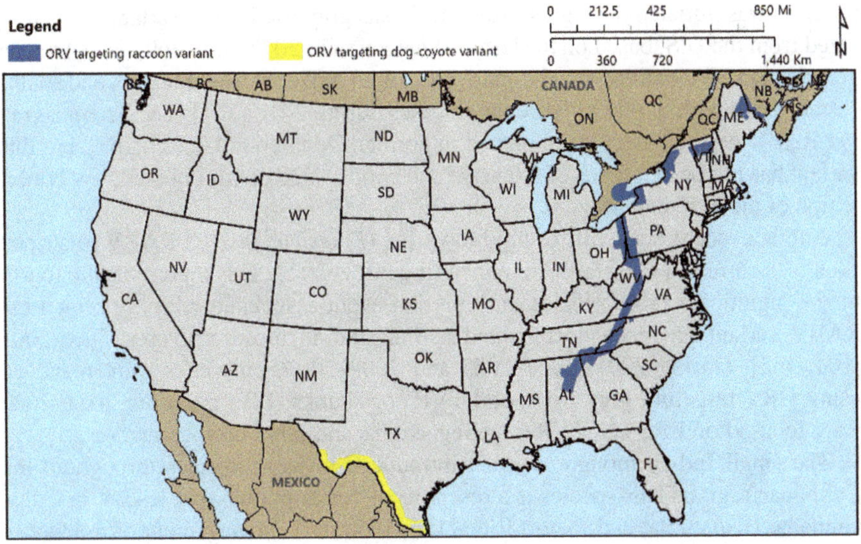

Fig. 7.2 Oral rabies vaccination (ORV) zones targeting two rabies virus variants in North America, 2021

detected raccoon RABV in 2014, after 12 years with no reported cases and ORV management was reinitiated [153]. Thirty-three cases have been reported in New Brunswick during 2014–2021 in skunks and raccoons. However, 24 of the total cases were reported during 2015, with the last reported case in July 2020 and no new cases confirmed during 2021 [154]. Raccoon RABV was once again reported in Ontario during late 2015, when a rabid raccoon was confirmed in the suburb of Stony Creek, outside the city of Hamilton [51]. Whole genome sequencing of raccoon RABV isolates collected in Canada and the U.S. suggest that this recent incursion in Ontario may be the result of a long-distance translocation of an infected raccoon from southeastern New York, more than 500 km away. However, continuous samples from southeastern NY to the Niagara Frontier, would be required to fully validate the origins of this translocation event. As a result of ORV management, TVR, and other wildlife rabies control strategies, the outbreak of raccoon RABV in southern Ontario has been contained and cases have continued to decline through 2022.

During 1988, an epizootic involving coyotes and domestic dogs began in South Texas along the Mexico-U.S. border [156–158]. During 1988–1994, 531 cases of the coyote-dog RABV variant were reported from 18 contiguous Texas counties and more than 2000 people received PEP [105, 157]. During 1995, the first coordinated large scale ORV program began in the U.S. in Texas. By 2004, the dog-coyote RABV variant was eliminated [29, 105, 159]. To date, the ORV program continues at the Mexico-U.S. border (Fig. 7.2). A RABV epizootic also emerged in gray fox populations in west Texas during 1988–1995, with 764 cases of Texas gray fox RABV variant confirmed in 46 contiguous counties [105]. A successful ORV program was initiated during 1996 and the Texas gray fox RABV variant was eliminated from the U.S. based on the last case detection in 2013 [159, 160]. The Arizona gray fox RABV continues to circulate throughout Arizona, New Mexico, and southeastern Utah with a total of 64 reported cases during 2018 [161]. The Arizona gray fox RABV has also been documented in northern Mexico [144, 162, 163], but this variant has not been a focus of intensive and coordinated control in the cross-border region of the U.S. and Mexico.

Another recent host shift of big brown bat (*Eptesicus fuscus*) RABV to striped skunks occurred in Flagstaff, Arizona during 2021–2022. This represents the fourth time a significant cross species transmission event associated with big brown bat RABV variant has been documented in Flagstaff in mesocarnivores since 2001 [164, 165]. During 2001, 2004–2005 and 2008–2009, outbreaks were managed using ORV targeting gray foxes and TVR for skunks. Efforts during 2021–2022 were focused on ERS and TVR targeting skunks and other mesocarnivores.

The small Indian mongoose is an introduced invasive species throughout the Caribbean region. This species is a reservoir of cosmopolitan dog RABV in Cuba, Grenada, Hispaniola, and Puerto Rico [125, 166, 167]. Management of mongoose rabies though population reduction was unsuccessful in terms of control or elimination [168, 169]. The small Indian mongoose is a potential candidate for management with ORV based on immunogenicity and effectiveness of candidate oral vaccines [170–174]. A recent placebo field trial in Puerto Rico showed promise in reaching a fraction of the mongoose population likely sufficient to control rabies with minimal interference from non-target species [175]. These and other research advances provide a potential path forward for planned ORV field trials that may demonstrate applicability to other Caribbean islands where mongoose rabies occurs [176].

Integrated Wildlife Rabies Management in the New World

Historic and contemporary management of RABV in wild carnivores, where ORV is the central tactic, has demonstrated that RABV can be prevented, controlled, and locally eliminated within target populations in the Americas [51, 98, 105, 159, 177, 178]. Several publications comprehensively summarize the history and detail various aspects of contemporary RABV management in carnivores in the New World [24, 34, 41, 46–48, 51, 52, 77, 106, 178, 179]. New World rabies researchers and program managers have shown global leadership in proof-of-concept testing of oral rabies vaccines and subsequent field testing of vaccine-baits, distribution strategies, and systems in the pursuit of effective management of one of the most neglected tropical diseases [88, 102, 104, 138, 180].

Sustained and timely communication, coordination and transparency among international and local partners has been the cornerstone of wildlife rabies management and has proven essential for broad-based support and for moving decision makers from skepticism to optimism to advocates and champions of wildlife rabies management. A focus on communication with stakeholders and partners has helped program managers address the "issue fatigue" that comes with long-term planning horizons and control programs that may take decades to meet management objectives. Frequent communication has also been essential for developing synergistic public-private partnerships between industry and government wildlife management agencies for research, development, and testing of vaccine-baits in support of cost-effective advances to address complex technical issues of efficacy and field effectiveness across an array of New World species [77]. Clearly, the benefits and costs of modern-day rabies control is inherently tied to expenditures associated with the purchase of vaccine-baits, required bait density on the landscape, and ORV distribution [181–183]. As a result, program transparency and an emphasis on collaboration between the private (vaccine manufacturer) and public sectors (customer) to find a cost-effective path forward for RABV control that is informed by science, experience, available funding and the competitive marketplace is a prerequisite for sustainability. Program effectiveness in meeting management goals in the context of the economics under various control scenarios will continue to be the primary drivers for future support by decision makers and the public [34, 47]. Historically, collaborative research in the U.S. described economics of ORV and the effect of management on reducing the cost for case investigations, pet and livestock vaccination and human PEP through comparison to the direct and indirect costs of living with mesocarnivore rabies in the absence of management [82–84, 101, 181, 182, 184–187]. This line of investigation remains invaluable in the context of integrated management scenarios to achieve local and regional elimination of RABV variants in developed landscapes. These data are essential to inform regional strategic planners as they relate to forecasting costs for ORV maintenance zones to prevent the re-incursion of RABV variants once locally eliminated and the dynamic process associated with low-risk ORV zone shifts toward RABV variant elimination.

As Rupprecht and others suggest [188], "...from a management standpoint, ORV strategies need to be dynamic and fulfill certain basic requirements including updated state-of-the-art use of science and technology, effectiveness under variable environmental conditions, cost-effectiveness (minimal expense but maximal success), and simplified principles to follow throughout a campaign..." To objectively assess ORV effectiveness in the field requires ERS and project monitoring at scale [77]. The development and continued refinement of regionally coordinated approaches to ERS as a complement to public health surveillance in Canada and the U.S. is remarkable given the scale and scope to support science-based decision making for rabies management [99, 141, 189–191]. As applied to wildlife rabies management in North America, ERS has been greatly bolstered through widespread use of the Direct Rapid Immunohistochemical Test (dRIT) as a counterpart to exposure-based public health surveillance since 2005 in the U.S. and 2010 in Quebec and Ontario [141, 167, 192–194]. Detection of rabid wildlife through a combination of public health surveillance and ERS is critical for documenting changes in rabies epizootiology and impacts of management [99, 111]. Estimation of vaccine-induced rabies virus neutralizing antibody (rVNA) seroprevalence as an index of population immunity provides another useful metric for strategy refinements and monitoring ORV management actions.

The coordinated management of RABV in raccoons and skunks with ORV has proven more challenging than with fox species. Despite significant progress to date in controlling the spread of raccoon RABV, periodic costly and often labor-intensive contingency actions involving intensive ERS and TVR still arise [48]. Projects during the late 1980s and 1990s in Canada and the U.S. informed the scaling up and development of more formal applied research on a variety of topics that would lead to enhanced understanding of vaccine-bait uptake by species, baiting strategy refinement, target bait densities and ultimately improved ORV [106]. For example, small scale pilot studies in Vermont, New York and Massachusetts focused on topics such as raccoon visitation of scent stations at different elevations, development of bait stations, multi-state vaccine bait disappearance studies using tracking stations, and telemetry studies documenting raccoon movement along linear shoreline habitat on the St. Lawrence River, aimed at improving vaccine distribution by fixed-wing aircraft (Fig. 7.3). Data shared regularly at international conferences on preliminary results of pilot studies shape the future direction and philosophy of RABV management and research in the Americas.

An unwavering commitment to applied research was established early on in North America to overcome ecological, technical, and logistical hurdles and cement the evidence-based nature of RABV management in mesocarnivores. Current control programs were built on a foundation of ideal principles and best practices first published by the World Health Organization for Animal Health (OIE) in 1989 and WHO in 1992 [195]. The global evolution of wildlife rabies management programs led to additional guidance published by the WHO Expert Consultation on Rabies in 2005, 2013, and most recently in 2018 [76, 196, 197]. The essential components of current wildlife RABV management programs incorporate six distinct emphasis areas including: (1) coordination, strategic planning and communication; (2) enhanced rabies surveillance; (3) integrated ORV management; (4) ORV monitoring; (5) contingency actions; and (6) applied research [34, 47].

Coordination, Strategic Planning, Communication

Critical building blocks for the development and implementation of wildlife rabies management in the New World include a strong foundation of coordination, strategic planning, and communication [34, 77, 195]. As management programs mature, continued emphasis on telegraphing future program direction, transparency and sharing technical and operational information among decision makers, collaborators, and stakeholders has proven to be a key ingredient for developing support and sustainability of programs in North America [80, 196, 198]. A diverse coalition includes state, provincial and federal government agencies; industry; universities; and non-government organizations (NGO). The strategic planning process helps identify a coalition of willing participants among agencies and organizations and

Fig. 7.3 Early pilot studies (Clockwise from top left). Early tracking stations (via animal tracks) and "hair salons" (via animal fur) with placebo baits placed in the center provided species presence data, preliminary raccoon density estimates, and bait attractiveness and fate information. A proof of concept bait station (5-gallon bucket full of baits) paired with a tracking station. Early GPS units to record trap locations were large and required a backpack to carry associated equipment through the woods. Radio telemetry provided raccoon home range and movement data along difficult to bait river shorelines. All work pictured was conducted in Vermont and New York between 1997–2003

outlines roles and responsibilities for implementing management, establishes a broad framework for coordination of research and management activities that target wildlife populations across diverse habitats, landscapes, and geo-political boundaries. This process has proven invaluable in strengthening working relationships and recruiting new partners representing subject matter experts who bring diverse viewpoints and perspectives that better inform strategic program direction [48]. Consensus building helps all partners to feel invested in short and long-term strategic goals and objectives, increasing the likelihood of program success locally and regionally. Highlighting important achievements and measurable successes across jurisdictions promotes advocacy and the political support to sustain funding to meet long-term rabies management objectives [77].

An interdisciplinary team of rabies experts has met regularly in the U.S. to conduct comprehensive strategic planning to facilitate the coordination of ORV among states and along the borders with Canada and Mexico [48]. The National Rabies Management Team has met 15 times since the first meeting in 1999, often involving more than 100 invited participants who participate in smaller teams, to discuss challenges and strategies for coordination and communication, rabies surveillance, ORV management and contingency actions, as well as program monitoring and applied research [34, 47]. This team-based approach facilitated detailed development of objectives and strategies that inform a U.S. National Plan for Wildlife Rabies Management [199]. Similarly, a Canadian Rabies Committee was formed in 2007, consisting of agency representatives from public health, agriculture, and wildlife agencies. This working group of experts developed the first Canadian Rabies Management Plan in 2009 [41]. The strategic planning and project development process has facilitated interjurisdictional collaboration at the municipal, county, state, provincial, federal, and international levels. These committees have also served as an important mechanism for regulatory compliance and permitting through interjurisdictional collaboration in support of wildlife rabies management [48, 76, 88, 200]. In addition, the U.S. Compendium of Animal Rabies Prevention and Control, from the National Association of State Public Health Veterinarians, represents a document by an independent but integral group that provides technical recommendations and guidance for rabies management based on available scientific data and the opinions of subject matter experts [39, 77].

The International Rabies in The Americas (RITA) conference provides an opportunity to bring together agency representatives from Canada, Mexico, the Navajo Nation, and the U.S. in a series of formal side meetings that led to the development of the NARMP [48, 77, 104, 106]. Four pillars that foster collaboration and coordination of activities outlined in the NARMP include: information transfer; rabies prevention and control; surveillance and monitoring; and applied research [201]. Since 2005, the Canada/Mexico/US Trilateral Committee for Wildlife and Ecosystem Conservation and Management has enhanced partnerships and supported activities under the NARMP among wildlife agencies and collaborative programs in support of One Health, science-based ORV programs targeting RABV variants of mutual interest in North America. Additional collaboration under the NARMP has led to: cross border ORV field trials; ORV program evaluation; modeling; economic analysis; and localized strategic planning [107, 108]. During 2011 and 2016, meetings of a diverse group of U.S. national and international rabies and wildlife experts were organized around the DELPHI forecasting process, to find consensus among experts to support decisions where data gaps existed [202]. During 2011, this process was used to better inform economic analysis of raccoon rabies management in the U.S. by seeking consensus expert opinion on likely rates and locations of raccoon rabies spread over a 22-year horizon in the absence of intervention with ORV and by exploring alternative rabies elimination scenarios

[183]. A second DELPHI meeting held March of 2016 solicited the opinions of three groups of experts to develop spatial-temporal elimination scenarios for raccoon RABV under specific budget and time frame scenarios and provided a framework for future development of a coordinated effort to collaborate among the provincial governments of Ontario, Quebec, and New Brunswick on raccoon rabies elimination in North America. The NARMP has also provided an up-to date framework for the donation of over 500,0000 ORV baits by the Ontario Ministry of Natural Resources and Forests (OMNRF) and the Quebec Ministry of Forests, Wildlife and Parks (QMFWP) to the U.S. Department of Agriculture for distribution along the U.S. side of the border to enhance management and prevent the northward spread of raccoon RABV variant [47]. Since 1997, over $2.7M USD has been donated to the U.S. by the Canadian Food Inspection Service, OMNRF and QMFWP, for bait purchase to help prevent raccoon RABV spread from the U.S. into Canada. Since the mid-1990's, Cornell University has led coordinated efforts in the Northeastern U.S. to secure state funding to purchase ORV baits for distribution in New York, including strategic areas along the border with Canada.

Rabies communication strategies at the local level for ORV may involve outreach and education campaigns to raise awareness of the risks associated with mesocarnivore rabies, scheduled ORV bait distribution, and other management actions. This type of communication can also build support and advocacy for local programs and may help to overcome reservations by segments of the public that may have concerns with large government programs, vaccination, and public health initiatives (Fig. 7.4). Easily accessible and personalized technical assistance by government agencies represents a value-added approach to reduce rabies risks to

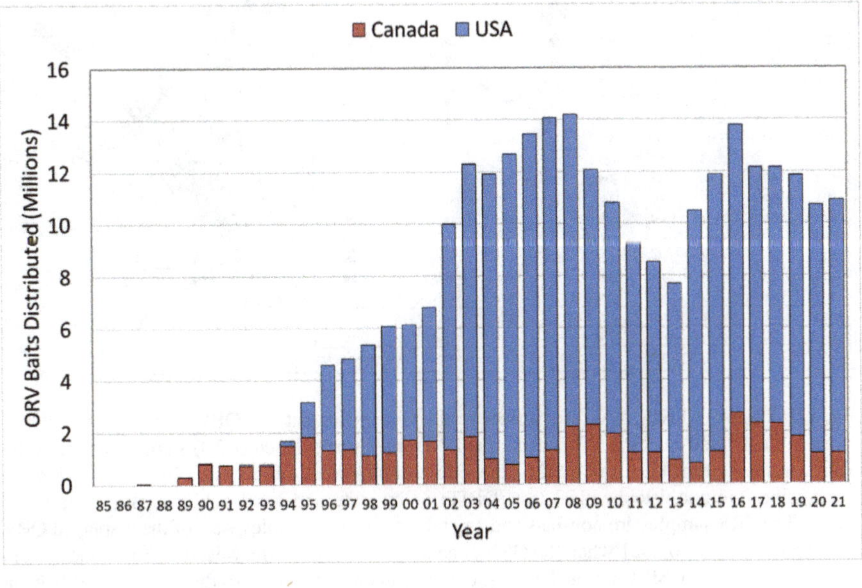

Fig. 7.4 Number of oral rabies vaccination (ORV) baits distributed in North America, 1985–2021

humans and animals and mitigate human-wildlife conflicts [203]. Specific messaging prior to and during ORV bait distribution may help to reduce human and pet vaccine-bait contacts [204]. Key public health messaging should include: ways to minimize RABV exposure to people and pets; steps to take when a human or pet exposure has occurred; mechanisms to easily report human or pet contacts with ORV baits; reminders to vaccinate pets according to recommended schedules; keeping wildlife at a healthy distance; and warns against the feeding or translocation of rabies vector species [39, 205].

Enhanced Rabies Surveillance (ERS)

When it comes to RABV surveillance in the context of informing science-based management decisions for wildlife, such as ORV, "…too much surveillance is never enough…" (Fig. 7.5). Laboratory-based public health surveillance is critical for case detection, reporting, and prevention of rabies in humans and animals. A public health "passive surveillance" approach is largely based on human and domestic animal exposures, with diagnostics conducted primarily by state public health

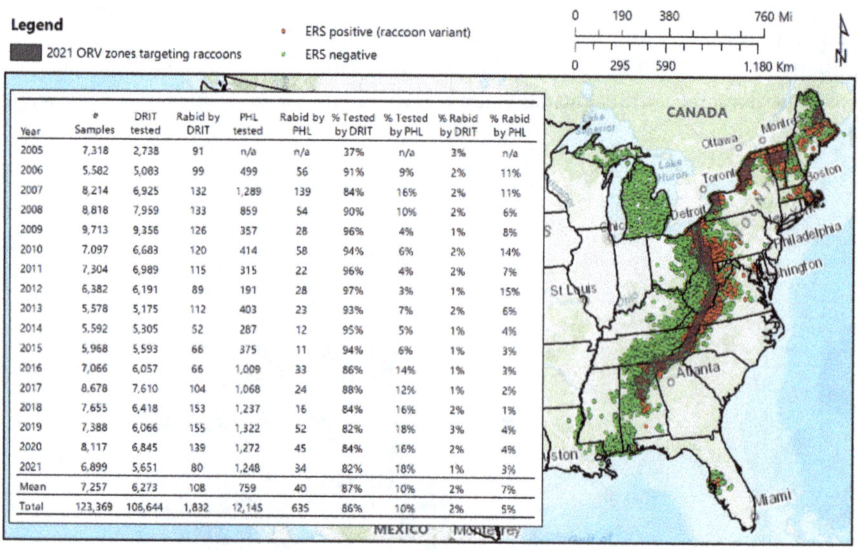

Year	# Samples	DRIT tested	Rabid by DRIT	PHL tested	Rabid by PHL	% Tested by DRIT	% Tested by PHL	% Rabid by DRIT	% Rabid by PHL
2005	7,318	2,738	91	n/a	n/a	37%	n/a	3%	n/a
2006	5,582	5,083	99	499	56	91%	9%	2%	11%
2007	8,214	6,925	132	1,289	139	84%	16%	2%	11%
2008	8,818	7,959	133	859	54	90%	10%	2%	6%
2009	9,713	9,356	126	357	28	96%	4%	1%	8%
2010	7,097	6,683	120	414	58	94%	6%	2%	14%
2011	7,304	6,989	115	315	22	96%	4%	2%	7%
2012	6,382	6,191	89	191	28	97%	3%	1%	15%
2013	5,578	5,175	112	403	23	93%	7%	2%	6%
2014	5,592	5,305	52	287	12	95%	5%	1%	4%
2015	5,968	5,593	66	375	11	94%	6%	1%	3%
2016	7,066	6,057	66	1,009	33	86%	14%	1%	3%
2017	8,678	7,610	104	1,068	24	88%	12%	1%	2%
2018	7,655	6,418	153	1,237	16	84%	16%	2%	1%
2019	7,388	6,066	155	1,322	52	82%	18%	3%	4%
2020	8,117	6,845	139	1,272	45	84%	16%	2%	4%
2021	6,899	5,651	80	1,248	34	82%	18%	1%	3%
Mean	7,257	6,273	108	759	40	87%	10%	2%	7%
Total	123,369	106,644	1,832	12,145	635	86%	10%	2%	5%

Fig. 7.5 USDA, APHIS, Wildlife Services oral rabies vaccination (ORV) zones targeting raccoons in the United States in 2021 and enhanced rabies surveillance (ERS) samples collected by USDA in the eastern US, 2005–2021. The ERS samples were either tested by USDA using the direct rapid immunohistochemical test (DRIT) or by state and local public health laboratories (PHL). The ERS samples are non-bats and any rabies-positive samples west of the displayed ORV zone occurred prior to 2021 when the ORV zone may have been to the west or north of the current zone. In addition to USDA national efforts, county and city officials also distributed ORV baits in Florida, Maryland, New Jersey, and New York City in 2021 (not reflected on map)

agencies [2]. In the U.S, over 130 state health, veterinary, and university laboratories test more than 100,000 samples each year from 50 states and Puerto Rico [30]. Laboratories perform the standard direct fluorescent antibody test (FAT) for rabies diagnosis in humans, domestic animals, and wildlife [195, 206]. Public health surveillance is invaluable for informing medical decisions regarding PEP after RABV exposure, as well as identifying outbreaks and trends in the risks to human and animal health. However, passive surveillance may be inadequate for timely detection of strategically important RABV variant cases in wildlife for management purposes [178]. A passive laboratory-based system may underestimate cases in an area by only testing samples associated with a human or domestic animal exposure [48, 189]. Public health surveillance may be biased toward areas with higher human population density, because strange acting animals are more likely to be reported and rates of human and pet exposure are more prevalent [48], where case detection is directly related to the level of public awareness and the ability of agencies to collect and transport samples for testing in a timely manner [76, 207]. The ability to document the epidemiology of RABV and evaluate management in mesocarnivore populations fundamentally relies on a robust system of national public health exposure-based surveillance and wildlife ERS that prioritizes continuous sampling in space and time to describe rabies dynamics in reservoir populations and delineate the geographic extent of RABV variants for management [30, 77, 99, 197, 206]. "Active surveillance" or ERS is considered complementary to public health surveillance to effectively improve case detection and provide real-time data to inform wildlife rabies management decisions [97–99, 189, 190]. Reporting both positive and negative cases provides a more complete picture of the intensity of sampling and improves confidence in monitoring systems for RABV variants and atypical events [76, 197]. The number and distribution of rabies cases in space and time, along with population levels of RABV antibodies as an index to vaccine-induced herd immunity represent the primary metrics used to measure ORV effectiveness in the U.S. and Canada [48, 189]. The typing of RABV variants is invaluable in geographic areas near ORV zones to define the leading edge of the epizootic front for targeted variants or to implement a contingency action in the event of an ORV zone breach [196, 197, 208, 209]. The timely sharing of case and variant typing information among all partners and other stakeholders is essential to successful management and eventual elimination of RABV variants across broad landscapes [34, 189, 195, 196].

Ultimately, the success of rabies management in mesocarnivores is directly related to a decrease in disease occurrence and a systematic approach to increasing ERS may be necessary to adequately detect changes in the disease burden at a landscape scale [99]. Intensive and real-time ERS is required to document the location and extent of rabies epizootic or enzootic foci to inform control actions, including delineating where and when ORV should be applied [141, 189]. Typically, public and enhanced rabies surveillance are risk-based and focus on suspect animals that have been involved with unprovoked bites, illness, or are found dead [197]. Sampling efforts must be sufficient in quality and quantity across space and time to detect a representative portion of both infected and healthy animals while working within the limits of time and available resources to meet rabies surveillance and management goals [190]. Surveillance should be maintained even in regional areas where mesocarnivore rabies has been eliminated [197]. A formalized strategy for ERS was implemented by USDA Wildlife Services in the U.S. in 2004 to improve decision making for ORV management [47, 189]. During 2005–2021 a total of 123,369 samples were collected by WS and cooperators in 26 states and Puerto Rico, resulting in 2467 additional confirmed cases to inform management [189]. The USDA Wildlife Services ERS system was updated in 2016 to focus surveillance toward high-value specimens by assigning points to different methods of sample collection, including strange-acting animals, roadkill animals, animals found dead other than roadkill, surveillance-trapped, nuisance animals collected by trapping ("nuisance wildlife control operator") and unknown [189]. This modification of the ERS system improved RABV detection probabilities and the quality of surveillance for raccoon rabies in the eastern U.S. [99].

In 2000, scientists at the U.S. Centers for Disease Control and Prevention (CDC) developed the dRIT as a rapid, reproducible, field-based alternative to the FAT that greatly increased throughput for ERS sample testing while maintaining high diagnostic sensitivity and specificity [141, 167, 192, 194]. Prior to 2003, ORV zones were almost universally delineated by public health surveillance [104]. The use of the dRIT by USDA Wildlife Services field staff has fundamentally changed the way ERS is implemented in North America [189, 193]. During 2018, USDA Wildlife Services tested 6602 wildlife samples by dRIT, accounting for 7% of all animals tested for rabies in the U.S. that year [47, 161]. Real time polymerase chain reaction tests have high sensitivity and specificity for lyssavirus diagnosis under laboratory conditions [193, 210, 211]. However, its use in the near term for field-based decentralized ERS is highly unlikely, given current equipment costs and laboratory capacity needs [78].

Conceptually, ERS provides more functional eyes on the landscape to increase the reporting of strange acting animals or roadkill and other sample sources to procure additional suspect animals for testing [207]. The majority of rabies cases reported in Quebec were sourced through public notification and both Quebec and Ontario confirmed raccoon rabies for the first time as a result of public reporting [190, 212]. Strange acting animals are often the primary focus of public reports and accounted for more than 70% of rabid animals detected through ERS in one study from the U.S. [189]. Rabies cases in Quebec are more likely to be documented in spring followed by summer and fall and less frequently during winter months, which may be influenced by the behavior of mesocarnivore populations but also human reporting bias and frequency during winter [190]. A TET approach to ERS involving the collection of apparently healthy animals not involved with a human or domestic animal exposure does not represent the highest quality samples for case detection and often produces no or an extremely low percentage of positives, which could be misleading about the presence of rabies locally [99]. However, intensive application of multiple ERS sampling methods may be appropriate, particularly in the context of strategically important areas and the need for critical data to inform decisions around the implementation of emergency contingency actions [76].

Population Reduction

Throughout history, a first line of defense in reducing the threat of disease associated with wild mesocarnivores often involved non-specific reduction of local populations thought to pose a threat to human or domestic animal health (Fig. 7.6). In the twentieth century, prior to the development of ORV, broad scale reduction of reservoir populations below an arbitrary threshold density was the only management technique available to attempt to reduce RABV in wildlife [24, 66, 213, 214]. In North America, canids, raccoons, skunks, and mongooses have all been historic targets of population reduction for rabies control [24, 38, 47]. No readily available published data exists for Central and South America regarding carnivore population reduction as a RABV management strategy. Toxicants, including strychnine, sodium fluoroacetate (1080), thallium sulfate, and zinc phosphide (for mongoose only) were historically used in poultry, egg and other baits to manage carnivore reservoirs [34, 213]. Zinc phosphide was used with little effect to control rabies in mongoose populations on Grenada in 1965 [168]. In the U.S., suppression of rabies was authorized and directed under the Animal Damage Control Act of 1931 (Fig. 7.7). This legislation was expanded to include other zoonotic diseases in the Act of December 1987 (101 Stat. 1329–1331, 7 USC 426c) [47].

During the mid-1950s to early 1960s, after fox rabies spread to the U.S. from Canada, strychnine-tallow baits were distributed by USDA and cooperators in several eastern states to control rabies [27, 47, 214, 215]. Efforts targeted coyotes, foxes, and other wild canids in Alberta in the early 1950s as part of an aggressive population reduction campaign to combat fox rabies from the Northwest Territories [216, 217]. Methods involved the use of professional trappers, strychnine and sodium fluoroacetate poison baits, and sodium cyanide 'coyote-getters.' These cumulative efforts reportedly removed some 105,000 coyotes, 50,000 foxes, 7500 lynx, 4200 wolves, and smaller numbers of other wild carnivores, which was reported to have kept Alberta free from rabies until 1971 [35]. Lethal control efforts later targeted striped skunks to limit the spread of rabies in southern Alberta [36–38, 218]. Factors including ecology of rabies reservoir species in northern climates, favorable geographic features facilitating control in the treatment area, lack of previous history of rabies in the area, and a well-organized control effort all likely contributed to the apparent effectiveness of the management action in Alberta [38].

Fig. 7.6 Government trappers removing predators for livestock protection in the western US, circa 1915–1930

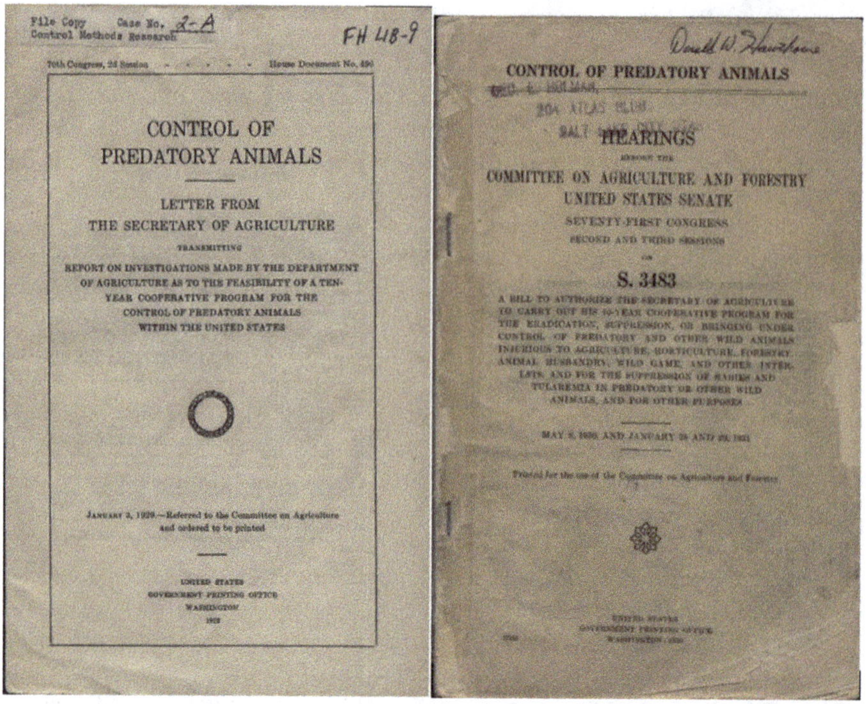

Fig. 7.7 USDA, APHIS, Wildlife Services program enabling legislation the Animal Damage Control Act of 1931 specifically authorizing the Secretary of Agriculture to "carry out this 10-year cooperative program for the…suppression of rabies and tularemia in predatory or other wild animals and for other purposes."

In general, the effectiveness of these intensive, single-strategy management programs has been poorly documented and failed to eliminate or prevent further disease spread [24, 46, 66, 195, 197]. More labor intensive trapping was more target specific compared to toxicants and the potential broad scale impact on non-target wildlife was significantly reduced [34]. During 1995–2003, management of coyote rabies and a unique variant of gray fox RABV in Texas using ORV as the primary tactic also included the removal of 963 coyotes (1 coyote removed/405 km^2) and 699 gray fox (1 gray fox removed/503 km^2) for surveillance purposes [24, 105]. Removal of foxes and coyotes by hunting was presumed to have a low impact on populations, yet may have been a factor in the overall effectiveness of the ORV program [105]. Short-term success in reducing wildlife rabies through population reduction was achieved under certain management scenarios, target species and management techniques. However, increased movements and immigration of incubating or rabid animals, enhanced fecundity of mesocarnivores associated with reduced densities and related rapid repopulation of treatment areas (potentially increasing the contact rates of conspecifics) makes this management approach unsuitable for most rabies control programs [46, 219, 220]. Effectiveness of this method for ecological generalists, growing sociocultural concerns, and cost effectiveness limits the use of population reduction as a stand-alone approach [46, 76]. The PIC technique used in Canada for raccoon rabies control was based on the integration of population reduction, ORV, and TVR around an index case, and included increased sampling via ERS [51, 52, 102]. Integrated management strategies like PIC to manage rabies in carnivores remain a potentially important contingency action in North America [196]. However, its application is likely limited to unique circumstances where there is relative certainty in the timely confirmation (early detection) of the index case and management can be implemented immediately if no effective oral vaccine is readily available or if there is no vaccine bait currently deemed effective for the target species.

Trap-Vaccinate-Release (TVR)

Urban-suburban areas and green space in developed landscapes, including urban parks, present unique challenges for wildlife rabies control [221, 222]. The live capture and parenteral vaccination of skunks and raccoons has proven to be an effective rabies control strategy on a limited scale, often in urbanized and developed landscapes, especially when conducted in an adaptive management framework with ERS and ORV [41, 106, 221]. Historically, TVR involved cage trapping and parenteral vaccination with an approved, off-label use of inactivated veterinary vaccine with release of target species at the site of capture [24, 197, 223, 224]. Injectable rabies vaccines, such as Imrab® (Boehringer Ingelheim, Ingelheim, Germany), produce nearly 100% rVNA response in raccoons and skunks after intramuscular injection [223, 225]. Generally, TVR has been used in smaller, more urbanized landscapes as part of contingency actions to address the spread of raccoon rabies and RABV variant spillover in skunks [47, 77, 102, 104, 106, 226]. However, TVR is labor and cost-intensive compared to ORV alone and has typically been limited to areas less than 2000 km^2 [84, 178, 212]. Both TVR and PIC methods are more expensive per unit area (approximately \$616/km^2) than ORV, regardless of target species [84].

Originally, TVR was developed as a vaccination strategy targeting skunks and raccoons prior to ORV, first implemented in metropolitan Toronto during 1984 [221]. Since the mid-1980s, successful application of TVR to manage RABV variants has been documented in North America, most often integrated with ORV, PIC, and other RABV management strategies [98, 102, 178, 223]. In Ontario, PIC operations resulted in more than 90,000 raccoons live-trapped and vaccinated during 1991–2008 representing the most intensive application of the TVR strategy to date [223]. A RABV management project with TVR as the central tactic was initiated in a 60 km^2 area of Toronto Canada in 1987 targeting raccoons and skunks [221, 227]. During 1987–1991, 2266 unique raccoons and 955 unique skunks were captured and vaccinated over a 5-year period [221, 227]. In addition, TVR was used extensively to eliminate raccoon RABV variant from Wolfe Island, the largest island in the 1000 Island region. on the border with Ontario and New York [177]. However, increased host densities may require additional trap nights to meet the recommended vaccination rate for target populations. Only two cases of rabies in raccoons have been reported during TVR campaigns in Ontario [226].

During 2001, a documented rabies outbreak involving 19 striped skunks with big brown bat RABV variant in Flagstaff, Arizona, was successfully controlled with a cooperative TVR program, despite evidence of skunk-to-skunk transmission [47, 48, 164]. However, emergence of the same RABV variant in skunks occurred during 2004, 2005, and 2008 and most recently in 2021 [48, 106]. Further research is required to understand cross species transmission events of RABV from bats to mesocarnivores, which may necessitate a timely management response (including TVR) to prevent the emergence of novel RABV variants [47, 164, 165]. A coordinated TVR campaign involving federal, state and city agencies and a NGO was initiated in three parks in Manhattan, New York City, U.S. after an outbreak of raccoon rabies was reported near the 843-acre Central Park [222]. Management with ORV was considered logistically difficult and population reduction was opposed by the public [106]. Three separate TVR events took place between February 2010 and December 2011 resulting in 3822 trap nights and a total of 484 unique raccoons vaccinated and released [222]. Using ERS, this timely TVR campaign in the midst of high raccoon mortality from an intensive rabies epizootic resulted in no reported cases in Central Park through 2021 [222]. In contrast, a 14-day TVR campaign on peninsular Cape Cod, Massachusetts was not effective in stopping spread of rabies beyond the established ORV barrier [228]. A total of 481 raccoons and 20 unique skunks were vaccinated in four towns in the contingency action zone as part of an integrated response involving ORV, TVR, and TET of suspect animals [228]. The control strategy was likely ineffective due to inadequate surveillance to define the size of the contingency action zone prior to the implementation of TVR [228]. Conversely, more aggressive TVR on Cape Cod was conducted in 2021 after a breach of a long-standing ORV zone. No additional RABV cases have been reported through early 2022 beyond the initial confirmed case.

The use of TVR will remain an important rabies management tool in mesocarnivores. However, future research should focus on a better understanding of the required scale, scope, and timing to reduce the risk of rabies control failures. The benefit to cost ratio associated with TVR as an effective component of control programs targeting developed areas and contingency actions, which may include ORV and other methods, also requires further evaluation.

ORV Management

Historically, ORV has been successfully employed on three continents to meet wildlife RABV management goals targeting a diversity of reservoir species and RABV variants across a variety of habitat types [24, 48, 66, 76, 88, 106, 188, 229]. Threats to public health, agriculture and wildlife populations led to the implementation of local, regional, and national wildlife management programs using ORV as the primary method for landscape level rabies control. From laboratory proof of concept in the U.S. (1970s), through field testing in Switzerland (1978), Germany (1983), and Italy (1984), and subsequent wide-scale distribution of millions of commercially produced vaccine baits in Europe, Canada, Republic of Korea (ROK), and the U.S., managers in both the Old and New Worlds have prevented, controlled, or locally eliminated RABV variants in red (Europe, Canada) and gray fox (U.S.), raccoons (Canada, locally U.S.), coyotes (U.S.), skunks (Canada), and raccoon dogs (Europe, ROK) [24, 43, 46, 48, 66, 76, 88, 188, 196].

Contemporary wildlife rabies management involves the strategic application of oral rabies vaccine-baits at prescribed bait densities and frequency in targeted habitats to achieve sufficient herd immunity to prevent, reduce or eliminate RABV variants in wild mesocarnivore populations [34, 46, 196, 197, 229]. The application of ORV as a wildlife disease management strategy is complex, challenging, and costly, requiring a long-term commitment of "time, talent and treasure." The effectiveness of ORV is influenced by target and non-target species, treatment, habitats, and landscape features such as bodies of water that may facilitate or impede successful implementation of RABV control strategies [104]. The ability to sustain multi-year ORV is essential to reach management objectives regardless of the species-specific approach or vaccine bait type deployed [65]. Intensive program monitoring provides empirical data to demonstrate progress and assists managers in refining implementation of coordinated ORV programs. However, the sustainability and success of ORV programs in North America over the last 25 years and associated lessons learned provide a template to facilitate the broader application of this management approach elsewhere, all of which began as small-scale applied research projects during the 1960s–1970s (Fig. 7.8).

In the Americas, the early field success in Southern Ontario, starting in the 1980s in red fox populations using baits containing the Evelyn-Rokitniki-Abelseth (ERA) oral rabies vaccine under experimental permit provided confidence in this approach and paved the way for future application of other vaccine-baits for research and management in Canada and the U.S [24, 51, 106, 138, 179, 230, 231]. Besides a RABV for wildlife ORV, also during the 1980s, joint studies focused upon the laboratory development and eventual field release of a genetically modified virus in the U.S. (Fig. 7.9).

Multidisciplinary collaboration among international partners resulted in the development of the vaccinia-rabies-glycoprotein recombinant virus vaccine RABORAL V-RG and to subsequent safety and efficacy field trials in multiple states in the US [40, 88, 159]. During 1990–1994, field trials conducted in Virginia, Pennsylvania and New Jersey demonstrated safety and field effectiveness of prototypic RABORAL V-RG vaccine-baits [40, 159]. In 1997, RABORAL V-RG® was licensed for use in raccoons and is currently the only licensed oral rabies vaccine for RABV management in raccoons and coyotes (added in 2002) in the U.S. [159]. The strategic use of RABORAL V-RG in Texas, in conjunction with free domestic dog vaccination clinics, led to the formal declaration of successful elimination of the canine RABV variant in the U.S. during 2007, and a unique gray fox RABV variant

Fig. 7.8 Applied research on oral vaccination during the 1970s (Clockwise from top left). Skunk and red fox pens at the NY State Department of Health (NYSDOH) Wadsworth Center housing animals for early oral rabies vaccine trials. Researchers hold the jaw open of an immobilized fox and skunk while oral vaccine is placed into the back of the mouth. The first attempt at a vehicle (or bait) to carry the oral vaccine was a straw filled with vaccine, placed into a hollowed-out sausage bait (i.e., Slim Jim®) and the ends plugged. All photos were taken between 1974–1976 and courtesy of the Wadsworth Center, NYS Department of Health

Fig. 7.9 Proposed field trial reports for live experimental vaccina vectored recombinant rabies vaccine

was eliminated from Texas in 2013. Currently, RABORAL V-RG comprises about 60% of all oral rabies vaccine distributed in the U.S.

During 2007, Ontario initiated experimental field projects involving the distribution of the human adenovirus-rabies glycoprotein vaccine-baits (ONRAB) to manage a persistent focus of Arctic fox RABV in red fox and frequent spillover into striped skunks [24, 232–234]. Since 2009, ONRAB is the only oral rabies vaccine-bait used in Canada and it is currently licensed for skunks, with distribution under experimental permit for RABV management in foxes and raccoons [24]. Since 2011, ONRAB vaccine-baits have been distributed in the U.S. under an experimental use permit for field trials, contingency actions and other research and novel management projects targeting raccoon RABV in West Virginia, Vermont, New Hampshire, New York, Pennsylvania, Ohio, and Tennessee [110, 235–237]. Results are promising in the U.S. for use with raccoons, and ONRAB is in the late stages of the U.S. licensing process [24]. An additional ONRAB field trial collaboratively conducted by the USDA WS and Texas Department of State Health Services is ongoing in coyotes.

Vaccine baits are distributed by four different methods in North America (Fig. 7.10). These included fixed wing aircraft, helicopter, hand baiting by ground vehicle, and bait stations [24, 41, 46, 88, 102, 106, 157, 238]. Logistical considerations during ORV operations involve: flight planning, staffing; and scheduling; safety and navigator-bait distribution training; communications and media support; and on-site management of ORV logistics, historically involving different aircraft (Fig. 7.11). Multiple vaccines, baits, target species, distribution methods, and strategies result in many sources of variation that may affect successful implementation of ORV programs at the landscape scale. Tailoring ORV strategies to different reservoir species involves a deliberative approach to determine bait density, bait distribution pattern by delivery method, timing of baiting, frequency of baiting (> once/year) and establishing an ORV zone of adequate size [34, 46, 47, 76, 196, 229, 239]. In North America, bait densities used regularly for management purposes range from 37.5 baits/km^2 to 300 baits/km^2, with 75 baits/km^2 for rural environments and

Fig. 7.10 Bait distribution techniques (Clockwise from left). Fixed wing aircraft are used to distribute oral rabies vaccine baits across rural landscapes, helicopters are used in suburban areas. Hand baiting by vehicle or walking is used in urban areas along with bait stations

Fig. 7.11 Bait distribution in the early years in the US was completed with the Ontario Ministry of Natural Resources de Havilland Canada DHC-6 Twin Otter fixed wing aircraft and belt machine designed for fishmeal polymer block baits (top row). Today Beechcraft King Air 90's, contracted by Dynamic Aviation (Virginia), are used with a belt that can accommodate multiple bait types

150 baits/km^2 for urban environments as the most common application densities regardless of bait type [4, 34, 110]. For RABV management in coyote and gray fox in Texas bait densities average 29 baits/km^2, which are aligned with species population densities and habitat [34]. The ORV distribution in North America generally occurs once/year typically during late summer to late fall depending on latitude. Baiting frequency may be increased to twice/year during spring and fall for contingency actions [24, 34, 48]. During ORV distribution with fixed-wing aircraft, GPS-planned flight lines are established and flown at 125, 250, 500 or 750 m apart depending on species and management objective [34].

Urban-suburban environments and associated changes to habitats and access to anthropogenic food subsidies may increase population density, reduce bait attractiveness, and alter movement patterns of wild mesocarnivores [100]. The intentional and accidental translocation of mesocarnivores is common in developed habitats, where artificially high population densities of rabies reservoir populations lead to an increase in human-wildlife conflicts [17, 240, 241]. The live capture and translocation of wildlife raise significant concerns regarding disease transmission risk to humans, domestic animals, and wildlife, as well as the potential spread of RABV from enzootic to naïve areas [19, 34, 88, 106, 242]. Reducing translocation remains a top priority for wildlife managers in North America due to its potential to negatively impact ORV successes, and to protect the significant commitment of public

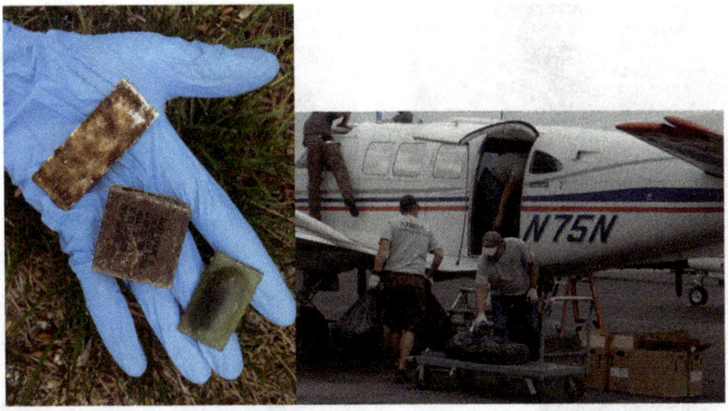

Fig. 7.12 Bait types (Left). Three oral rabies vaccine bait types are currently distributed in the US: the RABORAL V-RG® coated sachet (top), the RABORAL V-RG® fishmeal polymer (middle), and the ONRAB ultralite (bottom). (Right) Baits are loaded on a fixed wing airplane for distribution

resources to date [17, 240]. Undeveloped landscapes that may include features such as lakes, rivers, ocean shoreline, and inner coastal pine habitat may also require additional research and management focus to "right size" and tactically locate ORV zones after local RABV elimination to enhance the probability of cost-efficient management. Finally, the impact of other wildlife diseases on rabies epizootiology or control success (e.g., canine distemper, toxoplasmosis, leptospirosis, etc.) in populations of rabies reservoir species has not been evaluated (but see [243–247].

Clearly, ORV has revolutionized wildlife rabies management. However, the need to purchase oral rabies vaccine in quantities to strategically maintain ORV zones, conduct meaningful levels of ERS, move zones after local RABV elimination, or to conduct emergency contingency actions as they arise to meet long-term management objectives is essential for program sustainability. Future success demands continued innovation in wildlife rabies management programs and a persistent focus on applied research to enhance efficiency and effectiveness of ORV at the landscape scale, using selected vaccine-laden baits delivered to specific target species (Fig. 7.12).

ORV Monitoring

Intensive monitoring of RABV management in wild carnivores is key to understanding progress and increasing knowledge to refine control strategies (Fig. 7.13). Developing standardized metrics for measuring the impacts of rabies management actions allows comparison among projects across space and time [76]. Diagnostic support for ORV monitoring is needed and standards are also required to ensure consistency, accuracy, and comparability among results that can be provided by collaborating laboratories in a timely way at a reasonable cost [248]. The gold standard

Fig. 7.13 Wildlife monitoring (Top left). At least four weeks after baiting, target animals are live trapped and immobilized; (top right) a blood sample is collected from the jugular vein, spun into serum, and analyzed for rabies virus neutralizing antibodies; (bottom right) a small premolar tooth is pulled and analyzed for age and biomarker (when present in the bait); (bottom left) animals are ear tagged with a unique identifier before recovery from anesthesia and release at the capture site

for measuring ORV success is the substantial decrease or elimination of rabies cases in the ORV zone, based on adequate spatiotemporal ERS [46, 48, 76, 189]. Another metric for monitoring ORV campaigns as a complement to ERS involves serologic testing for rVNA in wildlife serum as an index to vaccine-induced population immunity [178, 179, 206]. The most common techniques to measure rVNA include the Rapid Fluorescent Focus Inhibition Test (RFFIT) and the Fluorescent Antibody Virus Neutralization Test (FAVN) [206, 249, 250]. Another common test for ORV monitoring is the Enzyme Linked Immunosorbent Assay (ELISA) [249, 251]. The ELISA has some advantages over RFFIT and FAVN including biosafety and increased standardization and comparability of results among laboratories for monitoring of ORV programs [188, 248, 249]. Canada employs a competitive ELISA calibrated against a 0.5 IU/mL neutralizing antibody level as evidence of adequate vaccine-induced seroprevalence [252]. The USDA WS has historically used a 0.05–0.06 IU/mL as minimum evidence of circulating rVNA in raccoon sera to evaluate ORV programs [34], based in part on recommendations from rabies experts and laboratorians participating on the USDA National Rabies Management Team meeting in 2007, as well as evidence that some raccoons exposed to ORV under field conditions with low titers (≥0.05 IU/mL–0.11 UI/mL) survived experimental challenge [253].

A blue-ribbon panel (BRP) was convened by USDA WS in 2014 to provide insight on factors that may affect the measurement and subsequent interpretation of serologic results in wild mesocarnivores in response to ORV. The BRP included experts from all laboratories conducting rabies serologic testing in the U.S. and Canada, as well as a select group of international experts. As a result of the BRP and the regular meeting and recommendations provided by a U.S. subcommittee on wildlife serology, strategies including partial validation studies for wildlife sera, routine cross-laboratory evaluations and comparisons, and regular proficiency testing have been developed and implemented. Future research and analysis regarding ORV monitoring should explore opportunities to optimize sampling by implementing a stratified sampling scheme for serologic testing to reduce costs for ORV monitoring while still addressing dynamic program needs. A more rapid turnaround in serologic results would be ideal to allow for adaptive management of wildlife RABV using ORV.

Biomarkers also provide an indirect measure of bait uptake but not necessarily vaccine exposure because they are commonly formulated into the bait matrix. Biomarkers are indispensable in evaluating new baits and ORV efforts. Understanding ORV effectiveness in North America will continue to rely on ERS, as well as monitoring biomarker and rVNA seroprevalence in target species populations pre- and post-ORV as indices of bait and vaccine uptake. Moving ORV zones to progress toward raccoon RABV elimination will require standardized metrics for program monitoring. For example, ORV zones with high quality spatial-temporal ERS demonstrating RABV absence and target population rVNA seroprevalence greater than 60% with adequate sampling across multiple years may represent lower risk areas to shift ORV zones to address other RABV enzootic areas. Applied research based on empirical field data and experimental use of an oral vaccine will still require comprehensive sampling to establish benchmarks and determine short and long-term trends in impact for target populations.

Contingency Actions

Contingency actions and emergency response preparedness are integral to successful management of rabies in mesocarnivores in North America (Fig. 7.14). These can be characterized as distinct management actions to prevent RABV variants from spreading to new areas [34, 46, 178]. Typically, management is focused on maintaining or reestablishing existing ORV zones that have an elevated risk of being compromised or that have been breached to prevent rabies from spreading and gaining a larger geographic footprint or areas where no ORV zone has been established [46, 48]. In the U.S., contingency actions for the raccoon RABV variant are typically considered when RABV is confirmed >80km beyond but still in close proximity to an ORV zone, active within the leading edge of an ORV zone, spreading toward an ORV zone, and spreading near where no ORV zone has been established [104]. These events trigger a risk assessment to evaluate contingency action options,

Fig. 7.14 Example of a contingency action conducted on Cape Cod, Massachusetts by USDA, APHIS, Wildlife Services, 2021. After a rabid raccoon was confirmed on the Cape (first case in >8 years), a trap-vaccinate-release (TVR) zone was established with 10-km2 cells approximately 10-25 km east of the index case (map). Sufficient trap coverage across the designated area and hand vaccinating at least 60% of the raccoon population is key to a successful TVR effort. During TVR, target species are trapped (bottom left), hand injected with rabies vaccine (bottom middle), ear tagged and released (bottom right) at the point of capture. In addition, oral rabies vaccination (ORV) after TVR is a critical component to most contingency actions and stopping the spread of raccoon rabies

including a no action alternative. A variety of factors (e.g., number and location of the RABV cases, habitat, raccoon density if known, time of year and resource availability) may inform the need and management strategy for implementing a contingency action. Focal TVR accompanied by spring and fall emergency ORV baiting operations often represent the preferred response [104, 106]. Intensified ERS radiating out from first confirmed case is an integral component of essentially all contingency actions. In specific instances, ERS may involve cage trapping, euthanasia and testing raccoons, skunks and other reservoirs to help define the geographic scope and intensity of rabies outbreaks and a new epizootic front may be considered for specific contingency actions [104]. During 1999–2022, USDA WS cooperatively conducted 16 major contingency actions as part of the integrated ORV rabies management strategy.

In the Canadian provinces of Quebec, Ontario and New Brunswick, cage trapping, euthanasia and testing has been used as an "early detection-early response" PIC, which represents a contingency action that may involve intensive population reduction, TVR and ORV [24, 212]. The PIC involves population reduction followed by TVR and ORV around detected cases until elimination is reached. A modified PIC that relies on TVR and ORV is now a common contingency action strategy in Canada [24, 46, 102]. While an essential component of wildlife rabies management, contingency actions are inherently complex, requiring a labor-intensive response that may cost at least 2.5 times more than ORV alone [212]. Therefore, the resources required to conduct a sustained emergency response over larger areas and longer time frames remain logistically and financially difficult [46].

During 2007–2011, three contingency actions were conducted that integrated ORV and TVR in Ohio, Vermont and New York. In Ohio, rabies cases were continuing to occur within an existing contingency action zone which had been in place since 2004. The TVR coverage averaged 157 km^2/year from 2007–2011. In Vermont and New York, contingency actions were in response to the first case of raccoon RABV variant confirmed in Quebec in 2006 near the Vermont border. The increased number of cases in the Vermont ORV zone (125 cases in 2007) and multiple high risk spread corridors in eastern New York were key risk factors of international concern. The TVR encompassed > 31,000 raccoons (as well as skunks) in these three states during a 5-year period. In addition, ORV operations were completed twice per year in Ohio (no spring baiting in 2010) and once per year in Vermont and New York using densities of 75–150 baits/km^2 across the three treatment areas and high-density baiting (150 baits/km^2) along the U.S.- Quebec border. Recent contingency actions in the U.S. and Canada are focused on containing the spread of raccoon RABV variant, including provinces of New Brunswick (2014), Ontario (2015), and Quebec (2015). The first raccoon RABV variant case detected in the most recent reemergence event in Hamilton, Ontario, Canada outbreak was in December of 2015 [51, 142]. This epizootic represented the largest documented raccoon RABV epizootic in Canada [142]. An integrated rabies management strategy included ERS, ORV distribution by hand, helicopter, fixed-wing aircraft, and bait stations, as well as public education and pet vaccination clinics. In 2022, TVR was also integrated into the rabies control strategy in the greater Hamilton, Ontario area and these collective actions resulted in a significant and consistent reduction in raccoon RABV cases [142].

During 2021, the USDA WS program and local cooperators coordinated TVR in response to the first rabid raccoon detected on peninsular Cape Cod, MA in 8 years (Fig. 7.14). Raccoon RABV management began in the 1990s, and has been conducted in MA by USDA in collaboration with local partners since 2002, involving ERS, ORV twice per year by helicopter, hand-baiting, bait stations, intensive program monitoring and applied research [228]. The confirmed rabid raccoon was located approximately 9 km (5.5 miles) east of the ORV zone, representing an immediate threat to spread throughout the peninsula. The USDA WS collaborated with the CDC, the MA Department of Health and the local Cape Cod and Southeast MA Rabies Task Force to develop a response plan, which included development and distribution of several press releases by the Commonwealth of MA and Barnstable County, increased ERS, TVR, and ORV baiting by helicopter and ground means. A timely contingency action was implemented six days after raccoon rabies was confirmed and TVR activities were conducted June 14 – July 2, 2021, involving 20 USDA WS specialist and biologists. A total of 716 animals (raccoons as well as skunks, coyotes, and fishers – *Pekania pennanti*) were trapped for parenteral vaccination using inactivated IMRAB® vaccine. Based on signs or behaviors indicative of rabies or that were found dead in traps, 6 animals were submitted for testing. None of these samples were confirmed positive for RABV antigens. Continuous ERS has not detected additional rabid animals on Cape Cod through the beginning of 2022. Contingency actions will remain integral to long-term rabies prevention, control, and elimination.

Applied Research

Wildlife accounts for >90% of rabies cases in North America, with mesocarnivores responsible for the majority of reported cases [111]. Most mesocarnivore species responsible for reported rabies cases are common along the human-wildlife interface (e.g. raccoons, striped skunks and foxes) and present unique and complex challenges for management [4, 224, 254–256]. Landscape scale rabies management requires a detailed understanding of how reservoir species use a variety of habitats and innovative applications of traditional methods, as well as development of novel tools and strategies to meet current RABV prevention and control objectives. Collaborative applied research to refine and improve the effectiveness of landscape ORV to control and eliminate RABV circulation in wild carnivores has been extensively reported and summarized from North America [34, 47, 77, 88, 104, 106, 188, 232, 257, 258]. Since the 1980s, applied research by CDC, the Canadian Food Inspection Agency (CFIA), the USDA WS National Wildlife Research Center, Ontario Ministry of Natural Resources and Provincial government agencies, State agencies, universities, NGOs and industry has focused on addressing ecological, technical, logistical, and economic data gaps in support of RABV management [79]. Rabies management programs typically begin with strategic planning to identify agency and organization roles and responsibility, gauging public support and identify funding streams for management and research [76]. Economic research to address foundational questions around management impacts, required resource commitments and the benefits and costs associated with ORV to stop the spread of

RABV variants or to maintain RABV variant elimination have helped to shape rabies prevention and control programs in North America [82–85, 101, 181, 182, 184–187]. The ecology of RABV hosts is generally poorly understood and has slowed the implementation of effective rabies management [47]. Research has focused by necessity on improving our understanding both of target species and RABV epizootiology, as well as vaccine and bait development and field-testing ORV products [138, 159, 259–262]. Research is also conducted to refine field operations, logistics and methods for monitoring management actions to inform adaptive management.

Research to improve our understanding of target and non-target species for rabies management have focused on population responses to ORV, movement patterns, habitat use and associated RABV spread [263–273]. The focus has largely been on species that are recognized reservoirs for RABV variants targeted for elimination, but non-target species (e.g., Virginia opossum, *Didelphis virginiana*) that compete with raccoons for ORV baits also have been the subject of studies [274]. Documenting population density of the raccoon and other host species is critical for defining ORV bait densities to control and eliminate specific RABV variants [4, 109, 268, 269]. Slate et al. [4] indexed raccoon population densities based on 253 standardized studies conducted by USDA WS from Maine to Alabama that were conservatively estimated to range from 0.0–56.9 raccoons/km^2 with lower densities confirmed in evergreen forests and higher densities in more developed landscapes. These and results from other studies assist managers in enhancing program effectiveness and efficiency by revealing opportunities for altering vaccine-bait density to improve uptake in a given reservoir species and potentially reduce non-target interference across habitat types [106].

Vaccine immunogenicity and safety trials have targeted multiple mesocarnivores (e.g., raccoons; striped skunks; Arctic, gray and red foxes; coyotes; and the small Indian mongoose). Studies from the 1970s to date have evaluated field application of ORV baits in Europe and the Americas. These studies originally focused on live RABV vaccines derived from the Street Alabama Dufferin (SAD) virus strain, including ERA (Evelyn-Rokitniki-Abelseth), SAD-B19, and SAD-Berne, or the second generation modified live RABV vaccines SAG1 and SAG2, which were selected under monoclonal antibody pressure [230, 275–277]. The ERA vaccine was the only attenuated RABV product used in the New World among the first and second generation rabies vaccines developed for wildlife [34]. Canada conducted ORV in the Americas during the mid-1980s using ERA vaccine to control arctic fox RABV variant in red foxes [278].

The recombinant vaccinia-rabies glycoprotein (RABORAL V-RG) was first applied in the field in Belgium in 1988 to address RABV in red foxes [159, 279], and later in the U.S. as part of experimental safety and efficacy trials on Paramore Island in Virginia in 1990, wildlife management areas near Williamsport Pennsylvania in 1991, and Cape May, New Jersey in 1992 and 1993 [87, 180, 280]. These experimental field trials with RABORAL V-RG focused on vaccine safety and control of raccoon RABV variant. They paved the way for a diversity of larger-scale management programs in the coyote and gray fox in Texas and raccoon in 17

US states and Canada [46, 88, 157, 281, 282]. The first broad scale use of RABORAL V-RG for management purposes occurred in 1995 in south Texas to address spill-over of canine RABV into coyotes and in 1996 to target a unique gray fox RABV variant in west Texas [105, 157]. Development and experimental field use of ONRAB occurred in Ontario, Canada in 2007 against Arctic fox RABV that had spilled over into red foxes and striped skunks [232, 233, 261, 283]. ONRAB has also been distributed for raccoon RABV variant management in Ontario (2006), Quebec (2007) and New Brunswick (2008), as well as under experimental use permit in Maine, New Hampshire, Vermont, New York, Pennsylvania, Ohio, West Virginia, Tennessee and Texas [107, 108, 110, 235–237, 261].

Vaccine-bait distribution methods using fixed-wing, helicopter, hand baiting and bait stations evolved over time and been used in varying degrees based on local management focus since the late 1990s [34, 47, 238, 284]. In 2020 in the U.S., > 9.5 million oral rabies vaccine-baits were distributed (67% RABORAL V-RG and 33% ONRAB) across 166,000 km^2 (64,000 mi^2), 76% by fixed-wing aircraft, 14% by helicopter, 9% by hand-baiting and about 1% by bait stations. Improving efficiency and effectiveness of ORV delivery systems and strategies remains a continuous priority for study [258]. New tools and technology are consistently being tested and incorporated into ORV programs including data collection, data management systems, and vaccine-bait dispensing equipment [258]. Field diagnostics and laboratory support are an integral part of rabies programs and research continues to discern temporal and spatial differences in ORV effectiveness, allowing for timely program changes through adaptive management [34].

Other research in North America has focused on vaccine bait immunogenicity and efficacy, as well as improvements including bait flavor preferences, bait handling, as well as bait enhancement and acceptance [171, 285–294]. Biomarkers have been used to assess ORV bait uptake in target and non-target species. Rhodamine B, tetracycline and iophenoxic acid (IPA) have been used to determine bait uptake for research and program monitoring [257, 295–297]. Biomarkers incorporated into the oral vaccine bait matrix provide an indirect measure of vaccine uptake in target and non-target wildlife. Captive and field studies in the U.S. and Canada have determined the potential of these products to mark carnivores and monitor vaccine-bait uptake [257, 295, 298–300]. To date, USDA Wildlife Services researchers have compared marking ability and serum decay of different types of IPA biomarkers at different dosages delivered orally to target mesocarnivores [175, 299]. Results are promising and collaborative field studies are ongoing in Vermont, Tennessee, South Carolina, and Texas including ORV baits containing IPA biomarkers for raccoons and coyotes.

Fertility control in mesocarnivores has been explored by scientists as a potential alternative to population reduction or as a potential adjunct to enhance rabies control [301, 302]. One primary advantage of a surgical, chemical or immunocontraception in contrast to population reduction is that fertility control may be less likely to alter animal movements and dispersal [106, 303, 304], yet effective application for wild mesocarnivores is not practical at this time.

Deterministic and stochastic predictive modeling approaches have been developed, refined, and used by managers to estimate rabies spread risk and inform rabies management recommendations [93–96, 149, 305–309]. Retrospective statistical modeling has helped refine and improve ERS sample collection and provide additional clarity on the status of raccoon rabies in the northeastern U.S. along the border with Canada [97–99]. Collaboration between the U.S. and Canada has facilitated data sharing to better predict and guard against breaches in the ORV zones and delineate high risk RABV spread corridors [106]. The application of theoretical models combined with empirical field data will continue to provide insight into terrestrial wildlife RABV management and future research questions [100].

Recent advances in methods to determine population and landscape genetics of RABV reservoir species has offered a new opportunity for research to enhance adaptive management. Genomic analysis of field derived tissue samples assists in identifying host dispersal patterns, population structure, demographics, range expansions, and local adaptation of pathogens [310]. Through a more refined understanding of subpopulations of related individuals and the typing of RABV variants managers gain insight into variant distribution, spillover, and transmission dynamics, which subsequently enhance the effectiveness of the management response to a translocation or the need for a contingency action [120, 124, 140, 150, 155, 165, 311].

Priorities for future research in the Americas should focus on the ecology of RABV reservoir species in Central and South America and the Caribbean, including white-nosed coati, crab-eating fox, hoary fox, and the small Indian mongoose. Management of RABV variants in urban environments has proven to be particularly challenging in North America. Additional research is necessary to efficiently estimate densities of target, non-target bait competitors, and RABV vector species; understand bait interaction opportunity for target species in the context of resource selection to improve ORV strategies; and bait preference studies to improve bait attractiveness and effectiveness. Emphasis on a multidisciplinary, collaborative approach among field biologists and researchers should provide the best available scientific data to inform large scale ORV and a future path forward toward attaining RABV management goals in mesocarnivores [80].

Lessons Learned and Future Directions

The focus on management strategies in the New World targeting rabies at its source in wild mesocarnivore populations continues to evolve and is further refined through applied research and adaptive management. Of the 35 countries in the Americas, only Canada and the U.S. are actively managing RABV in mesocarivores through coordinated programs at landscape scales. This will likely change as countries such as Mexico eliminate canine rabies, which may re-focus efforts on a reduction of RABV cases and associated exposures from free-ranging wildlife reservoirs that could spillover and become reestablished in dogs. In North America, specific mesocarnivore RABV variants are managed intensively to meet one of three primary objectives: (1) prevent reintroduction; (2) prevent the further spread; or (3) achieve local and regional RABV elimination. Despite the need for contingency actions to maintain integrity of ORV zones, these programs have demonstrated promising advances toward long-term goals. Managers of coordinated programs will continue to explore opportunities to enhance efficiency, effectiveness, and optimal use of resources. Collaboration across state, provincial and international borders is a key ingredient for success and will continue to expand, including strategic planning projects such as the "management without borders" initiative among USDA and provincial governments of Ontario and Quebec. New initiatives like this further foster synergism to improve ways to apply resources and expertise to manage rabies in wildlife populations towards a common goal. Nevertheless, many programs are at a critical juncture requiring objective examination of scientific evidence and long-term data in the context of program success, setbacks, and future initiatives. Retrospective analysis should focus on critical data gaps and near and long-term research prioritization schedules. Science-based planning and decision making will facilitate transitions from preventing spread of raccoon rabies and localized elimination to regional and national elimination.

Public health, agriculture and wildlife agency engagement protects human and animal health and reduces costs associated with rabies in North American wildlife populations, which sets a precedent for the further development of One Health concepts and strategies to address other zoonotic and non-zoonotic diseases, including diseases of global importance (e.g., SARS CoV-2). Implementation of management projects with supporting research to prevent, control or eliminate RABV at its source in coyotes, foxes, mongoose, raccoons, and skunks is the result of creative vision, development and implementation of the required tools and techniques for coordination, ERS, ORV, program monitoring and emergency contingency actions at a meaningful landscape scale to manage specific RABV variants. Program sustainability in the U.S. and Canada required nurturing of rabies management programs to maturity in the face of challenges such as nuanced ecological, logistical, political, and administrative complexities associated with regional RABV elimination. Management programs with ORV as a central tactic often face strong headwinds associated with sustained funding for long-term management priorities. Political willingness to embrace long-term planning horizons involving incremental implementation of local programs coordinated across geopolitical state, national and international boundaries requires measurable progress, which may not always be acutely visible. Strategic vision and scientific integrity fused to transparency, advocacy, and passion represent characteristics for coordinators of RABV management programs. Remaining flexible and open-minded to new ideas and engaging in proof-of-concept research are critical to improve upon management strategies to tackle issues such as RABV reemergence after regional elimination. Commitment to 15- to 30-year planning horizons to accomplish long-term goals requires patience and the ability to market successes and place setbacks into context. Shifting cultural ideals may impact rabies management over time. The role of government, as well as the public support of science and research, has been questioned by some segments of the public, including the use of safe and efficacious vaccines in disease prevention. Human dimensions research to better understand and bolster public support for rabies and other wildlife disease management initiatives is warranted along with economic research and data analysis associated with historic and future planned management. Reducing vaccine-bait contacts by non-target species and improving timely reporting and case resolution remains critical for regulatory compliance and to maintain support for wildlife rabies management programs. Continued transparency and public involvement through regulatory and non-regulatory activities, combined with a well-designed communication plan, are essential to address bait-vaccine contacts in a timely fashion even at currently low incidence levels. A formal collaborative process to update and refresh critical strategic planning documents is underway in Canada and the U.S., including the Canadian Rabies Management Plan, U.S. National Plan for Wildlife Rabies Management, and the NARMP, to incorporate recent scientific advances and to underscore new opportunities for partnership.

The continued quest for the safest, most effective, and least expensive ORV-bait remains a high priority, requiring new and existing industry partners to research and develop innovate refinements to ORV products in a competitive business environment. Transferrable or weakly transmissible rabies vaccines for wildlife remain an interesting concept to potentially increase vaccination rates of target animals through social contacts. Nevertheless, this approach may encounter significant regulatory obstacles and challenges for experimental field-use targeting mesocarnivores in North America [312, 313]. Understanding the impacts from potential non-target competitors (e.g., Virginia opossums, feral swine, small mammals, birds, invertebrates, etc.) on vaccine baits is important. A wide variety of such non-target species may reduce bait availability, palatability, and acceptance by target species, which is important for improving field effectiveness of ORV. Vaccine-bait development to increase uptake by target species while reducing attractiveness to competitors represents a priority focus of applied research. Development of biodegradable and matrix free vaccine-baits should continue in collaboration with industry. Initial proof of concept research is desirable to identify opportunities and challenges for RABV management in other New World carnivore species (e.g., crab-eating fox, hoary fox and the white-nosed coati).

Managing RABV in reservoirs involves a variety of landscapes, ranging from heavily managed forested habitats, monoculture agriculture, and more developed environments that can alter target and non-target species ecology, especially movement patterns and population density, resulting in the need to adjust and refine control strategies. Retrospective analyses of existing data are needed to explore habitat-based opportunities for reducing vaccine-bait densities or altering distribution patterns for target species to improve program efficiency. Similar analyses are required in developed landscapes, where higher vaccine-bait density or a paradigm shift in management strategy could be envisioned. Suburban-urban settings need additional bait-fate camera trap studies to explore alternative methods of bait delivery. Optimization of the integration and potential combinations of helicopter, hand baiting and bait stations may improve program effectiveness in developed areas.

Modeling to identify landscape patches and corridors with heightened RABV spread risk is valuable, particularly where it can incorporate target species behavior, movement patterns and resource use in more developed landscapes. Disposable bait stations or dispensing technology could exploit trash containers (with vaccine-bait dispersers), commonly accessed by wildlife in urban settings, designed to specifically attract mesocarnivores when strategically distributed in urban environments. As technology improves, exploring the use of unmanned aerial vehicles for distribution of oral vaccines in developed habitats should be considered.

Future ORV field trials for proof-of-concept evaluation to manage mongoose RABV should be prioritized to protect human and animal health in impacted and underserved communities of the Caribbean. Recent regional interest in surveillance, prevention and control of feral swine diseases (e.g., African Swine Fever) in the Caribbean may provide an opportunity to promote a One Health approach to eliminate mongoose rabies. Effective programs on certain islands could reduce the risk of translocating mongoose rabies to neighboring Caribbean islands with at-risk mongoose populations. Climate change may also alter the timing and availability of predictable food and water resources exacerbating the planning, timing, and frequency of management operations over time. Intensive ORV program monitoring through ERS and seroprevalence of rVNA as evidence of population immune responses should continue as a companion metric, but these two metrics should be analyzed in greater detail to characterize the relationship with estimation and measurement of landscape level disease risk to inform management decisions. Rigorous ERS and monitoring are key to understanding progress towards rabies management goals. Raccoon and skunk RABV management has proven particularly challenging compared to coyotes and foxes, necessitating a persistent commitment to documenting the effectiveness of ORV under various environmental conditions. Epidemiological modeling could assist managers in refining ORV monitoring but will require that spatial-temporal ERS sampling is adequate in remote areas and corridors representing a high risk of RABV spread. Diagnostics for both local case confirmation and measuring vaccine-induced antibodies need refinement and validation for application among a growing list of target species. For example, a commercial source for predictable and sustainable dRIT reagents will be essential to maintain the current level of ERS in North America and elsewhere. Reduction in the frequency, scope and financial costs of contingency actions require shortened response time to address the index case, improved vaccine-bait uptake by target species, and enhanced resilience of long-standing and future ORV zones to pressure from adjacent enzootic areas. Abundant opportunities exist for future rabies management successes in mesocarnivores throughout the New World, as evidenced by past results and sustained public support to date.

Conclusions

Wildlife rabies management in the Americas will continue to serve as a forward-leaning One Health model for effective zoonotic disease management to protect public health, agriculture, and pets, and to mitigate economic impacts of living with rabies. While the technique is often still used to manage spillover of wildlife rabies, widespread population reduction of mesocarnivores has met limited success in the control and elimination of wildlife rabies based on ecological, economical, and disease dynamics in these targeted populations. After >50 years of proof-of-concept studies, applied research, methods development, and adaptive management, RABV in mesocarnivore reservoirs is prevented, controlled, and eliminated from strategically important landscapes based upon herd immunity elicited via vaccination, where the fundamental economical approach relies on ORV. Prioritizing large-scale national and international initiatives based on integrated management strategies anchored by ERS, ORV and effective collaboration and coordination is a successful working model. The coyote and Texas gray fox RABV variants have been eliminated and the raccoon RABV variant has been prevented from gaining a significantly larger geographic footprint with targeted ORV and TVR in the U.S. and Canada. Most countries in the Americas remain focused on the laudable 2030 goal of global elimination of canine-mediated human rabies. Wildlife RABV management accomplishments in North America provide insight and science-based strategies that may be applied in mesocarnivores in developing countries once canine RABV elimination has been achieved. Minimizing the risk of reinfection where RABV has been eliminated from large landmasses will require a sustained commitment to ERS and development of robust models to identify high risk spread corridors and levels of management effort needed to prevent re-introduction [314]. Subsequently, informed risk taking with robust metrics to measure program success will be key for future progress. Translocation of RABV will remain a potential significant threat to long term progress and management success. More stringent regulation and enforcement of wildlife reservoir movement, combined with enhanced information and education campaigns, should help mitigate disease risks of translocation. A primary challenge for wildlife rabies management includes managing expectations and sustaining and improving maturing programs. Support for research is imperative and fundamental to addressing knowledge gaps to improve vaccine-baits and baiting strategies particularly in urban-suburban landscapes where progress in management of raccoon, skunk, fox and coyote populations has stalled. As ORV programs in North America enter middle age and as wildlife rabies management programs are initiated elsewhere in the region, a renewed interest in collaboration to concentrate collective energies is necessary to adapt management for the best chance to meet current and long-term goals. Program managers must experiment to succeed and not allow enthusiasm for short-term management successes outpace the science. Responsible management calls for implementing the most effective programs based on the best available scientific evidence and available resources against which meaningful metrics can be applied to evaluate progress toward well-defined programmatic goals.

References

1. Whitehouse ER et al (2021) Evaluating surveillance for and estimating administration of rabies postexposure prophylaxis in the United States, 2012–2018. PLoS Negl Trop Dis 15(10):e0009878
2. Pieracci EG et al (2019) Vital signs: trends in human rabies deaths and exposures – United States, 1938–2018. MMWR Morb Mortal Wkly Rep 68(23):524–528
3. Meske M et al (2021) Evolution of rabies in South America and inter-species dynamics (2009–2018). Trop Med Infect Dis 6:98
4. Slate D et al (2020) Rabies management implications based on raccoon population density indexes. J Wildl Manag 84(5):877–890
5. Tarantola A (2017) Four thousand years of concepts relating to rabies in animals and humans, its prevention and its cure. Trop Med Infect Dis 2:5
6. Baer GM (2007) The history of rabies. In: Jackson AC, Wunner WH (eds) Rabies, 2nd edn. Elsevier, pp 1–22
7. Vos A et al (2011) The occurrence of rabies in pre-Columbian Central America: an historical search. Epidemiol Infect 139(10):1445–1452
8. Rupprecht C, Kuzmin I, Meslin F (2017) Lyssaviruses and rabies: current conundrums, concerns, contradictions and controversies. F1000Res 6:184
9. Hayman DT et al (2016) The global phylogeography of lyssaviruses – Challenging the 'Out of Africa' hypothesis. PLoS Negl Trop Dis 10(12):e0005266
10. Fisher CR, Streicker DG, Schnell MJ (2018) The spread and evolution of rabies virus: conquering new frontiers. Nat Rev Microbiol 16(4):241–255
11. Streicker D, Biek R (2020) Evolution of rabies virus. In: Fooks AR, Jackson A (eds) Rabies (4th edition): scientific basis of the disease and its management. Academic Press, pp 83–101
12. Crandell RA (1991) Arctic fox rabies. In: Baer GM (ed) The natural history of rabies. CRC Press, Boca Raton, pp 291–306
13. Colan T (1881) The dog disease, or canine madness of the Arctic regions, viewed in connection with hydrophobia; together with the measures used and suggested for its extinction, from information collected, and observations made in the country. Vet J Annals Comp Pathol 13(11):324–325
14. Steele JH (1975) History of rabies. In: Baer GM (ed) The natural history of rabies. Academic Press, pp 1–29
15. Rupprecht CE et al (1995) The ascension of wildlife rabies – a cause for public-health concern or intervention. Emerg Infect Dis 1(4):107–114
16. Troupin C et al (2016) Large-scale phylogenomic analysis reveals the complex evolutionary history of rabies virus in multiple carnivore hosts. PLoS Pathog 12(12):e1006041
17. Chipman R et al (2008) Downside risk of wildlife translocation. Dev Biol 131:223–232
18. Singh AJ et al (2018) Translocation of a stray cat infected with rabies from North Carolina to a terrestrial rabies-free county in Ohio, 2017. MMWR Morb Mortal Wkly Rep 67(42):1174–1177
19. Rosatte R et al (2007) Human-assisted movements of Raccoons, Procyon lotor, and Opossums, Didelphis virginiana, between the United States and Canada. Can Field-Nat 121(2):212–213
20. Hampson K et al (2015) Estimating the global burden of endemic canine rabies. PLoS Negl Trop Dis 9(4):e0003709
21. Wallace RM, Blanton JD (2020) Epidemiology. In: Fooks AR, Jackson A (eds) Rabies (4th edition): scientific basis of the disease and its management. Academic Press, London, pp 103–142
22. Changalucha J et al (2021) Human rabies: prospects for elimination. CAB Rev 16:039
23. Beran GW (1991) Urban rabies. In: Baer GM (ed) The natural history of rabies. CRC Press, Boca Raton, pp 427–443
24. Fehlner-Gardiner C (2018) Rabies control in North America – past, present and future. Rev Sci Tech 37(2):421–437

25. Freire de Carvalho M et al (2018) Rabies in the Americas: 1998–2014. PLoS Negl Trop Dis 12(3):e0006271
26. Steele JH (1988) Rabies in the Americas and remarks on global aspects. Rev Infect Dis 10(Suppl 4):S585–S597
27. Tabel H et al (1974) History and epizootiology of rabies in Canada. Can Vet J 15(10):271–281
28. Smith J, Orciari LA, Yager PA (1995) Molecular epidemiology of rabies in the United States. Semin Virol 6:387–400
29. Blanton JD, Hanlon CA, Rupprecht CE (2007) Rabies surveillance in the United States during 2006. J Am Vet Med Assoc 231(4):540–556
30. Ma X et al (2021) Rabies surveillance in the United States during 2019. J Am Vet Med Assoc 258(11):1205–1220
31. Pieracci EG et al (2021) U.S. dog importations during the COVID-19 pandemic: do we have an erupting problem? PLoS One 16(9):e0254287
32. Hercules Y et al (2018) Rabies in a dog imported from Egypt – Connecticut, 2017. MMWR Morb Mortal Wkly Rep 67(50):1388–1391
33. Anderson RM et al (1981) Population dynamics of fox rabies in Europe. Nature 289(5800):765–771
34. Gilbert A, Chipman R (2020) Rabies control in wild carnivores. In: Fooks AR, Jackson A (eds) Rabies (4th edition): scientific basis of the disease and its management. Academic Press, pp 605–654
35. Aubert M (1994) Control of rabies in foxes: what are the appropriate measures? Vet Rec 134:55–59
36. Pybus MJ (1988) Rabies and rabies control in striped skunks (*Mephitis mephitis*) in three prairie regions of western North America. J Wildl Dis 24(3):434–449
37. Pybus MJ (2020) Striped skunks and rabies: ecology and epizootiology. In: Gregory DJ, Tinline RR (eds) Take the bite out of rabies: the evolution of rabies management in Canada. University of Toronto Press, pp 480–488
38. Rosatte RC, Pybus MJ, Gunson JR (1986) Population reduction as a factor in the control of skunk rabies in Alberta. J Wildl Dis 22(4):459–467
39. Brown CM et al (2016) Compendium of animal rabies prevention and control, 2016. J Am Vet Med Assoc 248(5):505–517
40. Hanlon CA, Childs JE, Nettles VF (1999) Recommendations of a national working group on prevention and control of rabies in the United States. Article III: rabies in wildlife. national working group on rabies prevention and control. J Am Vet Med Assoc 215(11):1612–1618
41. Rosatte R (2011) Evolution of wildlife rabies control tactics. In: Jackson A (ed) Advances in virus research: research advances in rabies. Academic Press, London, pp 397–419
42. Wandeler AI et al (1988) Oral immunization of wildlife against rabies: concept and first field experiments. Rev Infect Dis 10(Suppl 4):S649–S653
43. Baer GM, Abelseth MK, Debbie JG (1971) Oral vaccination of foxes against rabies. Am J Epidemiol 93(6):487–490
44. Tinline RR, Gregory DJ (2020) Assessing Canada's contribution to rabies management and control. In: Gregory DJ, Tinline RR (eds) Taking the bite out of rabies: the evolution of rabies management in Canada. University of Toronto Press, pp 631–642
45. Wandeler A et al (1974) Rabies in wild carnivores in Central Europe. I. Epidemiological studies. Zentralbl Veterinarmed B 21(10):735–756
46. Slate D, Rupprecht CE (2012) Rabies management in wild carnivores. In: Miller RE, Fowler ME (eds) Fowler's zoo and wild animal medicine: current therapy. Elsevier Sanders, St. Louis, pp 366–375
47. Slate D, Chipman RB (2020) The role of the U.S. Department of Agriculture (USDA), Wildlife services in wildlife rabies management. In: Gregory DJ, Tinline RR (eds) Taking the bite out of rabies: the evolution of rabies management in Canada. University of Toronto Press, pp 615–630

48. Slate D et al (2009) Oral rabies vaccination in north America: opportunities, complexities, and challenges. PLoS Negl Trop Dis 3(12):e549
49. Giles RH (1978) Wildlife management. W. H. Freeman and Company, San Francisco, p 416
50. Krausman PR, Cain JW (eds) (2013) Wildlife management and conservation: contemporary principles and practices. The Johns Hopkins University Press, Baltimore, p 342
51. Tinline RR, Rosatte R (2020) Ontario. In: Gregory DJ, Tinline RR (eds) Taking the bite out of rabies: the evolution of rabies management in Canada. University of Toronto Press, pp 125–156
52. Belanger D et al (2020) Quebec. In: Gregory DJ, Tinline RR (eds) Taking the bite out of rabies: the evolution of rabies management in Canada. University of Toronto Press, pp 157–178
53. Organ JF et al (2012) The North American model of wildlife conservation. Wildl Soc Tech Rev 12(4):1–60
54. Fuller AK et al (2020) Ratcheting up rigor in wildlife management decision making. Wildl Soc Bull 44(1):29–41
55. Follmann E et al (2011) Preliminary evaluation of Raboral V-RG(R) oral rabies vaccine in Arctic foxes (Vulpes lagopus). J Wildl Dis 47(4):1032–1035
56. Ramey AM et al (2022) Highly pathogenic avian influenza is an emerging disease threat to wild birds in North America. J Wildl Manag 86(2):e22171
57. Gilbert A (2018) Rabies virus vectors and reservoir species. Rev Sci Tech Off Int Epiz 37(2):371–384
58. Wobeser G (2002) Disease management strategies for wildlife. Revue Scientifique Et Technique-Office International Des Epizooties 21(1):159–178
59. Artois M et al (2001) Control of infectious diseases of wildlife in Europe. Vet J 162(2):141–152
60. Rabinowitz PM et al (2013) Toward proof of concept of a one health approach to disease prediction and control. Emerg Infect Dis 19(12):e130265
61. Cliquet F et al (2018) Oral vaccination of dogs: a well-studied and undervalued tool for achieving human and dog rabies elimination. Vet Res 49(1):61
62. Chanachai K et al (2021) Feasibility and effectiveness studies with oral vaccination of free-roaming dogs against rabies in Thailand. Viruses-Basel 13(4):571
63. Yale G et al (2022) Review of oral rabies vaccination of dogs and its application in India. Viruses-Basel 14(1):155
64. Wallace RM et al (2020) Role of oral rabies vaccines in the elimination of dog-mediated human rabies deaths. Emerg Infect Dis 26(12):1–9
65. Müller TF et al (2015) Spatio-temporal use of oral rabies vaccines in fox rabies elimination programmes in Europe. PLoS Negl Trop Dis 9(8):e0003953
66. Müller TF, Freuling CM (2018) Rabies control in Europe: an overview of past, current and future strategies. Rev Sci Tech 37(2):409–419
67. Lockwood, J.L., M.F. Hoopes, and M.P. Marchetti, Global climate change and invasive species, in Invasion ecology (2nd edition), J.L. Lockwood, M.F. Hoopes, and M.P. Marchetti, Editors. 2013, Wiley-Blackwell. p. 356-392.
68. Ricciardi A et al (2017) Invasion science: a horizon scan of emerging challenges and opportunities. Trends Ecol Evol 32(6):464–474
69. Crosby AW (2003) The Columbian exchange: biological and cultural consequences of 1492. Greenwood Publishing Group, Westport, p 283
70. Smith JA et al (2018) Human activity reduces niche partitioning among three widespread mesocarnivores. Oikos 127(6):890–901
71. Barrett LP, Stanton LA, Benson-Amram S (2019) The cognition of ‘nuisance’ species. Anim Behav 147:167–177
72. Schell CJ et al (2021) The evolutionary consequences of human-wildlife conflict in cities. Evol Appl 14(1):178–197
73. Kim BI et al (2014) A conceptual model for the impact of climate change on fox rabies in Alaska, 1980–2010. Zoonoses Public Health 61(1):72–80

74. Simon A et al (2021) Ecology of Arctic rabies: 60 years of disease surveillance in the warming climate of northern Canada. Zoonoses Public Health 68(6):601–608
75. Van de Vuurst P et al (2022) Climate change has affected the spillover risk of bat-borne pathogens. BioRxiv:1–15
76. WHO (2018) WHO expert consultation on rabies, third report, WHO technical report series, vol 195. World Health Organization, Geneva
77. Rupprecht CE, Slate D (2012) Rabies prevention and control: advances and challenges. In: Dietzgen RG, Kuzmin I (eds) Rhabdoviruses: molecular taxonomy, evolution, genomics, ecology, host-vector interactions, cytopathology and control. Caister Academic Press, Norfolk, pp 215–252
78. Rupprecht CE et al (2018) Global rabies management: perspectives on regional strategies for prevention and control. Revue Scientifique Et Tech-Off Int Des Epizooties 37(2):711–727
79. Slate D, Decker T (2003) Oral rabies vaccination: unresolved issues and data gaps. In: Proceedings of 10th wildlife damage management conference
80. Vercauteren K et al (2014) Rabies in North America: need and call for a one health approach. In: Fourteenth wildlife damage management conference. Nebraska City
81. Rupprecht CE et al (1987) Epidemiology of rabies virus variants. Differentiation using monoclonal antibodies and discriminant analysis. Am J Epidemiol 126(2):298–309
82. Kemere P et al (2000) Economic analysis of a large scale oral vaccination program to control raccoon rabies. In: Clark L et al (eds) Human conflicts with wildlife: economic considerations. National Wildlife Research Center, Fort Collins, pp 109–115
83. Shwiff SA, Kirkpatrick KN, Sterner RT (2008) Economic evaluation of an oral rabies vaccination program for control of a domestic dog-coyote rabies epizootic: 1995–2006. J Am Vet Med Assoc 233(11):1736–1741
84. Sterner RT et al (2009) Tactics and economics of wildlife oral rabies vaccination, Canada and the United States. Emerg Infect Dis 15(8):1176–1184
85. Chipman RB et al (2013) Costs of raccoon rabies incidents in cattle herds in Hampshire County, West Virginia, and Guernsey County, Ohio. J Am Vet Med Assoc 243(11):1561–1567
86. Nettles VF et al (1979) Rabies in translocated raccoons. Am J Public Health 69(6):601–602
87. Hanlon CA, Rupprecht CE (1998) The reemergence of rabies. In: Scheld WM, Armstrong D, Hughes JM (eds) Emerging infections 1. American Society for Microbiology Press, Washington, D.C., pp 59–80
88. Slate D et al (2005) Status of oral rabies vaccination in wild carnivores in the United States. Virus Res 111(1):68–76
89. Vos A et al (2017) Oral vaccination of wildlife against rabies: differences among host species in vaccine uptake efficiency. Vaccine 35(32):3938–3944
90. Te Kamp V et al (2020) Responsiveness of various reservoir species to oral rabies vaccination correlates with differences in vaccine uptake of mucosa associated lymphoid tissues. Sci Rep 10(1):2919
91. TWS (2012) Final position statement: wildlife disease. The Wildlife Society
92. Fontaine JJ (2011) Improving our legacy: incorporation of adaptive management into state wildlife action plans. J Environ Manag 92(5):1403–1408
93. Rees EE et al (2008) Assessing a landscape barrier using genetic simulation modelling: implications for raccoon rabies management. Prev Vet Med 86(1-2):107–123
94. Rees EE et al (2011) Understanding effects of barriers on the spread and control of rabies. Adv Virus Res 79:421–447
95. Rees EE et al (2013) Modelling the effect of landscape heterogeneity on the efficacy of vaccination for wildlife infectious disease control. J Appl Ecol 50(4):881–891
96. McClure KM et al (2020) Variation in host home range size decreases rabies vaccination effectiveness by increasing the spatial spread of rabies virus. J Anim Ecol 89(6):1375–1386
97. Davis AJ et al (2019) Not all surveillance data are created equal – prioritizing methods for estimation of rabies virus elimination from wildlife. J Appl Ecol 56:2551–2561

98. Davis AJ et al (2019) Rabies surveillance identifies potential risk corridors and enables management evaluation. Viruses 11(11):1006
99. Davis AJ et al (2021) Data-driven management-a dynamic occupancy approach to enhanced rabies surveillance prioritization. Viruses-Basel 13(9):1795
100. McClure KM et al (2022) Accounting for animal movement improves vaccination strategies against wildlife disease in heterogeneous landscapes. Ecol Appl 32(4):e2568
101. Sterner RT, Smith GC (2006) Modelling wildlife rabies: transmission, economics, conservation. Biol Conserv 131:163–179
102. Rosatte RC et al (2009) The control of raccoon rabies in Ontario Canada: proactive and reactive tactics, 1994–2007. J Wildl Dis 45(3):772–784
103. Morters MK et al (2013) Evidence-based control of canine rabies: a critical review of population density reduction. J Anim Ecol 82(1):6–14
104. Slate D et al (2008) Attaining raccoon rabies management goals: history and challenges. Dev Biol 131:439–447
105. Sidwa TJ et al (2005) Evaluation of oral rabies vaccination programs for control of rabies epizootics in coyotes and gray foxes: 1995–2003. J Am Vet Med Assoc 227(5):785–792
106. Elmore SA et al (2017) Management and modeling approaches for controlling raccoon rabies: the road to elimination. PLoS Negl Trop Dis 11(3):e0005249
107. Mainguy J et al (2013) Oral rabies vaccination in raccoons: comparison of ONRAB® and RABORAL V-RG® vaccine-bait field performance in Quebec, Canada and Vermont, USA. J Wildl Dis 49(1):190–193
108. Fehlner-Gardiner C et al (2012) Comparing ONRAB® AND RABORAL V-RG® oral rabies vaccine field performance in raccoons and striped skunks, New Brunswick, Canada, and Maine, USA. J Wildl Dis 48(1):157–167
109. Sattler AC et al (2009) Influence of oral rabies vaccine bait density on rabies seroprevalence in wild raccoons. Vaccine 27(51):7187–7193
110. Johnson SR et al (2021) Serological responses of raccoons and striped skunks to ontario rabies vaccine bait in West Virginia during 2012–2016. Viruses 13(2):157
111. Ma X et al (2022) Rabies surveillance in the United States during 2020. J Am Vet Med Assoc 215:1157–1165
112. Rupprecht CE, Turmelle A, Kuzmin IV (2011) A perspective on lyssavirus emergence and perpetuation. Curr Opin Virol 1(6):662–670
113. Charlton KM, Webster WA, Casey GA (1991) Skunk rabies. In: Baer GM (ed) The natural history of rabies. CRC Press, Inc., Boca Raton, pp 307–324
114. Kuzmina NA et al (2013) A reassessment of the evolutionary timescale of bat rabies viruses based upon glycoprotein gene sequences. Virus Genes 47(2):305–310
115. Winkler WG (1975) Fox rabies. In: Baer GM (ed) The natural history of rabies. Academic Press, Inc., New York, NY, pp 3–22
116. Parker RL (1975) Rabies in skunks. In: Baer G (ed) The natural history of rabies. Academic Press, pp 41–51
117. Müller T, Freuling C (2020) Rabies in terrestrial animals. In: Fooks AR, Jackson AC (eds) Rabies: scientific basis of the disease and its management. Acadmic Press, pp 195–230
118. Worsley-Tonks KEL et al (2020) Using host traits to predict reservoir host species of rabies virus. PLoS Negl Trop Dis 14(12):e0008940
119. Mollentze N, Biek R, Streicker DG (2014) The role of viral evolution in rabies host shifts and emergence. Curr Opin Virol 8:68–72
120. Nadin-Davis SA, Fehlner-Gardiner C (2019) Origins of the arctic fox variant rabies viruses responsible for recent cases of the disease in southern Ontario. PLoS Negl Trop Dis 13(9):e0007699
121. Wallace RM et al (2014) Right place, wrong species: a 20-year review of rabies virus cross species transmission among terrestrial mammals in the United States. PLoS One 9(10):e107539
122. Streicker DG et al (2010) Host phylogeny constrains cross-species emergence and establishment of rabies virus in bats. Science 329(5992):676–679

123. Mollentze N et al (2020) Virulence mismatches in index hosts shape the outcomes of cross-species transmission. Proc Natl Acad Sci U S A 117(46):28859–28866
124. Nadin-Davis S (2020) Molecular epidemiology. In: Fooks AR, Jackson A (eds) Rabies: scientific basis of the disease and its management. Academic Press, pp 143–193
125. Seetahal JFR et al (2018) Rabies in the Caribbean: a situational analysis and historic review. Trop Med Infect Dis 3(3):89
126. Arechiga-Ceballos N et al (2010) New rabies virus variant found during an epizootic in white-nosed coatis from the Yucatan Peninsula. Epidemiol Infect 138(11):1586–1589
127. Paez A et al (2005) Molecular epidemiology of rabies in northern Colombia 1994–2003. Evidence for human and fox rabies associated with dogs. Epidemiol Infect 133(3):529–536
128. Cordeiro Rde A et al (2016) The importance of wild canids in the epidemiology of rabies in northeast Brazil: a retrospective study. Zoonoses Public Health 63(6):486–493
129. Caraballo DA et al (2021) A novel terrestrial rabies virus lineage occurring in South America: origin, diversification, and evidence of contact between wild and domestic cycles. Viruses 13(12):2484
130. Carnieli P Jr et al (2009) Molecular characterization of rabies virus isolates from dogs and crab-eating foxes in Northeastern Brazil. Virus Res 141(1):81–89
131. de Souza DN et al (2017) Phylogenetic analysis of rabies virus isolated from canids in North and Northeast Brazil. Arch Virol 162(1):71–77
132. Carnieli P Jr et al (2008) Characterization of Rabies virus isolated from canids and identification of the main wild canid host in Northeastern Brazil. Virus Res 131(1):33–46
133. Favoretto SR et al (2013) The emergence of wildlife species as a source of human rabies infection in Brazil. Epidemiol Infect 141(7):1552–1561
134. Kotait I et al (2019) Non-human primates as a reservoir for rabies virus in Brazil. Zoonoses Public Health 66(1):47–59
135. Tierkel ES (1975) Control of urban rabies. In: Baer GM (ed) Natual history of rabies. Routledge, New York, pp 189–201
136. Williams RB (1949) Rabies in Alaska. Can J Comp Med Vet Sci 13(6):136–143
137. Plummer PJG (1947) Preliminary note on arctic dog disease and its relationship to rabies. Can J Comp Med 11:154–160
138. MacInnes CD et al (2001) Elimination of rabies from red foxes in eastern Ontario. J Wildl Dis 37(1):119–132
139. Johnston DH, Beauregard M (1969) Rabies epidemiology in Ontario. Wildl Dis 5(3):357–370
140. Nadin-Davis SA, Muldoon F, Wandeler AI (2006) Persistence of genetic variants of the arctic fox strain of Rabies virus in southern Ontario. Can J Vet Res 70(1):11–19
141. Middel K et al (2017) Incorporating direct rapid immunohistochemical testing into large-scale wildlife rabies surveillance. Trop Med Infect Dis 2(3):21
142. Lobo D et al (2018) Raccoon rabies outbreak in Hamilton, Ontario: a progress report. Can Commun Dis Rep 44(5):116–121
143. Kuzmina NA et al (2013) The phylogeography and spatiotemporal spread of South-central skunk rabies virus. PLoS One 8(12):e82348
144. Jaramillo-Reyna E et al (2020) Rabies virus variants identified in Nuevo Leon State, Mexico, from 2008 to 2015. Javma-J Am Vet Med Assoc 256(4):438–443
145. Garcés-Ayala F et al (2017) Molecular characterization of atypical antigenic variants of canine rabies virus reveals its reintroduction by wildlife vectors in southeastern Mexico. Arch Virol 162(12):3629–3637
146. McLean RG (1975) Raccoon rabies. In: Baer GM (ed) The natural history of rabies. Academic Press, New York, NY, pp 53–77
147. Bigler WJ, McLean RG, Trevino HA (1973) Epizootiologic aspects of raccoon rabies in Florida. Am J Epidemiol 98(5):326–335
148. Rupprecht CE, Smith JS (1994) Raccoon rabies – The reemergence of an epizootic in a densely populated area. Semin Virol 5(2):155–164

149. Childs JE et al (2000) Predicting the local dynamics of epizootic rabies among raccoons in the United States. Proc Natl Acad Sci U S A 97(25):13666–13671
150. Biek R et al (2007) A high-resolution genetic signature of demographic and spatial expansion in epizootic rabies virus. Proc Natl Acad Sci U S A 104(19):7993–7998
151. Trewby H et al (2017) Processes underlying rabies virus incursions across US-Canada border as revealed by whole-genome phylogeography. Emerg Infect Dis 23(9):1454–1461
152. Stevenson B, Goltz J, Masse A (2016) Preparing for and responding to recent incursions of raccoon rabies variant into Canada. Can Commun Dis Rep 42(6):125–129
153. Goltz J, Badcock J, Tinline R (2020) Maratime provinces: Nova Scotia, Prince Edward Island, New Brunswick. In: Gregory DJ, Tinline RR (eds) Taking the bite out of rabies. University of Toronto Press, pp 157–178
154. CFIA (2021) Rabies in Canada. Available from: https://inspection.canada.ca/animal-health/terrestrial-animals/diseases/reportable/rabies/rabies-cases-in-canada-2021/eng/1613407237949/1613407238418
155. Nadin-Davis S et al (2020) A long-distance translocation initiated an outbreak of raccoon rabies in Hamilton, Ontario, Canada. PLoS Negl Trop Dis 14(3):e0008113
156. Clark KA et al (1994) Epizootic canine rabies transmitted by coyotes in south Texas. J Am Vet Med Assoc 204(4):536–540
157. Fearneyhough MG et al (1998) Results of an oral rabies vaccination program for coyotes. J Am Vet Med Assoc 212(4):498–502
158. Clark KA, Wilson PJ (1995) Canine and gray fox rabies epizootics in Texas. In: Great plains wildlife damage control workshop proceedings. Noble Foundation, Ardmore
159. Maki J et al (2017) Oral vaccination of wildlife using a vaccinia-rabies-glycoprotein recombinant virus vaccine (RABORAL V-RG®): a global review. Vet Res 48(1):57
160. Ma X et al (2018) Rabies surveillance in the United States during 2017. J Am Vet Med Assoc 253(12):1555–1568
161. Ma X et al (2020) Public veterinary medicine: public health: rabies surveillance in the United States during 2018. J Am Vet Med Assoc 256(2):195–208
162. Garces-Ayala F et al (2022) Rabies virus variants detected from Cougar (Puma concolor) in Mexico 2000–2021. Pathogens 11(2):265
163. Ramírez CVH, García DCS (2022) Determinants for the persistence of canine rabies in the Sierra Madre Occidental of Sinaloa. Braz J Anim Environ Res 5(1):424–437
164. Leslie MJ et al (2006) Bat-associated rabies virus in skunks. Emerg Infect Dis 12(8):1274–1277
165. Kuzmin IV et al (2012) Molecular inferences suggest multiple host shifts of rabies viruses from bats to mesocarnivores in Arizona during 2001–2009. PLoS Pathog 8(6):e1002786
166. Zieger U et al (2014) The phylogeography of rabies in Grenada, West Indies, and implications for control. PLoS Negl Trop Dis 8(10):e3251
167. Rupprecht CE et al (2018) Additional progress in the development and application of a direct, rapid immunohistochemical test for rabies diagnosis. Vet Sci 5(2):59
168. Everard CO, Everard JD (1992) Mongoose rabies in the Caribbean. Ann N Y Acad Sci 653:356–366
169. Everard COR, Everard JD (1985) Mongoose rabies in Grenada. In: Bacon PJ (ed) Population dynamics of rabies in wildlife. Academic Press, London, pp 43–69
170. Berentsen A et al (2020) Immunogenicity of Ontario rabies vaccine for small Indian mongooses (Herpestes auropunctatus). J Wildl Dis 56(1):000–000
171. Berentsen AR et al (2021) Oral rabies vaccination of small Indian Mongooses (Urva auropunctata) with ONRAB via Ultralite Baits. Viruses 13(5):734
172. Ortmann S et al (2018) Safety studies with the oral rabies virus vaccine strain SPBN GASGAS in the small Indian mongoose (Herpestes auropunctatus). BMC Vet Res 14(1):90
173. Vos A et al (2013) Oral vaccination of captive small Indian mongoose (Herpestes auropunctatus) against rabies. J Wildl Dis 49(4):1033–1036
174. Blanton JD et al (2006) Vaccination of small Asian mongoose (*Herpestes javanicus*) against rabies. J Wildl Dis 42(3):663–666

175. Berentsen A et al (2020) Placebo oral rabies vaccine uptake by small Indian mongooses (Herpestes auropunctatus) in southwestern Puerto Rico. J Wildl Dis 56(2):000–000
176. Sauve CC et al (2021) Modeling mongoose rabies in the Caribbean: a model-guided fieldwork approach to identify research priorities. Viruses 13(2):323
177. Rosatte R et al (2007) The elimination of raccoon rabies from Wolfe Island, Ontario: animal density and movements. J Wildl Dis 43(2):242–250
178. Rosatte R (2013) Rabies control in wild carnivores. In: Jackson A (ed) Rabies, 3rd edn. Academic Press, London, pp 617–670
179. Rosatte RC, Tinline RR, Johnston DH (2007) Rabies control in wild carnivores. In: Jackson AC, Wunner WH (eds) Rabies, 2nd edn. Academic Press, London, pp 595–634
180. Hanlon CA et al (1998) First North American field release of a vaccinia-rabies glycoprotein recombinant virus. J Wildl Dis 34(2):228–239
181. Sterner RT et al (2008) Skunk rabies in California (1992–2003)–implications for oral rabies vaccination. J Wildl Dis 44(4):1008–1013
182. Shwiff SA et al (2009) Benefit cost scenarios of potential oral rabies vaccination for skunks in California. J Wildl Dis 45(1):227–233
183. Anderson A et al (2014) Forecasting the spread of raccoon rabies using a purpose-specific group decision-making process. Hum Wildl Interact 8(1):130–138
184. Meltzer MI, Rupprecht CE (1998) A review of the economics of the prevention and control of rabies. Part 2: rabies in dogs, livestock and wildlife. PharmacoEconomics 14(5):481–498
185. Shwiff SA et al (2007) Direct and indirect costs of rabies exposure: a retrospective study in southern California (1998–2002). J Wildl Dis 43(2):251–257
186. Shwiff SA et al (2011) A retrospective economic analysis of the Ontario red fox oral rabies vaccination programme. Zoonoses Public Health 58(3):169–177
187. Elser JL et al (2016) The economics of a successful raccoon rabies elimination program on long island, New York. PLoS Negl Trop Dis 10(12):e0005062
188. Rupprecht CE et al (2008) Can rabies be eradicated? Dev Biol 131:95–121
189. Kirby JD et al (2017) Enhanced surveillance to support effective oral rabies vaccination of raccoons in the eastern United States. Trop Med Infect Dis 2(3):34
190. Rees EE et al (2011) Targeted surveillance of raccoon rabies in Québec, Canada. J Wildl Manag 75(6):1406–1416
191. Slate D et al (2017) Cost and relative value of road kill surveys for enhanced rabies surveillance in raccoon rabies management. Trop Med Infect Dis 2(2):13
192. Lembo T et al (2006) Evaluation of a direct, rapid immunohistochemical test for rabies diagnosis. Emerg Infect Dis 12(2):310–313
193. Rupprecht C et al (2014) Progress in the development of a direct rapid immunohistochemical test for diagnosing rabies. OIE Bull 3:87–95
194. Patrick EM et al (2019) Enhanced rabies surveillance using a direct rapid immunohistochemical test. J Vis Exp 146:e59416
195. WHO (1992) WHO expert committee on rabies: 8th report. WHO, Geneva, p 84
196. WHO (2005) WHO expert consultation on rabies, first report, World health organization technical report series, vol 121, Geneva
197. WHO (2013) WHO expert consultation on rabies, second report, WHO technical report series, vol 139, Geneva
198. Octaria R et al (2018) From recognition to action: a strategic approach to foster sustainable collaborations for rabies elimination. PLoS Negl Trop Dis 12(10):e0006756
199. USDA (2008) In: U.D.o. Agriculture (ed) US national plan for wildlife rabies management. US Department of Agriculture, Concord, 43 pages
200. MacInnes CD, LeBer CA (2000) Wildlife management agencies should participate in rabies control. Wildl Soc Bull 28:1156–1167
201. USDA (2008) In: U.D.o. Agriculture (ed) North American rabies management plan. US Department of Agriculture, Concord, 76 pages

202. Dalkey N, Helmer O (1963) An experimental application of the Delphi method to the use of experts. Manag Sci 9(3):458–467
203. Chipman R et al (1993) Rabies and problem wildlife: a toll-free management approach in Vermont and Maryland. Northeast Wildl Trans Northeast Section 50:63–68
204. Kellogg F et al (2013) Human contacts with oral rabies vaccine baits distributed for wildlife rabies management – Ohio, 2012. Mmwr-Morb Mortal Wkly Rep 62(14):267–269
205. Rosatte R et al (1997) A proactive prevention strategy for raccoon rabies in Ontario, Canada. Wildl Soc Bull 25(1):110–116
206. OIE (2018) Rabies (infection with rabies virus and other lyssaviruses). In: Manual of diagnostic tests and vaccines for terrestrial animals. World Animal Health Organization, pp 578–612
207. Pepin KM et al (2017) Predicting spatial spread of rabies in skunk populations using surveillance data reported by the public. PLoS Negl Trop Dis 11(7):e0005822
208. Pieracci EG et al (2020) Evaluation of rabies virus characterization to enhance early detection of important rabies epizootic events in the United States. J Am Vet Med Assoc 256(1):66–76
209. Pieracci EG et al (2020) Evaluation of species identification and rabies virus characterization among bat rabies cases in the United States. J Am Vet Med Assoc 256(1):77–84
210. Szanto AG et al (2011) Re-assessment of direct fluorescent antibody negative brain tissues with a real-time PCR assay to detect the presence of raccoon rabies virus RNA. J Virol Methods 174(1–2):110–116
211. Wadhwa A et al (2017) A Pan-Lyssavirus Taqman Real-Time RT-PCR assay for the detection of highly variable rabies virus and other lyssaviruses. PLoS Negl Trop Dis 11(1):e0005258
212. Rosatte R et al (2001) Emergency response to raccoon rabies introduction into Ontario. J Wildl Dis 37(2):265–279
213. Debbie JG (1991) Rabies control of terrestrial wildlife by population reduction. In: Baer GM (ed) The natural history of rabies, 2nd edn. CRC Press, Inc., Boca Raton, pp 477–484
214. Lewis JC (1975) Control of rabies among terrestrial wildlife by population reduction. In: Baer GM (ed) The natural history of rabies. Academic Press, Inc., New York, NY, pp 243–259
215. Blancou J, Aubert MF, Artois M (1991) Fox rabies. In: Baer G (ed) The natural history of rabies. CRC Press, Boca Raton, pp 257–290
216. Ballantyne EE, O'Donoghue JG (1954) Rabies control in Alberta. J Am Vet Med Assoc 125(931):316–326
217. Ballantyne EE (1956) Rabies control program in Alberta. Can J Comp Med Vet Sci 20(1):21–30
218. Gunson JR, Dorward WJ, Schowalter DB (1978) An evaluation of rabies control in skunks in Alberta. Can Vet J 19(8):214–220
219. Bögel K et al (1974) Recovery of fox populations in rabies control 1. Zentralbl Veterinarmed B 21(6):401–412
220. Müller T et al (2015) Terrestrial rabies control in the European Union: historical achievements and challenges ahead. Vet J 203(1):10–17
221. Rosatte RC et al (1992) Trap-vaccinate-release and oral vaccination for rabies control in urban skunks, raccoons and foxes. J Wildl Dis 28(4):562–571
222. Slavinski S et al (2012) Trap-vaccinate-release program to control raccoon rabies, New York, USA. Emerg Infect Dis 18(7):1170–1172
223. Sobey KG et al (2010) Field evaluation of an inactivated vaccine to control raccoon rabies in Ontario, Canada. J Wildl Dis 46(3):818–831
224. Wohlers A et al (2018) Challenges to controlling rabies in skunk populations using oral rabies vaccination: a review. Zoonoses Public Health 65(4):373–385
225. Rosatte RC et al (1990) Intramuscular vaccination of skunks and raccoons against rabies. J Wildl Dis 26(2):225–230
226. Rosatte R et al (2007) Rabies in vaccinated raccoons from Ontario, Canada. J Wildl Dis 43(2):300–301
227. Rosatte RC et al (1993) Tactics for the control of wildlife rabies in Ontario (Canada). Rev Sci Tech 12(1):95–98

228. Algeo T et al (2008) Anatomy of the Cape Cod oral rabies vaccination program. In: 23rd vertebrate pest conference. University of California Davis, San Diego
229. Müller T et al (2012) Elimination of terrestrial rabies in Germany using oral vaccination of foxes. Berl Munch Tierarztl Wochenschr 125(5–6):178–190
230. Lawson KF, Crawley JF (1972) The ERA strain of rabies vaccine. Can J Comp Med 36(4):339–344
231. MacInnes CD et al (1988) Planning for rabies control in Ontario. Rev Infect Dis 10(Suppl 4):S665–S669
232. Campbell JB, Prevec L (2020) Oral vaccine development in Canada. Oral vaccines. In: Gregory DJ, Tinline R (eds) Taking the bite out of rabies. The evolution of rabies management in Canada. University of Toronto Press, pp 281–286
233. Lawson K, MacInnes C (2020) Oral vaccine development in Canada. Bait development. In: Gregory DJ, Tinline R (eds) Taking the bite out of rabies. University of Toronto Press, pp 287–290
234. Artemis Technologies I (2020) Oral vaccine development in Canada. Bait production. In: Gregory DJ, Tinline R (eds) Taking the bite out of rabies. University of Toronto Press, pp 291–298
235. Slate D et al (2014) Safety and immunogenicity of ontario rabies vaccine bait (ONRAB) in the first us field trial in raccoons (*Procyon lotor*). J Wildl Dis 50(3):582–595
236. Gilbert AT et al (2018) Field trials of Ontario rabies vaccine bait in the northeastern USA, 2012–14. J Wildl Dis 54:790–801
237. Pedersen K et al (2019) Raccoon (*Procyon lotor*) response to Ontario Rabies Vaccine Baits (ONRAB) in St. Lawrence County, New York, USA. J Wildl Dis 55(3):645–653
238. Boulanger JR et al (2008) Comparison of suburban vaccine distribution strategies to control raccoon rabies. J Wildl Dis 44(4):1014–1023
239. Selhorst T, Thulke HH, Müller T (2001) Cost-efficient vaccination of foxes (*Vulpes vulpes*) against rabies and the need for a new baiting strategy. Prev Vet Med 51(1–2):95–109
240. Craven S, Barnes T, Kania G (1998) Toward a professional position on the translocation of problem wildlife. Wildl Soc Bull 26(1):171–177
241. Griffith B et al (1989) Translocation as a species conservation tool: status and strategy. Science 245(4917):477–480
242. Group, F.W. Recommended best management practices to address possession, transportation and disposition of rabies vector species. 2016 [cited 2019 6/14/2019]; Available from: http://www.seafwa.org/Documents%20and%20Settings/46/Site%20Documents/Resources/Committee%20Resources/SEAFWA_BMPs%20for%20Rabies%20Vector%20Species_AS%20ADOPTED.pdf
243. Jardine CM et al (2018) Frequency of virus coinfection in raccoons (*Procyon lotor*) and Striped Skunks (*Mephitis mephitis*) during a concurrent rabies and canine distemper outbreak. J Wildl Dis 54(3):622–625
244. Wostenberg DJ et al (2018) Evidence of two cocirculating canine distemper virus strains in mesocarnivores from Northern Colorado, USA. J Wildl Dis 54(3):534–543
245. Elmore SA et al (2016) Estimating Toxoplasma Gondii exposure in arctic foxes (Vulpes Lagopus) while navigating the imperfect world of wildlife serology. J Wildl Dis 52(1):47–56
246. Bouchard E et al (2019) Transmission dynamics of toxoplasma gondii in arctic foxes (Vulpes Lagopus): a long-term mark-recapture serologic study at karrak lake, Nunavut, Canada. J Wildl Dis 55(3):619–626
247. Cranford HM et al (2021) Mongooses (Urva auropunctata) as reservoir hosts of Leptospira species in the United States Virgin Islands, 2019–2020. PLoS Negl Trop Dis 15(11):e0009859
248. Wasniewski M et al (2016) First international collaborative study to evaluate rabies antibody detection method for use in monitoring the effectiveness of oral vaccination programmes in fox and raccoon dog in Europe. J Virol Methods 238:77–85
249. Moore SM et al (2017) Rabies virus antibodies from oral vaccination as a correlate of protection against lethal infection in wildlife. Trop Med Infect Dis 2(31):1–24

250. Moore SM, Gordon CR (2020) Measures of rabies immunity. In: Fooks AR, Jackson A (eds) Rabies (4th edition): scientific basis of the disease and its management. Academic Press, pp 445–479
251. Rosatte R et al (2008) Prevalence of tetracycline and rabies virus antibody in raccoons, skunks, and foxes following aerial distribution of V-RG baits to control raccoon rabies in Ontario, Canada. J Wildl Dis 44(4):946–964
252. Fehlner-Gardiner C, Wandeler AI (2014) Detection of rabies virus antibodies by competitive enzyme-linked immunosorbent assay. In: Rupprecht C, Nagarajan T (eds) Current laboratory techniques in rabies diagnosis, research and prevention. Elsevier, London, pp 145–155
253. Blanton JD et al (2018) Evaluation of oral rabies vaccination: protection against rabies in wild caught raccoons (*Procyon Lotor*). J Wildl Dis 54(3):520–527
254. Riley SPD, Hadidian J, Manski DA (1998) Population density, survival, and rabies in raccoons in an urban national park. Can J Zool 76:1153–1164
255. Prange S, Gehrt SD, Wiggers EP (2003) Demographic factors contributing to high raccoon densities in urban landscapes. J Wildl Manag 67(2):324–333
256. Prange S, Gehrt SD, Wiggers EP (2004) Influences of anthropogenic resources on raccoon (*Procyon lotor*) movements and spatial distribution. J Mammal 85(3):483–490
257. Bachmann P, Tinline R (2020) Biomarkers. In: Gregory DJ, Tinline R (eds) Taking the bite out of rabies: the evolution of rabies management in Canada. University of Toronto Press, pp 384–392
258. Bachmann P, Brown LJ, Ayers NR (2020) The development of aerial baiting. In: Gregory DJ, Tinline R (eds) Taking the bite out of rabies: the evolution of rabies management in Canada. University of Toronto Press, pp 311–330
259. Bachmann P et al (1990) Wild carnivore acceptance of baits for delivery of liquid rabies vaccine. J Wildl Dis 26(4):486–501
260. Boyer JP et al (2011) Oral vaccination against raccoon rabies: landscape heterogeneity and timing of distribution influence wildlife contact rates with the ONRAB vaccine bait. J Wildl Dis 47(3):593–602
261. Rosatte RC et al (2009) Aerial distribution of ONRAB[(R)] baits as a tactic to control rabies in raccoons and striped skunks in Ontario, Canada. J Wildl Dis 45(2):363–374
262. Rosatte RC et al (2011) High-density baiting with ONRAB[(R)] rabies vaccine baits to control Arctic-variant rabies in striped skunks in Ontario, Canada. J Wildl Dis 47(2):459–465
263. Rosatte R et al (2010) Density, movements, and survival of raccoons in Ontario, Canada: implications for disease spread and management. J Mammal 91(1):122–135
264. Johnson SR et al (2016) Estimates of small Indian mongoose densities: implications for rabies management. J Wildl Manag 80(1):37–47
265. Berentsen A et al (2013) Spatial ecology of urban raccoons in Northeastern Ohio: implications for oral rabies vaccination. Prairie Nat 45:39–45
266. Berentsen AR et al (2018) Seroconversion of raccoons following two oral rabies vaccination baiting strategies. J Wildl Manag 82(1):226–231
267. Atwood TC, Fry T, Leland BR (2011) Partitioning of a limited resource by sympatric carnivores in the Chihuahuan Desert and the implications for disease transmission. J Wildl Manag 75:1609–1615
268. Ramey PC et al (2008) Oral rabies vaccination of a northern Ohio raccoon population: relevance of population density and prebait serology. J Wildl Dis 44(3):553–568
269. Blackwell BF et al (2004) Exposure time of oral rabies vaccine baits relative to baiting density and raccoon population density. J Wildl Dis 40(2):222–229
270. Rosatte R, Kelly P, Power M (2011) Home range, movements, and habitat utilization of striped skunk (*Mephitis mephitis*) in Scarborough, Ontario, Canada: disease management implications. Can Field-Nat 125:27–33
271. Arjo WM et al (2008) Effects of natural barriers and habitat on the western spread of raccoon rabies in Alabama. J Wildl Manag 72(8):1725–1735

272. Puskas RB et al (2010) Raccoon (*Procyon lotor*) movements and dispersal associated with ridges and valleys of Pennsylvania: implications for rabies management. Vector-Borne Zoonotic Dis 10(10):1043–1048
273. Tardy O et al (2014) Density-dependent functional responses in habitat selection by two hosts of the raccoon rabies virus variant. Ecosphere 5(10):art132
274. Bernasconi DA et al (2022) Influence of landscape attributes on Virginia opossum density. J Wildl Manag 86(7):e22280
275. Schumacher CL et al (1993) SAG-2 oral rabies vaccine. Onderstepoort J Vet Res 60(4):459–462
276. Lafay F et al (1994) Vaccination against rabies: construction and characterization of SAG2, a double avirulent derivative of SADBern. Vaccine 12(4):317–320
277. Le Blois H et al (1990) Oral immunization of foxes with avirulent rabies virus mutants. Vet Microbiol 23(1-4):259–266
278. Fehlner-Gardiner C et al (2008) ERA vaccine-derived cases of rabies in wildlife and domestic animals in Ontario, Canada, 1989–2004. J Wildl Dis 44(1):71–85
279. Kieny MP et al (1984) Expression of rabies virus glycoprotein from a recombinant vaccinia virus. Nature 312(5990):163–166
280. Roscoe DE et al (1998) Efficacy of an oral vaccinia-rabies glycoprotein recombinant vaccine in controlling epidemic raccoon rabies in New Jersey. J Wildl Dis 34(4):752–763
281. Robbins AH et al (1998) Prevention of the spread of rabies to wildlife by oral vaccination of raccoons in Massachusetts. J Am Vet Med Assoc 213(10):1407–1412
282. Olson CA, Mitchell KD, Werner PA (2000) Bait ingestion by free-ranging raccoons and nontarget species in an oral rabies vaccine field trial in Florida. J Wildl Dis 36(4):734–743
283. Prevec L et al (1990) A recombinant human adenovirus vaccine against rabies. J Infect Dis 161(1):27–30
284. Bigler LL et al (2021) Oral rabies vaccination strategies toward raccoon (*Procyon Lotor*) rabies elimination on Suburban Long Island, New York, USA. J Wildl Dis 57(1):132–144
285. Jojola SM, Robinson SJ, VerCauteren KC (2007) Oral rabies vaccine (ORV) bait uptake by captive striped skunks. J Wildl Dis 43(1):97–106
286. Gilbert A et al (2018) Flavor preference and efficacy of variable dose Ontario Rabies Vaccine Bait (ONRAB) delivery in striped skunks (*Mephitis mephitis*). J Wildl Dis 54(1):122–132
287. Brown LJ et al (2014) Oral vaccination and protection of red foxes (*Vulpes vulpes*) against rabies using ONRAB, an adenovirus-rabies recombinant vaccine. Vaccine 32(8):984–989
288. Brown LJ et al (2012) Immune response and protection in raccoons (*Procyon lotor*) following consumption of baits containing ONRAB(R), a human adenovirus rabies glycoprotein recombinant vaccine. J Wildl Dis 48(4):1010–1020
289. Brown LJ et al (2011) Immunogenicity and efficacy of two rabies vaccines in wild-caught, captive raccoons. J Wildl Dis 47(1):182–194
290. Brown LJ et al (2014) Oral vaccination and protection of striped skunks (*Mephitis mephitis*) against rabies using ONRAB(R). Vaccine 32(29):3675–3679
291. Gilbert A et al (2018) Efficacy of Ontario Rabies Vaccine Baits (ONRAB) against rabies infection in raccoons. Vaccine 36(32 Pt B):4919–4926
292. Steelman HG, Henke SE, Moore GM (2000) Bait delivery for oral rabies vaccine to gray foxes. J Wildl Dis 36(4):744–751
293. Linhart SB et al (2002) A new flavor-coated sachet bait for delivering oral rabies vaccine to raccoons and coyotes. J Wildl Dis 38(2):363–377
294. Linhart SB, Kappeler A, Windberg LA (1993) A review of baits and bait delivery systems for free-ranging carnivores and ungulates. United States Department of Agriculture, pp 69–132
295. Algeo TP et al (2013) Oral rabies vaccination variation in tetracycline biomarking among Ohio raccoons. J Wildl Dis 49(2):332–337
296. Ballesteros C et al (2013) Iophenoxic acid as a bait marker for wild mammals: efficacy and safety considerations. Mammal Rev 43(2):156–166

297. Fisher P (1999) Review of using Rhodamine B as a marker for wildlife studies. Wildl Soc Bull 27(2):318–329
298. Fry T, Atwood T, Dunbar M (2010) Utility of rhodamine B as a biomarker in raccoons. Hum-Wildl Interact 4:275–282
299. Berentsen AR et al (2019) Analysis of iophenoxic acid analogues in small Indian mongoose (Herpestes Auropunctatus) Sera for use as an oral rabies vaccination biological marker. J Vis Exp 147:e59373
300. Johnston JJ et al (2005) Evaluation and significance of tetracycline stability in rabies vaccine baits. J Wildl Dis 41(3):549–558
301. Stohr K, Meslin FX (1997) Zoonoses and fertility control in wildlife–requirements for vaccines. Reprod Fertil Dev 9(1):149–155
302. WHO (1993) WHO informal consultation on reproductive control of carnivores. World Health Organization, Geneva, pp 1–11
303. Gese EM, Terletzky PA (2021) Space use in free-ranging canids: are gonadal hormones required for territory maintenance? Can J Zool 99(10):939–947
304. Saunders G et al (2002) The effects of induced sterility on the territorial behaviour and survival of foxes. J Appl Ecol 39(1):56–66
305. Real LA et al (2005) Unifying the spatial population dynamics and molecular evolution of epidemic rabies virus. Proc Natl Acad Sci U S A 102(34):12107–12111
306. Real LA, Biek R (2007) Spatial dynamics and genetics of infectious diseases on heterogeneous landscapes. J R Soc Interface 4(16):935–948
307. Smith DL et al (2002) Predicting the spatial dynamics of rabies epidemics on heterogeneous landscapes. Proc Natl Acad Sci U S A 99(6):3668–3672
308. Smith DL et al (2005) Assessing the role of long-distance translocation and spatial heterogeneity in the raccoon rabies epidemic in Connecticut. Prev Vet Med 71(3-4):225–240
309. Reynolds JJ et al (2015) Raccoon contact networks predict seasonal susceptibility to rabies outbreaks and limitations of vaccination. J Anim Ecol 84(6):1720–1731
310. Brunker K, Nadin-Davis S, Biek R (2018) Genomic sequencing, evolution and molecular epidemiology of rabies virus. Rev Sci Tech 37(2):401–408
311. Borucki MK et al (2013) Ultra-deep sequencing of intra-host rabies virus populations during cross-species transmission. PLoS Negl Trop Dis 7(11):e2555
312. Lentzos F et al (2022) Eroding norms over release of self-spreading viruses. Science 375(6576):31–33
313. Murphy AA, Redwood AJ, Jarvis MA (2016) Self-disseminating vaccines for emerging infectious diseases. Expert Rev Vaccines 15(1):31–39
314. Jeon S et al (2019) Determining the post-elimination level of vaccination needed to prevent re-establishment of dog rabies. PLoS Negl Trop Dis 13(12):e0007869

Chapter 8
Management of Vampire Bats and Rabies: Past, Present, and Future

Tonie Rocke, Daniel Streicker, and Ariel Leon

Abstract Rabies virus transmitted via the bite of common vampire bats (*Desmodus rotundus*) has surpassed canine-associated cases as the predominant cause of human rabies in Latin America. Cattle, the preferred prey of *D. rotundus*, suffer extensive mortality from vampire bat associated rabies, with annual financial losses estimated in the tens of millions of dollars. Organized attempts to manage or curtail vampire bat populations and rabies virus transmission have been conducted since the early 1900s, when vampire bat-associated rabies cases in humans and livestock were first recognized. However, these attempts largely failed, as the distribution of vampire bat populations expanded geographically with the intensification of livestock production, and the incidence of vampire bat rabies (VBR) increased. Current methods of control rely primarily on culling vampire bat populations using poisons (vampiricides) that are transferred from bat to bat after topical application. Despite widespread use of vampiricides for the last 50 years, little evidence exists to demonstrate their effectiveness in reducing the incidence of VBR. Culling may further result in dispersion of bats, which could have an unintended consequence of spreading VBR. New methods to manage VBR are being developed or considered, including topical rabies vaccine that transfer among bats, much like vampiricides or a transmissible vaccine that would spread naturally among bats. Vaccination of vampire bats against rabies could lower the incidence of VBR and prevent viral transmission to cattle and humans without the animal welfare concerns and potential negative effects of culling. However, this approach would not deter vampire bat bites, and some form of population reduction (e.g., fertility control) would likely also be needed. An integrated strategy to reduce both the incidence of VBR and the abundance of vampire bats would be ideal for protecting both human and animal health.

T. Rocke (✉) · A. Leon
U.S. Geological Survey National Wildlife Health Center, Madison, WI, USA
e-mail: trocke@usgs.gov; aleon@usgs.gov

D. Streicker
Institute of Biodiversity, Animal Health and Comparative Medicine & MRC-University of Glasgow Centre for Virus Research, University of Glasgow, Glasgow, UK
e-mail: daniel.streicker@glasgow.ac.uk

C. E. Rupprecht (ed.), *History of Rabies in the Americas: From the Pre-Columbian to the Present, Volume I*, Fascinating Life Sciences,
https://doi.org/10.1007/978-3-031-25052-1_8

Keywords Anti-coagulants · Bats · Blood · Control · Culling · Livestock · Vaccination · Vampire bats

Introduction

As the prevalence of canine rabies has declined in Latin America, the relative significance of rabies virus transmission via common vampire bats (*Desmodus rotundus*) has increased, causing mortality in both humans and livestock, and severe economic losses for livestock owners [1–3]. Due to extensive control programs instituted by the Pan American Health Organization (PAHO) and others during the early 1980s, canine rabies has been nearly eliminated in most Latin American countries, whereas concurrent efforts to manage vampire bat rabies (VBR) have been less effective [3]. During 2005, the number of human rabies cases associated with vampire bats and reported to PAHO exceeded canine-associated cases for the first time [4]. Bites from rabid vampire bats have now become the leading cause of human rabies in Latin America. [3] In fact, cases are likely under-reported due to the geographic isolation of many affected communities [5]. It has been argued that VBR should be considered an emerging public health issue [6]. Cattle, the preferred prey of *D. rotundus*, suffer extensive losses from VBR-related outbreaks, in the tens to hundreds of thousands, annually [2, 7]. The economic fallout can be devastating for family farms and small producers. Financial losses of approximately 30 million USD annually have been estimated [8]. These figures do not include the costs of surveillance, diagnostics, and preventative measures, such as livestock vaccination. Unfortunately, this financial burden is not likely to decrease in the short term, as reports of VBR have increased in several countries and also expanded geographically (Fig. 8.1). Recent reports of vampire bat bites of humans in less traditional locales, such as urban areas and fishing communities, represent new contexts for human risk [9, 10]. Finally, evidence that vampire bats feed on and transmit rabies virus to dogs [11, 12] raises the prospect that VBR may also complicate the worldwide effort to eliminate canine rabies, requiring genetic or antigenic typing of viruses to differentiate new or unrecognized foci of dog-to-dog transmission from sporadic spillover of VBR. Moreover, given relatively sparse understanding of historical bat-to-carnivore host shifts, a risk for onward transmission in unvaccinated dog populations following spillover from bats cannot be ruled out [13]. In this chapter, we review the history of vampire bat management in the context of VBR occurrence in humans and livestock, and we discuss the effectiveness of current management programs and probable reasons for failure. We also explore new options, strategies and methodologies that could be used in future management of vampire bats and rabies.

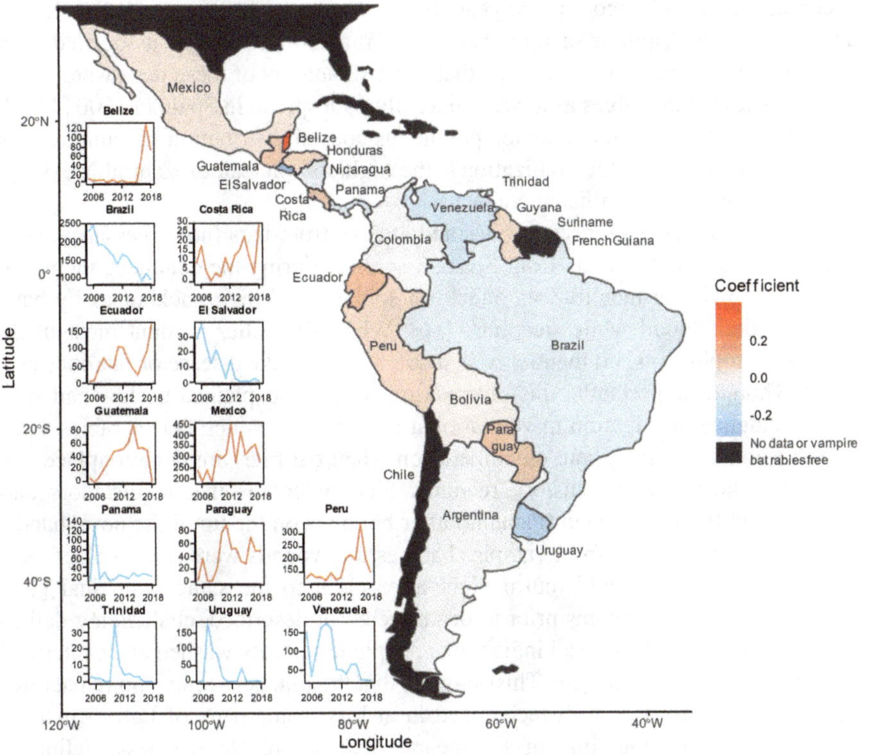

Fig. 8.1 Reprinted from Benavides et al. [6] Geographic trends in reported rabies outbreaks across Latin America. Countries are colored according to the coefficient estimated from a Poisson-distributed generalized linear model, relating year to the number of reported rabies cases in livestock (N = 40,287) and *Desmodus rotundus* (N = 214). Reds indicate increases and blues indicate decreases in rabies from 2005 to 2018. The inset panels show time series for countries with statistically significant trends ($p < 0.05$). Countries without reports of rabies transmitted by *D. rotundus* or where data were unavailable are colored in dark gray. (Data from the World Organization for Animal Health (OIE))

Historical Perspective on Vampire Bat Rabies and Management

The fossil record indicates that relatives of contemporary vampire bat species have occupied the Americas since at least the Pleistocene period, 2.4 million years ago, so indigenous civilizations likely had contact with these species [14, 15]. Vos et al. reviewed available resources and the few remaining codices (most were destroyed) from the Mayan people that occupied Mesoamerica (southern Mexico and Central America) from 1800 BC to about the ninth century AD [16]. This material indicated the Mayans were familiar with vampire bats, but no clear reference was found to link rabies or any other infectious disease associated with bats to bat bites. Likewise, in early manuscripts, written about the life and customs of Aztec people that occupied Central Mexico for 200 years before the Spanish conquest (1300–1521), no definitive descriptions or references to a "rabies-like" disease associated with bats was mentioned, despite the fact that Spanish authors of these texts were likely familiar with canine rabies as it was reportedly widespread in Spain in 1500 [16]. A lack of overlap between the contemporaneous coastal distribution of vampire bats and the heart of the Aztec civilization – the highland plateau of Central Mexico – may have precluded significant contact.

Several indirect references to bites and attacks from bats (not necessarily vampire bats) are found in texts from Spanish soldiers during the conquest, including one from 1527 that mentions an attack on soldiers and their packhorses by bats, "sucking their blood while sleeping" [16]. While this latter account most likely refers to vampire bats, no mention was made of rabies-like disease or fatalities as a result. Though, importantly, this observation could be explained by low and sporadic incidence of infection in vampire bat populations, which makes rabies virus transmission a relatively rare outcome, even when bat bites are commonplace. An early treatise on natural history recounts fatal outcomes in conquistadors and Christians in the Darien Gap (Panama) after bat bites on the tip of the nose, hands, and feet, most certainly from vampire bats, as the wounds were said to bleed profusely and the bats would return night after night to the same individual [17]. Although disease symptoms prior to death were not described, challenging definitive attribution to rabies, local indigenous people reportedly washed and cauterized these wounds to avoid disease. This suggests that the link between vampire bat bites and human disease was well recognized in at least some parts of Latin America, before and during the time of European colonization. Nevertheless, definitive accounts of rabies in Latin America were not reported until 200 years later in Mexico, and these cases were attributed to canine rabies [18]. It is quite possible that early explorers failed to recognize the more paralytic or "dumb" form of rabies, which is more typically associated with VBR, compared to the more familiar, "furious" form transmitted by dogs. Alternatively, victims may not have realized they had been exposed to a bat, especially if bitten while exhausted during deep sleep with limited pain and small peripheral lesions, and a link to vampire bats was unnoticed [19].

Initial phylogenetic analyses suggested that bat rabies virus lineages in the Americas arose about the time of European colonization, most likely due to introduction of rabies virus from domestic animals [20]. However, more recent studies and analyses provide evidence that rabies virus lineages of New World bats may have been present earlier [21, 22]. Estimating the exact origin of bat rabies viruses from molecular sequence data is challenging because substitution saturation in viral genomes can lead to systematic under-estimation of time until the most recent common ancestor [23]. As such, the origins of VBR and more broadly bat rabies in the Americas remains an unresolved mystery. Despite this uncertainty, European colonists played a major role in the expansion of vampire bat populations, and consequently, the spread of VBR, due to deforestation of the Americas and intensified livestock production. The first documented reports of rabies most likely associated with vampire bats occurred in southern Brazil from 1906–1908, resulting in the death of thousands of cattle, horses and mules [24]. However, the outbreak was not linked to vampire bats until Haupt and Rehaag demonstrated the presence of Negri bodies indicative of rabies virus in the brains of affected bats [25]. Following rabies outbreaks in livestock in 1925, and in humans in 1929 in Trinidad, experimental studies established the first definitive evidence of rabies virus transmission by vampire bats [26, 27]. In the late 1940s, the highly fatal paralytic disease of cattle (derriengue), prevalent along the Mexican coast since the 1900s, was determined to be rabies virus transmitted from vampire bats [28]. Thus, it appears that transmission of VBR to humans and livestock was first recognized about the same time (early 1900s) in Brazil, Mexico and the Caribbean island of Trinidad. This synchrony may reflect increased contact between humans, livestock, and vampire bats as all of their populations in these regions increased, improved inter-regional communication once the threat of VBR was initially identified, or coincidence.

By 1965, annual losses of approximately half a million head of cattle in Latin America from VBR were reported, and more than 2.7 million cattle were vaccinated yearly [29]. Early attempts to control the disease included vaccination of livestock with a fixed virus developed in the brain of calves or horses (phenolized emulsion of 20% brain tissue in glycerol and water) [30]. Other attempts included bat-proofing stables and keeping lanterns burning at night [31]. For humans, protection against bat bites included use of lights at night, barriers of spiky branches, wire screening, and mosquito nets over beds. Bats (vampire and other species) were destroyed by poisoning or fumigating roosts, applying strychnine to bitten animals so as to poison returning bats, capture and destruction of bats in traps or nets, and shooting. The 1939 mural by Pedro Martinez (which can be viewed at Bellas Artes, San Miguel de Allende, Mexico) and the "Corrido del Derriengue" written during the 1963 outbreak in Oaxaca epitomizes the public's fear of and battles with vampire bats during this period (Fig. 8.2) [32]. Destroying roosts and other tactics employed to control vampire bats were not very effective in the long term, given their fairly rapid expansion throughout Latin America and the increased incidence of VBR in both humans and livestock in the 100 years hence.

Current Management Practices

Current management of the common vampire bat relies predominantly on efforts to reduce population sizes (Figs. 8.3 and 8.4). The immediate management objective is to reduce bat bites on humans and domesticated animals with the secondary objective of reducing either the prevalence of rabies in local bat populations or the incidence of rabies-induced mortality in other species.

Methods for vampire bat population control span both efforts carried out by affected communities and government-led or sanctioned programs. Informal efforts include direct capture and killing of bats, typically using improvised bat killing devices, such as torches, wooden poles, or branches wrapped in thorns that are used to swat bats as they exit roosts (Fig. 8.3). Informal efforts also include activities that

CORRIDO DEL DERRIENGUE

A corrido I'm going to sing
Hope nobody gets offended
I only want to tell you, about that "derriengue"
This "derriengue" business is really quite a thing
Believe me at the rate it's going, we'll be left with less than nothing.
With this sad death, you see no money flow
It's already done in the cattle of Ajal and Rincon Vaquero.
The death moves, killing cow and horse, just see it go
Around Almoloya, the Haciendita, and the Barrio
When you pass the plum orchard you see the buzzards low.
They're picking away at the cows of Don Waldo Cabrero.
The cows are dying, both calves and old cattle
In the place called Collolapa.
They die straight on. Not even the tail's for you.
They're already dying by Paso Falso
And now it's at Chigola.
Two big cows died, the hides couldn't be used, oh no,
One died at Reforma, the other at Sardinero
Look, dear God, you've killed enough it seems.
Gone is Uncle Martiniano Rueda's prize ox team.
And Mr. Angel Dolores, well, he talked to me right straight.
With that damned derriengue, ten of his cows met their fate.
We're lamenting it now, it's not even the exception but the ru.
It's killed Uncle Pedro's ox, and old Lolo's bull.

Many are complaining, no action, just words again.
What we need is the fame of some great veterinarian.
When they go out to sell meat,
Man, don't even turn it over.
They say it's "that derriengue"
Who'd eat that now-or ever.
With that I take my leave, but not from this place.
I want a stiff battle against this worldwide plague.

Fig. 8.2 (**a**) Vampire bat mural by Pedro Martinez (1939) at the Bellas Artes in San Miguel de Allende, Mexico. (Photo credit Tonie Rocke). (**b**) Lyrics to Corrido del Derriengue written by local rancher Rosendo Reyes during the 1963 rabies outbreak in cattle in Oaxaca, Mexico. (From Ref. [32])

destroy or damage bat roosts, such as lighting fires within, dynamiting or gassing caves, sealing off exits with bats inside, or cutting down trees that are known to harbor bats [33–35]. These informal efforts are typically illegal and discouraged by governments due to associated animal welfare issues, the high likelihood that implementation by untrained individuals will kill non-target and beneficial bat species, and their general environmental destructiveness. While more extreme methods such as dynamiting and gassing bats are thought to be infrequent, roost burning and improvised bat killing remain widely and frequently used until the present day [36].

State-sanctioned efforts to reduce vampire bat populations by poisoning have been in place at least since the early 1970s and have used a variety of slow acting toxicants including chlorophacinone, diphenadione, brodifacoum, and warfarin [37–39]. Vampiricides are either topically applied to bats captured during foraging, enabling spread to conspecifics during the frequent social grooming characteristic of *D. rotundus* (Fig. 8.4), or injected systemically to cattle at concentrations that are innocuous to cattle but lethal to bats upon consumption of blood meals [40]. Numerous studies have demonstrated that both methods dramatically reduce vampire bat populations, with estimates of 10 to 20 bats killed per treated individual and up to 90% reductions in the intensity of bat bites on livestock [36, 41]. Given that vampire bats tend to roost separately from beneficial bats, risks to non-target bat

Fig. 8.3 Destruction of vampire bats by: (**a**) roost destruction; (**b**) capture in nets; (**c**) improvised swatting implement; (**d**) beating; and (**e**) burning of roosts

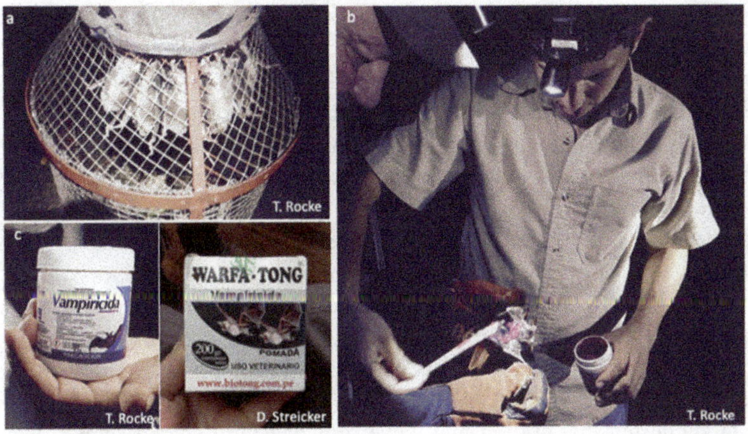

Fig. 8.4 State sanctioned culling of vampire bats. (**a**) Captured bats waiting treatment; (**b**) application to vampire bats; and (**c**) two different kinds of "vampiricida"

species are generally considered low and significant secondary poisoning has not been reported [42]. Nevertheless, given recurrent evidence of occasional virus transmission between vampire bats and other bat species, studies of inter-species contacts in additional environmental contexts are encouraged [43–46]. Moreover, effects of consumption of poisoned bat carcasses on scavengers has received relatively little attention, but evidence from anticoagulant rodenticides indicates this could be an important source of wildlife mortality across taxa [47–50]. As a result of the apparently low risk to other species, and combined with the slow reproductive and population growth rates of *D. rotundus*, vampiricides were initially praised as a promising solution for vampire bat control that were envisioned to require applications only every 3 to 8 years [51, 52].

Limitations of Current Vampire Bat Management Strategies

Vampiricide application initially seemed very promising, but evidence accumulated over the last few decades has shown that poisoning is insufficient as a singular tool for vampire bat management and rabies prevention. Due to the remote, dispersed and often undocumented locations of vampire bat colonies, many populations are effectively unreachable by poisons. Perhaps as a consequence of these unmanaged populations, range expansions of either vampire bats or variants of rabies virus associated with vampire bats have been observed in both the northern (Mexico) and southern (Peru, Argentina, Uruguay) limits of the reservoir species range [53–57]. Similarly, while some countries have reported reductions in reports of rabies outbreaks in livestock over the past decade, others, including Peru, Belize, Mexico and Costa Rica, have seen increased burdens despite national bat culling programs and substantial public or private investments in livestock vaccination. [6, 58] Reliance on state-led culling programs also introduces socioeconomic conflicts between agriculture and bat conservation because real or perceived neglect of locally affected communities may increase the use of illegal efforts to kill bats (Fig. 8.3). Thus, while culling vampire bats has the potential to protect other bat species from such vigilante efforts, the practical inability of governments to achieve levels of control sufficient to appease agricultural communities using existing technologies jeopardizes beneficial bats.

Despite the widespread use of vampiricide for at least 50 years, astonishingly few investigators have tested whether culling vampire bats reduces rabies virus cases in livestock. In one rare example, Fornes et al. exterminated vampire bats by gassing roosts with cyanide in a 15,000 km^2 area of Argentina that was in the expected path of a southbound rabies epizootic [35]. Rabies cases occurred in cattle, both in the culled area and in a surrounding (non-culled) surveillance area, though the data suggested that culling may have impeded viral spatial spread. However, even if validated, it is dubious that the massive reductions in vampire bat populations achieved using cyanide (95–100%) are ethically acceptable or could be applied on sufficient geographic scale to be considered for management. While we are not aware of other experiments that have evaluated lower levels of population reduction on livestock rabies burden, mathematical models have indicated that near eradication of vampire bats over large geographic areas may be necessary to substantially reduce risk [59, 60].

A major challenge in interpreting existing data on the effects of culling vampire bats on VBR incidence in livestock is that spatial or temporal replications have not been attempted. Consequently, hypothesized effects of culling cannot be separated from other factors that could impede viral spatial spread (i.e., landscape features), influence detection (e.g., variation in livestock density or vaccination coverage) or change occurrence (i.e., natural fluctuations that are unrelated to culling). For example, in the Fornes et al. study, the greatest reduction of bovine rabies occurred in the eastern part of the culled area, but cases were also absent from the non-culled area farther east, meaning the apparent reduction could have at least in part reflected the natural course of the epizootic [35]. Clearly, replicated experimental interventions or analyses carried out at sufficiently large spatial and temporal scales to separate effects of culling from natural variation in epidemic cycles would be beneficial. Such studies would offer a secondary benefit of enabling the development of evidence-based recommendations regarding the level of culling necessary to achieve rabies prevention targets. These recommendations are currently lacking. Without such knowledge, cost-benefit analyses of culling, in relation to other emerging management strategies (see below), will be largely futile.

Ecological Complexities of Current Management Strategies for Vampire Bat Rabies

Although culling vampire bat populations is intended to reduce the local incidence of VBR in bat populations, its true impact remains poorly understood. One study carried out over four years in Peru found that culled colonies (both sanctioned and ad hoc) had similar or higher anti-rabies virus seroprevalence compared to colonies that were not culled, suggesting that culling did not reduce rabies virus circulation and, controversially, may have increased rabies virus exposures [61]. However, the observational nature of that study precluded concrete conclusions on the value of culling, and to our knowledge, no subsequent field studies have attempted to relate rabies virus prevalence or seroprevalence in bats to culling efforts.

Theory and examples from other wildlife disease systems provide enlightening context for whether culling may be expected to reduce or increase VBR. At the most fundamental level, population reduction is expected to be most successful for disease control when prevalence is tightly linked to population density (i.e., density dependent transmission) [62]. While density dependence commonly influences the prevalence of human infections, a variety of factors specific to wildlife, such as spatial and social structure, preclude strict thresholds relating host density to pathogen extinction, and they rarely occur [63]. Consistent with this expectation, a metagenomic study showed that the richness of viruses in vampire bat colonies was unrelated to the size of colonies [64]. In the specific case of rabies, the relationship between population density and transmission is further diminished by bite-mediated viral transmission: a rabid animal may bite a similar number of conspecifics regardless of the number available. This phenomenon allows for viral perpetuation or invasion even at low population densities and decouples bat population size from the transmission rate of rabies virus. Indeed, for both dog and vampire bat rabies, relationships between host density and incidence appear tenuous [61, 65].

The ecology of vampire bats and the nature of disease maintenance may further complicate the relationship between culling and rabies virus transmission. Based upon field studies in wild bats, epidemiological studies in livestock, and rabies virus phylodynamics, VBR is maintained through recurrent extinction and re-invasion dynamics at the local level (i.e. metapopulation dynamics) [54, 59, 66]. As a result of these spatially asynchronous dynamics, culls are unlikely to affect the right populations at the right time, akin to a game of 'whack-a-mole' [67]. The dependence of long-term rabies virus maintenance on bat dispersal further supports the view that increased mixing among colonies, due to culling or dispersal of bats, may increase rabies virus incidence, as observed in other wildlife disease systems, most notably bovine tuberculosis in badgers [59, 68]. Whether such counterproductive dispersal effects occur for vampire bats is currently untested but could be reasonably expected following activities such as roost destruction or attempts to kill bats within their roosts.

New Approaches to Managing Vampire Bat Populations and Rabies

The continued increase in both the occurrence and distribution of VBR, despite ongoing control programs, underscores the need for new approaches to curtail both viral transmission and the density of vampire bat populations. The challenge, however, is to develop tools or strategies that are both financially and practically sustainable in the long-term, that are ecologically sound, and that balance conservation goals. Methods of population control are possible but would need to be targeted specifically to vampire bats, and not bats in general, as this approach could be harmful to the many beneficial bats species that reduce insects, disperse seeds, and pollinate plants in Latin America. Integrated strategies to reduce rabies virus incidence in vampire bats, as well as their livestock and human targets, could potentially curtail viral transmission, much as rabies vaccination of wild canids has reduced disease in domestic dogs, providing significant public health benefits [69]. Several such strategies have been proposed, as discussed below, and research is progressing on a number of fronts.

Oral Rabies Vaccines for Bats For several wildlife diseases that affect either humans or domestic animals, oral vaccination has superseded the use of culling for disease management, as culling is often counterproductive or otherwise unacceptable. Wild carnivores have been vaccinated against rabies since the 1980s in both Europe and North America by distributing attractive baits that contain attenuated rabies viruses or recombinant vaccinia virus expressing the rabies virus glycoprotein G (V-RG) [69–71]. As a direct result of these efforts, rabies in both domestic animals and humans declined [69]. Modeling this approach, other orally delivered wildlife vaccines have since been developed, including chimeric viruses for controlling classical swine fever in wild boar and vaccinia-vectored vaccines against bovine tuberculosis in European badgers [72, 73]. Critical to these efforts has been the use of vaccine constructs that can be safely, practically, and efficiently delivered to large numbers of wildlife, and that are effective via the oral route.

Vaccination could be especially effective for managing rabies in bats. The strategy behind vaccination is to induce sufficient population immunity to reduce viral transmission rates to unsustainable levels. Vampire bats have slower population growth rates than many carnivores, with females giving birth to one pup per year, which would preserve population level immunity [74]. Sero-surveys and experimental infection studies have demonstrated that naturally acquired immunity in bats, due to previous rabies virus exposure, exists in some populations already and is not uncommon [61, 75, 76]. Furthermore, the estimated natural reproductive number of VBR transmission (R_0) among bats is low (0.06–1.03) [59]. Given that vaccination efforts required for elimination are inversely proportional to R_0, even relatively low levels of vaccination have the potential to drive rabies virus occurrence in vampire bats to the point of local extirpation [77].

So far, rabies vaccines have not been tested in free-ranging bats, but several laboratory studies in captive bats have demonstrated that vaccination might have some utility in managing VBR. Vampire bats, orally vaccinated with V-RG (10^8 CCID50/ml cow blood), were protected when vaccinated 18 or 30 days prior to lethal rabies challenge, but, expectedly, not when vaccinated on the same day of challenge or 5 days post challenge [78]. Other routes of vaccine administration, including intramuscular, scarification, and aerosolization, were also found to protect vampire bats against rabies virus challenge [79]. However, the use of vaccinia as a vaccine vector for vampire bats is not ideal, as it could potentially cause poxvirus disease in humans and livestock if bitten by recently vaccinated bats [80, 81].

More recently, raccoon poxvirus (RCN), which is considered a safer alternative to vaccinia virus for humans and other animals, was utilized to develop a new rabies vaccine expressing a mosaic glycoprotein (MoG) [82]. Vaccination with RCN-MoG (5×10^7 pfu), orally or topically in glycerin jelly, was shown to protect big brown bats (*Eptesicus fuscus*) from rabies virus challenge. The MoG protein was developed with an *in silico* antigen designer tool using sequences available from the rabies virus Phylogroup I G protein, and thus may represent viral variants more likely to occur in bats (vampire bats as well as other American bat species), though its improvement over traditional glycoprotein G vaccines has not yet been demonstrated. In vampire bats, RCN-MoG, delivered orally shortly after bats started dying from a natural rabies virus exposure, did not improve survival, however rabies virus shedding in saliva (detected by PCR) was significantly reduced in vaccinated bats compared to controls [75]. The same phenomenon of reduced viral shedding was also observed in rabid vampire bats that had been vaccinated several months before challenge with a coyote rabies virus variant [83]. Although more research is needed to confirm these findings and determine the mechanism, reduced viral shedding in saliva as an outcome of vaccination would presumably reduce rabies virus transmission among vampire bats, and from bats to humans and other animals. Pre-exposure prophylaxis, if it occurs functionally in *D. rotundus* as in vaccinated *E. fuscus*, could create population level-immunity that further reduces viral transmission [82].

Delivery of Vaccines to Bats Before field testing to assess effectiveness of vaccination as a method to control VBR can proceed, additional studies are needed to confirm vaccine safety in non-target species and livestock, to determine duration of immunity in vaccinated bats, and to meet other regulatory requirements. Such assessments could likely be completed within the next decade. In the meantime, practical and efficient methods and media for delivering vaccines and possibly other pharmaceuticals (i.e., contraceptives) to large numbers of vampire bats are being tested and optimized. Like culling with vampiricides, the most feasible option is topical application resulting in oral delivery as the bats groom themselves and each other. In two studies in captive vampire bats, highly concentrated suspensions of V-RG mixed in petroleum jelly paste were topically applied to the backs of bats that were then returned to their cages with other untreated bats [84, 85]. Several of the untreated cage-mates survived rabies virus challenge, indicating viral vaccine transfer between bats. These studies provided initial proof of concept for delivering vaccines to bats, but the exact dosage of vaccine virus was not reported, and no details were provided on the survivability of the vaccine in paste over time.

Ideally, the medium used for delivering vaccines and other oral pharmaceuticals would need to adhere well to the fur of the bats but remain viscous enough to ingest easily and transfer among roost-mates, while maintaining integrity of the vaccine for as long as possible. Glycerin jelly has been tested as a possible delivery medium, as its viscosity is easily adjusted and vaccine titer has been shown to last longer than in other vehicles [86]. Other media, such as chitosan, are also being considered. Preliminary studies using glycerin jelly mixed with a biomarker Rhodamine B (RB), but without vaccine, were conducted in Peru to assess uptake in treated bats and transfer to untreated bats [60]. Upon ingestion by mammals, RB is detected easily in hair and nails using ultraviolet microscopy, making it a useful tool to assess rates of transfer, to compare different methods of application, and to evaluate uptake by non-target species [87]. In two colonies, where 43% and 50% of vampire bats (estimated populations sizes of 257 and 207, respectively) were treated topically with RB-marked glycerin jelly, the biomarker was later detected in the hair of 72% and 83% of untreated bats in each colony, indicating the jelly was successfully transferred among bats [60]. Different media, volumes, rates of application, and methods (hand versus automatic sprayers) would need to be tested in future studies to optimize vaccine distribution strategies. These biomarker studies would be useful for predicting the effectiveness, economic, and logistical feasibility of vaccinating vampire bats in future rabies control programs. Modelling studies indicated that use of an oro-topical, transferrable rabies vaccine (like RCN-MoG) in vampire bats could potentially increase population-level immunity up to 2.6 times beyond the same effort using injectable, protein-based vaccines, thereby reducing the probability, size, and duration of rabies outbreaks, even at low, but feasible, levels of vaccine application [60]. These models also favored vaccination over culling for reducing rabies in vampire bats.

Transmissible Vaccines As with any infectious disease in free ranging wildlife, delivering vaccines to sufficient numbers of inaccessible individuals to dampen transmission is an outstanding challenge for vampire bats and rabies. Encouragingly, a growing body of theoretical work suggests that even relatively small amounts of vaccine transfer among individuals can have considerable benefits for reducing pathogen transmission in reservoir hosts and consequent risks to other species [88, 89]. However, the relatively sedentary nature of vampire bats implies that such 'one-step' vaccines that exclusively spread from treated individuals (e.g., gel-based vaccines) or rely on simultaneous treatment of many individuals (e.g., aerosolized vaccines) are unlikely to benefit untreated populations. Efforts targeting regional elimination would therefore require repeated, geographically dispersed delivery campaigns which are likely to be financially costly or impractical in areas where populations are inaccessible. A provocative potential solution to reduce the need for recurring vaccination campaigns would be 'multi-step' vaccines that spread not only from treated bats, but also from untreated bats through an infectious process. This capability for true vaccine transmission, rather than transfer, would enable vaccines to spread beyond the initial population treated, and in principle, could be sustained in vampire bat populations over long timescales, reducing the need for re-deployments (Fig. 8.5). Indeed, when such vaccines transmit more efficiently than the virus they target, they are predicted to displace the target virus, creating an inexpensive route to rabies elimination [90]. In the case of vampire bat rabies, where estimated R_0 values range from 0.06–1.03 and the virus shows evidence of frequent natural extirpations in bat populations, a relatively low amount of vaccine transmission might reasonably be expected to eliminate rabies [59, 66].

As with transferrable vaccines, viral vectors are the most promising platform for generating transmissible vampire bat rabies vaccines. Yet, existing poxvirus vectors are unlikely to meet the differing safety and performance requirements that arise from vaccine transmission. From a safety perspective, poxviruses have a wide host range and must be attenuated genetically to reduce pathogenicity. Consequently, sustained transmission would open the door to reversion to wild type virulence, which might affect both vampire bats and non-target species. At a more fundamental level, available evidence indicates that existing RCN vectors are unlikely to transmit among bats, making them safe vectors for transferable vaccines but unsuitable to vector transmissible vaccines. Ideal vectors for transmissible vaccines would be intrinsically benign to reduce risks of reversion to virulence sometimes observed in live attenuated vaccines, such as oral polio vaccine [91]. For inherently benign viral vectors, the most likely evolutionary consequence of vaccine transmission is dropout of the antigenic insert (i.e., reversion to wild type), which is a far lesser concern [92]. A second requisite for vector design is high host specificity, which would reduce safety risks for non-target species. Finally, amenability to efficiently express foreign proteins from rabies virus while sustaining the capacity for

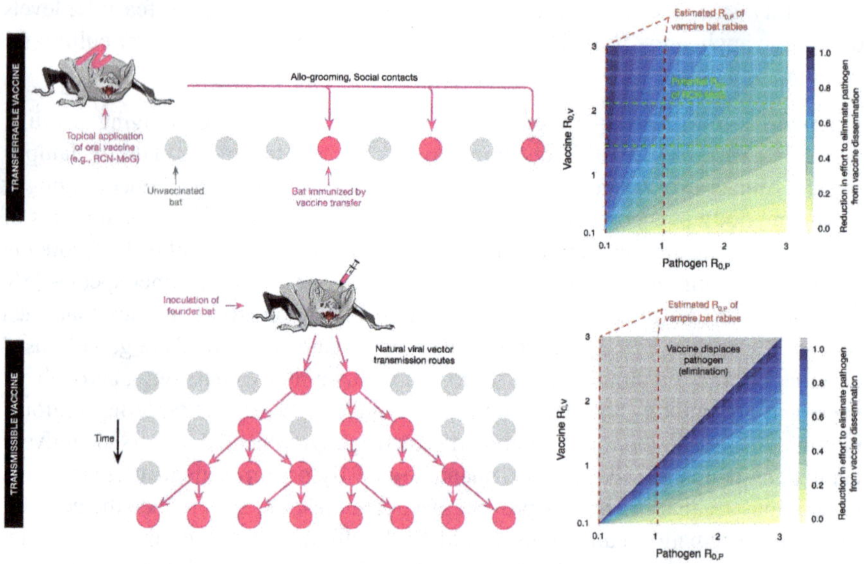

Fig. 8.5 Potential self-disseminating vaccine strategies for vampire bats. The top row illustrates how transferable oral vaccines immunize treated bats and those with social contacts that lead to vaccine ingestion. Bats immunized by vaccine transfer are assumed not to spread vaccines further. The right panel shows the predicted reduction in vaccination effort required to eliminate a pathogen from its host population, relative to a conventional (non-disseminating) vaccine. Darker colors indicate less effort to eliminate the pathogen. Red and green dashed lines indicate the approximate pathogen reproductive number ($R_{0,P}$) of VBR and the vaccine reproductive number ($R_{0,V}$) of RCN-MoG from Blackwood et al. [59] and Bakker et al. [60], implying that observed, low levels of vaccine transfer could reduce rabies virus transmission substantially compared to a conventional vaccine. The bottom row illustrates how transmissible vaccines spread infectiously from initially inoculated bats and those infected through vaccine transmission. The right panel shows predicted effects of vaccine transmission on the vaccination effort required for pathogen elimination. The gray area indicates the values of $R_{0,P}$ and $R_{0,V}$ where a single introduction of a transmissible vaccine would be predicted to eliminate a hypothetical pathogen. Panels showing reduction of vaccination effort compared to conventional vaccines were created using the equations of Nuismer and Bull [90]. Vampire bat illustration by K. Sutliff/Science. (From Hayman Science 354: 1099, 2016. https://doi.org/10.1126/science.aaj1818. Reprinted with permission from AAAS Ref. [133])

bat-to-bat spread would be vital. Betaherpesviruses are increasingly recognized to meet these criteria, while also offering an additional benefit of establishing lifelong persistent infections, such that reactivation of latent vaccine strains might naturally boost protection [88]. Considering the long lifespan of vampire bats (up to 17 years in the wild), conservation of this feature of betaherpesvirus biology in vaccine strains would have obvious benefits in reducing booster requirements [93]. Moreover, betaherpesviruses super-infect individual hosts, potentially allowing vaccine strains to invade populations where wild-type betaherpesviruses are prevalent.

Excitingly, recent work has identified a betaherpesvirus in Peruvian vampire bats that so far appears to meet the above prerequisites for a safe and effective transmissible vaccine [43]. Samples of this virus, termed *Desmodus rotundus* betaherpesvirus (DrBHV), formed a nearly vampire bat-exclusive monophyletic clade (a single *Artibeus literatus* was putatively infected), suggesting a vaccine version would have limited capability to infect non-vampire bats, or given the typical host-specificity of betaherpesviruses, would be extremely unlikely to infect non-bat species. In vampire bat populations across Peru, DrBHV infected >95% of individual vampire bats, indicating potential for high vaccine coverage if recombinant vaccines transmit similarly to wild type. Multi-strain infections within individual vampire bats were also observed, suggesting that a DrBHV-vectored rabies virus vaccine could superinfect wild-type DrBHV-infected individuals. Work spanning vaccine design, experimental verification of biological properties and efficacy in preventing rabies virus transmission, host restriction, and safety profiles will be required before transmissible vaccines can be realistically considered for real-world management of vampire bat rabies. In addition, given that only low levels of vaccine transmission are likely to be required to meet rabies management goals, methods to diminish DrBHV transmission or engineer programmable dropout of antigenic inserts could be explored [92, 94].

Public Acceptance of Rabies Vaccination of Bats As with other wildlife vaccination efforts, public support for rabies vaccination of vampire bats would be critical and highly dependent on educating farmers and other stakeholders of its benefits to both livestock and public health. This may prove difficult given the hostility many farmers feel towards vampire bats, as even in the absence of rabies, repeated bat bites cause anemia, secondary infections, lowered milk production, and even death of affected livestock [95]. Concerns have been raised that by protecting bats against rabies, vaccination would increase bat populations, although no evidence exists to support this assumption. Modeling indicates the probability of developing a lethal rabies virus infection is quite low for vampire bats (~10%), and most exposures are subclinical and potentially immunizing [59]. Thus, it is unlikely that rabies vaccination of bats would result in substantially increased survival and abundance. Even so, to win public acceptance, some means of controlling vampire bat populations will likely be needed through targeted fertility control or other strategies.

Use of replicating, genetically modified viruses as vectors adds a further public relations challenge for transmissible vaccines as viral mutations and spread may be perceived to carry hypothetical risks that are difficult to quantify [96]. Should they be demonstrated safe and effective in controlled settings, major communication efforts would be needed to help regulators and the general public to understand the benefits of releasing transmissible vaccines to wild bats. Involvement of social scientists from early in the vaccine development pipeline would be vital. Fortunately, transmissible disease control technologies and those that rely on genetic engineering are becoming mainstream. Examples span efforts to use CRISPR to engineer resistance to the causative agent of Lyme disease in wild mice, establishment of *Wolbachia* symbionts in mosquitoes to block arbovirus transmission, implementation of gene drives to manage malaria mosquito populations, and use of transgenic cotton and sterile insect releases to eliminate pink bollworm, an historically important crop pest [97–100]. These efforts provide frameworks to successfully engage stakeholders from across sectors. Given the potentially revolutionary benefits for management of VBR, similar investments in transmissible vaccines warrant consideration. Indeed, such work is already happening for other wildlife diseases with human health or agricultural effects, including vaccines targeting the rodent hosts of Lassa fever virus, intermediate ape hosts of ebolavirus, and bovine tuberculosis in European badgers [90, 101]. Yet, in all of these host-pathogen systems, success is contingent on achieving the level of biological understanding of pathogen reservoirs and transmission dynamics that is already well-established for rabies. As such, VBR has been singled out as an exceptionally promising system within which to refine transmissible vaccine technologies [90, 102]. Following on the initially controversial, but ultimately successful use of oral vaccines for control of rabies in carnivores, if regulatory challenges can be overcome, VBR may similarly pave the way for the next generation of self-disseminating wildlife vaccines [103].

Fertility Control of Vampire Bats A promising tool for managing vampire bat populations is fertility or reproductive control through the use of oral contraception [104–106]. Interest in the use of contraceptive methods for managing wildlife has increased, yet bias is apparent in the current literature in addressing individual-level rather than population-level effects of fertility control [107]. This bias limits our understanding of potential consequences of fertility management on populations. Reproductive control can possibly alter disease dynamics, genetic selection, reproductive behaviors, group dynamics, dispersion, and survival [107–116]. The directionality of these responses appears highly variable across species and systems, often leading to unintended consequences. For example, in brushtail possums, fertility control can increase male immigration, equaling the rates of birth loss from fertility control [111]. Population-level effects of fertility control in vampire bats would need to be explicitly studied in relation to rabies virus dynamics, because immigration plays a key role in the perpetuation of rabies virus among vampire bat populations [59]. However, current methods in development for vampire bats are promising in that they can reduce fecundity for both males and females, increasing the likelihood of reduced birth rates [105]. From an epidemiological standpoint, this control method would bias the age structure towards adults that are more likely to harbor immunity from previous rabies virus exposure, potentially reducing the prevalence and spread of the virus [6, 59]. This outcome differs from culling because the dependence of culling on social grooming (more common among adults) for vampiricide spread and the naturally adult-biased age structure of vampire bats would tend to promote relatively higher proportions of young, rabies-susceptible bats. Indeed, culled vampire bat colonies have been observed to have higher proportions of juveniles, the age class thought to be most important for viral spatial spread [12, 61]. By preventing births and lowering the proportion of rabies susceptible bats, fecundity reduction could alleviate problems with bat bites, while also improving prospects for rabies control more effectively than culling.

Integrated Strategies Strategies for controlling VBR that do not target the dispersal or reduction of vampire bat populations are varied. Efforts aimed at reducing VBR are often focused on livestock, where substantial economic losses are incurred every year [2]. A known effective measure of controlling rabies in livestock is vaccination [117]. While several bovine rabies vaccines are commercially available, only a minority of livestock are vaccinated each year [1, 7, 117, 118]. Several limitations to vaccinating all susceptible livestock exist: vaccines can be costly, require strict storage methods (i.e. low temperatures), as well as training for safe delivery into animals. Most vaccines currently available also require two doses or yearly boosters to elicit protective immunity, depending on the age of the animal [119]. Because many of the most affected populations are in remote geographic regions, these limitations represent significant challenges to widespread livestock vaccination, despite the economic efficiency of vaccination compared to bat control programs [1, 2]. Other methods of protecting livestock include bat exclusion or deterrent devices. Housing cattle overnight in enclosed spaces has been found to be effective in reducing bat bites but requires significant investment in infrastructure, which small-scale farmers may be unable to make [120]. Vampire bat deterrents, such as lighting of corrals overnight, or the use of ultrasonic noise producers have also been proposed as potential methods of protecting livestock. (reviewed in [6]) Additionally, a non-biological sound lure for the selective capture of rabid bats has been tested in California [121]. These methods present issues associated with cost, access to electricity, potentially short-lived benefits, and unintended consequences on both target and non-target bat populations [6]. Lastly, land-use and livestock management have been hypothesized to significantly alter risk of bites from vampire bats. Unfortunately, the current evidence is mixed as to whether grazing cattle versus corralling them overnight reduces their risk of depredation by vampire bats [122, 123]. The evidence may be conflicting because the risk to livestock can be dependent on the extent of surrounding forest fragmentation, precipitation seasonality, land-use changes, and proximity to vampire bat roosts, as incidence of vampire bat attacks is directly correlated with bat abundance [120, 123–127]. Deforestation and the intensification of livestock rearing in previously forested areas warrant consideration in the management of VBR [117, 127, 128].

Livestock use and management can also profoundly alter the risk of VBR in humans. The sudden removal of livestock from an area can lead to humans being targeted at significantly higher rates, and the presence of livestock may conversely reduce the risk of depredation of humans [129, 130]. The addition of livestock as an intervention method to reduce vampire bats targeting humans is not necessarily advisable, however, as the increase in available resources may ultimately increase vampire bat populations [6]. Preventing bat entry through structural changes to houses and the use of mosquito nets may provide more effective methods of protecting humans from bat bites [131]. As discussed in Benavides et al., these methods may be financially prohibitive or undesirable in hot humid environments, but netting could potentially reduce not only rabies disease burdens but other vector-borne illnesses [6]. Vaccination and post-exposure prophylaxis remain the most effective measure of protecting humans from rabies virus, yet they continue to be costly in certain countries, inaccessible in remote areas, and resisted in some affected communities [132, 133].

Models of vampire bat population dynamics in response to a variety of proposed mitigation tactics, in combination with other tactics to protect humans and livestock, can help elucidate which approaches are most likely to be effective. By evaluating social and economic tradeoffs in tandem with expected population level effects, optimal strategies that take into account objectives for both humans and bats can be developed.

Disclaimer: The use of trade, firm, or product names is for descriptive purposes only and does not imply endorsement by the U.S. Government.

References

1. Anderson A, Shwiff S, Gebhardt K et al (2014) Economic evaluation of vampire bat (Desmodus rotundus) rabies prevention in Mexico. Transbound Emerg Dis 61:140–146. https://doi.org/10.1111/tbed.12007
2. Benavides JA, Rojas Paniagua E, Hampson K et al (2017) Quantifying the burden of vampire bat rabies in Peruvian livestock. PLoS Negl Trop Dis 11:e0006105. https://doi.org/10.1371/journal.pntd.0006105
3. Vigilato MAN, Clavijo A, Knobl T et al (2013) Progress towards eliminating canine rabies: policies and perspectives from Latin America and the Caribbean. Philos Trans R Soc Lond B Biol Sci 368:20120143. https://doi.org/10.1098/rstb.2012.0143
4. Schneider MC, Romijn PC, Uieda W et al (2009) Rabies transmitted by vampire bats to humans: an emerging zoonotic disease in Latin America? Rev Panam Salud Pública 25:260–269. https://doi.org/10.1590/S1020-49892009000300010
5. Fenton MB, Streicker DG, Racey PA et al (2020) Knowledge gaps about rabies transmission from vampire bats to humans. Nat Ecol Evol 4:517–518. https://doi.org/10.1038/s41559-020-1144-3
6. Benavides JA, Valderrama W, Recuenco S et al (2020) Defining new pathways to manage the ongoing emergence of bat rabies in Latin America. Viruses 12:1002. https://doi.org/10.3390/v12091002
7. Rodrigues da Silva AC, Caporale GM, Gonçalves CA et al (2000) Antibody response in cattle after vaccination with inactivated and attenuated rabies vaccines. Rev Inst Med Trop Sao Paulo 42:95–98. https://doi.org/10.1590/s0036-46652000000200006
8. World Health Organization (2013) WHO expert consultation on rabies. Second report. World Health Organ Tech Rep Ser
9. de Paula Silva N, Andrade EA, Cardoso D et al (2019) Crab fisherman communities in North Brazil: a new high risk population for vampire bat rabies. bioRxiv
10. Ferraz C, Achkar SM, Kotait I (2007) First report of rabies in vampire bats (Desmodus rotundus) in an urban area, Ubatuba, São Paulo state, Brazil. Rev Inst Med Trop Sao Paulo 49:389–390. https://doi.org/10.1590/S0036-46652007000600010
11. Castilho JG, de Souza DN, Oliveira RN et al (2017) The epidemiological importance of bats in the transmission of rabies to dogs and cats in the state of São Paulo, Brazil, between 2005 and 2014. Zoonoses Public Health 64:423–430. https://doi.org/10.1111/zph.12320

12. Streicker DG, Winternitz JC, Satterfield DA et al (2016) Host–pathogen evolutionary signatures reveal dynamics and future invasions of vampire bat rabies. Proc Natl Acad Sci 113:10926–10931. https://doi.org/10.1073/pnas.1606587113
13. Kuzmin IV, Shi M, Orciari LA et al (2012) Molecular inferences suggest multiple host shifts of rabies viruses from bats to mesocarnivores in Arizona during 2001–2009. PLoS Pathog 8:e1002786. https://doi.org/10.1371/journal.ppat.1002786
14. Gut HJ (1959) A Pleistocene vampire bat from Florida. J Mammal 40:534. https://doi.org/10.2307/1376270
15. Koopan K (1958) A fossil vampire bat from Cuba. Breviera Museum Comp Zool 90:1–5
16. Vos A, Nunan C, Bolles D et al (2011) The occurrence of rabies in pre-Columbian Central America: an historical search. Epidemiol Infect 139:1445–1452. https://doi.org/10.1017/S0950268811001440
17. Fernández de Oviedo G (1526) Sumario de la Natural Historia de las Indias. Iberoamericana Vervuert, Orlando
18. Tellez G (1978) Apuntes para la Historia de la Rabia en México. Rev Vet Mex 9:37–48
19. Bernardes Filho F, Martins G, Luchi GS, Kac BK, Nery JA, Azulay-Abulafia L, Azulay DR (2014) Multiple lesions by vampire bat bites in a patient in Niterói, Brazil – case report. An Bras Dermatol 89(2):340–343. https://doi.org/10.1590/abd1806-4841.20142996
20. Hughes GJ, Orciari LA, Rupprecht CE (2005) Evolutionary timescale of rabies virus adaptation to north American bats inferred from the substitution rate of the nucleoprotein gene. J Gen Virol 86:1467–1474. https://doi.org/10.1099/vir.0.80710-0
21. Kuzmina NA, Kuzmin IV, Ellison JA et al (2013) A reassessment of the evolutionary timescale of bat rabies viruses based upon glycoprotein gene sequences. Virus Genes 47:305–310. https://doi.org/10.1007/s11262-013-0952-9
22. Rupprecht C, Kuzmin I, Meslin F (2017) Lyssaviruses and rabies: current conundrums, concerns, contradictions and controversies. F1000Res 6:184. https://doi.org/10.12688/f1000research.10416.1
23. Duchêne S, Holmes EC, Ho SYW (2014) Analyses of evolutionary dynamics in viruses are hindered by a time-dependent bias in rate estimates. Proc Biol Sci 281:20140732. https://doi.org/10.1098/rspb.2014.0732
24. Carini A (1911) Sur une grande epizootie de rage. Ann Inst Pasteur (Paris) 25:843
25. Haupt H, Rehaag H (1921) Durch Fledermäuse verbreitete seuchenhafte Tollwut unter Viehbeständen in Santa Catharina (Süd-Brasilien). Zeitschrift für Infekt der Haustiere
26. Hurst EW, Pawan JL (1959) An outbreak of rabies in Trinidad without history of bites, and with the symptoms of acute ascending myelitis. Caribb Med J 21:11–24
27. Pawan JL (1936) The transmission of paralytic rabies in Trinidad by the vampire bat (Desmodus Rotundus Murinus Wagner, 1840). Ann Trop Med Parasitol 30:101–130. https://doi.org/10.1080/00034983.1936.11684921
28. Johnson HN (1948) Derriengue; vampire bat rabies in Mexico. Am J Hyg 47:189–204. https://doi.org/10.1093/oxfordjournals.aje.a119196
29. Acha PN (1967) Epidemiology of paralytic bovine rabies and bat rabies. Bull Off Int Epizoot 67:343–382
30. Carneiro V (1954) Transmission of rabies by bats in Latin America. Bull World Health Organ 10:775–780
31. De Verteuil E, Urich FW (1959) The study and control of paralytic rabies transmitted by bats in Trinidad, British West Indies. Caribb Med J 21:85–109
32. Baer GM (1991) The natural history of rabies, 2nd edn. CRC, Boca Raton
33. Lewin R (1986) Why dynamiting vampire bats is wrong. Science 232:24–26. https://doi.org/10.1126/science.232.4746.24
34. Greenhall AM (1970) Vampire bat control: a review and proposed research programme for Latin America. In: Proceedings of the 4th vertebrate pest conference (1970)
35. Fornes A, Lord RD, Kuns ML et al (1974) Control of bovine rabies through vampire bat control. J Wildl Dis 10:310–316. https://doi.org/10.7589/0090-3558-10.4.310
36. O'Shea TJ, Cryan PM, Hayman DTS et al (2016) Multiple mortality events in bats: a global review. Mamm Rev 46:175–190. https://doi.org/10.1111/mam.12064

37. Linhart SB, Flores-Crespo R, Mitchell GC (1972) Control of vampire bats by topical application of an anticoagulant, Chlorophacinone. Boletín la OSP
38. Thompson RD, Mitchell GC, Burns RJ (1972) Vampire bat control by systemic treatment of livestock with an anticoagulant. Science 177:806–808. https://doi.org/10.1126/science.177.4051.806
39. Arellano-Sota C (1988) Biology, ecology, and control of the vampire bat. Clin Infect Dis 10:S615–S619. https://doi.org/10.1093/clinids/10.Supplement_4.S615
40. Carter G, Leffer L (2015) Social grooming in bats: are vampire bats exceptional? PLoS One 10:e0138430. https://doi.org/10.1371/journal.pone.0138430
41. Gomes MN, Uieda W, Latorre M do RD de O (2006) Influência do sexo de indivíduos da mesma colônia no controle químico das populações do morcego hematófago Desmodus rotundus (Phyllostomidae) no estado de São Paulo. Pesqui Veterinária Bras 26:38–43. https://doi.org/10.1590/S0100-736X2006000100008
42. Wohlgenant TJ (1994) Roost interactions between the common vampire bat (Desmodus rotundus) and two frugivorous bats (Phyllostomus discolor and Sturnira lilium) in Guanacaste, Costa Rica. Biotropica 26:344. https://doi.org/10.2307/2388857
43. Griffiths ME, Bergner LM, Broos A et al (2020) Epidemiology and biology of a herpesvirus in rabies endemic vampire bat populations. Nat Commun 11:5951. https://doi.org/10.1038/s41467-020-19832-4
44. Bergner LM, Orton RJ, Broos A et al (2021) Diversification of mammalian deltaviruses by host shifting. Proc Natl Acad Sci U S A 118:e2019907118. https://doi.org/10.1073/pnas.2019907118
45. Shoji Y, Kobayashi Y, Sato G et al (2004) Genetic characterization of rabies viruses isolated from frugivorous bat (Artibeus spp.) in Brazil. J Vet Med Sci 66:1271–1273. https://doi.org/10.1292/jvms.66.1271
46. Calderón A, Guzmán C, Mattar S et al (2019) Frugivorous bats in the Colombian Caribbean region are reservoirs of the rabies virus. Ann Clin Microbiol Antimicrob 18:11. https://doi.org/10.1186/s12941-019-0308-y
47. Elliott JE, Hindmarch S, Albert CA et al (2014) Exposure pathways of anticoagulant rodenticides to nontarget wildlife. Environ Monit Assess 186:895–906. https://doi.org/10.1007/s10661-013-3422-x
48. Lohr MT, Davis RA (2018) Anticoagulant rodenticide use, non-target impacts and regulation: a case study from Australia. Sci Total Environ 634:1372–1384
49. Rattner BA, Lazarus RS, Elliott JE et al (2014) Adverse outcome pathway and risks of anticoagulant rodenticides to predatory wildlife. Environ Sci Technol 48:8433–8445
50. Dennis GC, Gartrell BD (2015) Nontarget mortality of New Zealand lesser short-tailed bats (Mystacina tuberculata) caused by diphacinone. J Wildl Dis 51:177–186. https://doi.org/10.7589/2013-07-160
51. Greenhall AM, Joermann G, Schmidt U, Seidel MR (1983) Desmodus rotundus. Mamm Species 1:1. https://doi.org/10.2307/3503895
52. C.H. (1972) Doom Spelled for Vampires. Science 176:495–495. https://doi.org/10.1126/science.176.4034.495
53. Guarino H, Castilho JG, Souto J et al (2013) Antigenic and genetic characterization of rabies virus isolates from Uruguay. Virus Res 173:415–420. https://doi.org/10.1016/j.virusres.2012.12.013
54. Benavides JA, Valderrama W, Streicker DG (2016) Spatial expansions and travelling waves of rabies in vampire bats. Proc R Soc B Biol Sci 283:20160328. https://doi.org/10.1098/rspb.2016.0328
55. Delpietro HA, Nader AJ (1989) Rabies of herbivores transmitted by vampire bats in northeastern Argentina. Rev Sci Tech 8(1):177–198
56. Johnson N, Aréchiga-Ceballos N, Aguilar-Setien A (2014) Vampire bat rabies: ecology, epidemiology and control. Viruses 6:1911–1928. https://doi.org/10.3390/v6051911
57. Piaggio AJ, Russell AL, Osorio IA et al (2017) Genetic demography at the leading edge of the distribution of a rabies virus vector. Ecol Evol 7:5343–5351. https://doi.org/10.1002/ece3.3087

58. Becker DJ, Broos A, Bergner LM et al (2020) Temporal patterns of vampire bat rabies and host connectivity in Belize. Transbound Emerg Dis 68:870–879. https://doi.org/10.1111/tbed.13754
59. Blackwood JC, Streicker DG, Altizer S, Rohani P (2013) Resolving the roles of immunity, pathogenesis, and immigration for rabies persistence in vampire bats. Proc Natl Acad Sci 110:20837–20842. https://doi.org/10.1073/pnas.1308817110
60. Bakker KM, Rocke TE, Osorio JE et al (2019) Fluorescent biomarkers demonstrate prospects for spreadable vaccines to control disease transmission in wild bats. Nat Ecol Evol 3:1697–1704. https://doi.org/10.1038/s41559-019-1032-x
61. Streicker DG, Recuenco S, Valderrama W et al (2012) Ecological and anthropogenic drivers of rabies exposure in vampire bats: implications for transmission and control. Proc R Soc B Biol Sci 279:3384–3392. https://doi.org/10.1098/rspb.2012.0538
62. McCallum H (2001) How should pathogen transmission be modelled? Trends Ecol Evol 16:295–300. https://doi.org/10.1016/S0169-5347(01)02144-9
63. Lloyd-Smith JO, Cross PC, Briggs CJ et al (2005) Should we expect population thresholds for wildlife disease? Trends Ecol Evol 20:511–519. https://doi.org/10.1016/j.tree.2005.07.004
64. Bergner LM, Orton RJ, Benavides JA et al (2020) Demographic and environmental drivers of metagenomic viral diversity in vampire bats. Mol Ecol 29:26–39. https://doi.org/10.1111/mec.15250
65. Morters MK, Restif O, Hampson K et al (2013) Evidence-based control of canine rabies: a critical review of population density reduction. J Anim Ecol 82:6–14. https://doi.org/10.1111/j.1365-2656.2012.02033.x
66. Streicker DG, Fallas González SL, Luconi G et al (2019) Phylodynamics reveals extinction-recolonization dynamics underpin apparently endemic vampire bat rabies in Costa Rica. Proc Biol Sci 286:20191527. https://doi.org/10.1098/rspb.2019.1527
67. Blower S (2008) Modelling infectious diseases in humans and animals. Lancet Infect Dis 8:415. https://doi.org/10.1016/S1473-3099(08)70147-6
68. Bielby J, Donnelly CA, Pope LC et al (2014) Badger responses to small-scale culling may compromise targeted control of bovine tuberculosis. Proc Natl Acad Sci 111:9193–9198. https://doi.org/10.1073/pnas.1401503111
69. Maki J, Guiot A-L, Aubert M et al (2017) Oral vaccination of wildlife using a vaccinia–rabies-glycoprotein recombinant virus vaccine (RABORAL V-RG®): a global review. Vet Res 48:57. https://doi.org/10.1186/s13567-017-0459-9
70. Freuling CM, Hampson K, Selhorst T et al (2013) The elimination of fox rabies from Europe: determinants of success and lessons for the future. Philos Trans R Soc Lond B Biol Sci 368:20120142. https://doi.org/10.1098/rstb.2012.0142
71. Slate D, Algeo TP, Nelson KM et al (2009) Oral rabies vaccination in North America: opportunities, complexities, and challenges. PLoS Negl Trop Dis 3:e549. https://doi.org/10.1371/journal.pntd.0000549
72. Murphy D, Costello E, Aldwell FE et al (2014) Oral vaccination of badgers (Meles meles) against tuberculosis: comparison of the protection generated by BCG vaccine strains Pasteur and Danish. Vet J 200:362–367. https://doi.org/10.1016/j.tvjl.2014.02.031
73. Rossi S, Staubach C, Blome S et al (2015) Controlling of CSFV in European wild boar using oral vaccination: a review. Front Microbiol 6:1141. https://doi.org/10.3389/fmicb.2015.01141
74. Wilkinson GS (1985) The social organization of the common vampire bat – II. Mating system, genetic structure, and relatedness. Behav Ecol Sociobiol 17:123–134. https://doi.org/10.1007/BF00299243
75. Cárdenas-Canales EM, Gigante CM, Greenberg L et al (2020) Clinical presentation and serologic response during a rabies epizootic in captive common vampire bats (Desmodus rotundus). Trop Med Infect Dis 5:34. https://doi.org/10.3390/tropicalmed5010034
76. Seetahal JFR, Greenberg L, Satheshkumar PS et al (2020) The serological prevalence of rabies virus-neutralizing antibodies in the bat population on the Caribbean Island of Trinidad. Viruses 12:178. https://doi.org/10.3390/v12020178
77. Anderson RM, May RM (1985) Vaccination and herd immunity to infectious diseases. Nature 318:323–329. https://doi.org/10.1038/318323a0

78. Sétien AA, Brochier B, Tordo N et al (1998) Experimental rabies infection and oral vaccination in vampire bats (Desmodus rotundus). Vaccine 16:1122–1126. https://doi.org/10.1016/S0264-410X(98)80108-4
79. Aguilar-Setién A, Leon YC, Tesoro EC et al (2002) Vaccination of vampire bats using recombinant vaccinia-rabies virus. J Wildl Dis 38:539–544. https://doi.org/10.7589/0090-3558-38.3.539
80. Centers for Disease Control and Prevention (CDC) (2013) Human contacts with oral rabies vaccine baits distributed for wildlife rabies management – Ohio, 2012. MMWR Morb Mortal Wkly Rep 62:267–269
81. Baxby D (1991) Safety of recombinant vaccinia vaccines. Lancet 337:913. https://doi.org/10.1016/0140-6736(91)90241-G
82. Stading B, Ellison JA, Carson WC et al (2017) Protection of bats (Eptesicus fuscus) against rabies following topical or oronasal exposure to a recombinant raccoon poxvirus vaccine. PLoS Negl Trop Dis 11:e0005958. https://doi.org/10.1371/journal.pntd.0005958
83. Cárdenas-Canales EM, Velasco-Villa A, Ellison JA et al (2022) A recombinant rabies vaccine that prevents viral shedding in rabid common vampire bats (Desmodus rotundus). PLoS Negl Trop Dis 16(8):e0010699. https://doi.org/10.1371/journal.pntd.0010699
84. Almeida MF, Martorelli LFA, Aires CC et al (2005) Indirect oral immunization of captive vampires, Desmodus rotundus. Virus Res 111:77–82. https://doi.org/10.1016/j.virusres.2005.03.013
85. Almeida MF, Martorelli LFA, Aires CC et al (2008) Vaccinating the vampire bat Desmodus rotundus against rabies. Virus Res 137:275–277. https://doi.org/10.1016/j.virusres.2008.07.024
86. Stading BR, Osorio JE, Velasco-Villa A et al (2016) Infectivity of attenuated poxvirus vaccine vectors and immunogenicity of a raccoonpox vectored rabies vaccine in the Brazilian Free-tailed bat (Tadarida brasiliensis). Vaccine 34:5352–5358. https://doi.org/10.1016/j.vaccine.2016.08.088
87. Fernandez JR-R, Rocke TE (2011) Use of rhodamine B as a biomarker for Oral plague vaccination of prairie dogs. J Wildl Dis 47:765–768. https://doi.org/10.7589/0090-3558-47.3.765
88. Basinski AJ, Nuismer SL, Remien CH (2019) A little goes a long way: weak vaccine transmission facilitates oral vaccination campaigns against zoonotic pathogens. PLoS Negl Trop Dis 13:e0007251. https://doi.org/10.1371/journal.pntd.0007251
89. Nuismer SL, Althouse BM, May R et al (2016) Eradicating infectious disease using weakly transmissible vaccines. Proc R Soc B Biol Sci 283:20161903. https://doi.org/10.1098/rspb.2016.1903
90. Nuismer SL, Bull JJ (2020) Self-disseminating vaccines to suppress zoonoses. Nat Ecol Evol 4:1168–1173. https://doi.org/10.1038/s41559-020-1254-y
91. Macklin GR, O'Reilly KM, Grassly NC et al (2020) Evolving epidemiology of poliovirus serotype 2 following withdrawal of the serotype 2 oral poliovirus vaccine. Science 368:401–405. https://doi.org/10.1126/science.aba1238
92. Bull JJ, Smithson MW, Nuismer SL (2018) Transmissible Viral Vaccines. Trends Microbiol 26:6–15. https://doi.org/10.1016/j.tim.2017.09.007
93. Delpietro HA, Russo RG, Carter GG et al (2017) Reproductive seasonality, sex ratio and philopatry in Argentina's common vampire bats. R Soc Open Sci 4:160959. https://doi.org/10.1098/rsos.160959
94. Bull JJ (2015) Evolutionary reversion of live viral vaccines: can genetic engineering subdue it? Virus Evol 1:vev005. https://doi.org/10.1093/ve/vev005
95. Stoner-Duncan B, Streicker DG, Tedeschi CM (2014) Vampire bats and rabies: toward an ecological solution to a public health problem. PLoS Negl Trop Dis 8:e2867. https://doi.org/10.1371/journal.pntd.0002867
96. Sandbrink JB, Watson MC, Hebbeler AM, Esvelt KM (2021) Safety and security concerns regarding transmissible vaccines. Nat Ecol Evol 5:405–406. https://doi.org/10.1038/s41559-021-01394-3

97. Tabashnik BE, Liesner LR, Ellsworth PC et al (2021) Transgenic cotton and sterile insect releases synergize eradication of pink bollworm a century after it invaded the United States. Proc Natl Acad Sci 118:e2019115118. https://doi.org/10.1073/pnas.2019115118
98. Buchthal J, Evans SW, Lunshof J et al (2019) Mice against ticks: an experimental community-guided effort to prevent tick-borne disease by altering the shared environment. Philos Trans R Soc Lond B Biol Sci 374:20180105. https://doi.org/10.1098/rstb.2018.0105
99. Nazni WA, Hoffmann AA, NoorAfizah A et al (2019) Establishment of Wolbachia strain wAlbB in Malaysian populations of Aedes aegypti for dengue control. Curr Biol 29:4241–4248.e5. https://doi.org/10.1016/j.cub.2019.11.007
100. Gantz VM, Jasinskiene N, Tatarenkova O et al (2015) Highly efficient Cas9-mediated gene drive for population modification of the malaria vector mosquito Anopheles stephensi. Proc Natl Acad Sci 112:E6736–E6743. https://doi.org/10.1073/pnas.1521077112
101. Nuismer SL, Remien CH, Basinski AJ et al (2020) Bayesian estimation of Lassa virus epidemiological parameters: implications for spillover prevention using wildlife vaccination. PLoS Negl Trop Dis 14:e0007920. https://doi.org/10.1371/journal.pntd.0007920
102. Streicker DG, Gilbert AT (2020) Contextualizing bats as viral reservoirs. Science 370:172–173. https://doi.org/10.1126/science.abd4559
103. Flies AS, Flies EJ, Fox S et al (2020) An oral bait vaccination approach for the Tasmanian devil facial tumor diseases. Expert Rev Vaccines 19:1–10. https://doi.org/10.1080/14760584.2020.1711058
104. Pérez-Rivero J, Aguilar-Setién A, Villa-Godoy A, Serrano H (2005) Detection of ß estrogen receptors (ßER) in Desmodus rotundus testis using coumestrol. Vet México 36:469–474
105. Serrano H, Pérez-Rivero J-J, Aguilar-Setién A et al (2007) Vampire bat reproductive control by a naturally occurring phytooestrogen. Reprod Fertil Dev 19:470–472. https://doi.org/10.1071/RD06100
106. Pérez-Rivero JJ, Pérez-Martínez M, Aguilar-Setién A (2014) Histometric analysis of vampire bat (Desmodus rotundus) testicles treated with coumestrol by oral route. J Appl Anim Res 42:208–212. https://doi.org/10.1080/09712119.2013.827578
107. Ransom JI, Powers JG, Thompson Hobbs N, Baker DL (2014) Ecological feedbacks can reduce population-level efficacy of wildlife fertility control. J Appl Ecol 51:259–269. https://doi.org/10.1111/1365-2664.12166
108. Caley P, Ramsey D (2001) Estimating disease transmission in wildlife, with emphasis on leptospirosis and bovine tuberculosis in possums, and effects of fertility control. J Appl Ecol 38:1362–1370. https://doi.org/10.1046/j.0021-8901.2001.00676.x
109. Kirkpatrick JF, Turner JW (1991) Reversible contraception in nondomestic animals. Source J Zoo Wildl Med 22:392–408
110. Chambers LK, Singleton GR, Hinds LA (1999) Fertility control of wild mouse populations: the effects of hormonal competence and an imposed level of sterility. Wildl Res 26:579. https://doi.org/10.1071/WR98093
111. Ramsey D (2005) Population dynamics of brushtail possums subject to fertility control. J Appl Ecol 42:348–360. https://doi.org/10.1111/j.1365-2664.2005.01006.x
112. Harrenstien LA, Munson L, Chassy LM et al (2004) Effects of porcine zona pellucida immunocontraceptives in zoo felids. J Zoo Wildl Med 35:271–279. https://doi.org/10.1638/03-082
113. Madosky JM, Rubenstein DI, Howard JJ, Stuska S (2010) The effects of immunocontraception on harem fidelity in a feral horse (Equus caballus) population. Appl Anim Behav Sci 128:50–56. https://doi.org/10.1016/j.applanim.2010.09.013
114. Nuñez CMV, Adelman JS, Mason C, Rubenstein DI (2009) Immunocontraception decreases group fidelity in a feral horse population during the non-breeding season. Appl Anim Behav Sci 117:74–83. https://doi.org/10.1016/j.applanim.2008.12.001
115. Turner A, Kirkpatrick JF (2002) Effects of immunocontraception on population, longevity and body condition in wild mares (Equus caballus). Reprod Suppl 60:187–195
116. Twigg LE, Lowe TJ, Martin GR et al (2000) Effects of surgically imposed sterility on free-ranging rabbit populations. J Appl Ecol 37:16–39. https://doi.org/10.1046/j.1365-2664.2000.00471.x

117. Mayen F (2003) Haematophagous bats in Brazil, their role in rabies transmission, impact on public health, livestock industry and alternatives to an indiscriminate reduction of bat population. J Vet Med Ser B 50:469–472. https://doi.org/10.1046/j.1439-0450.2003.00713.x
118. Netto AR, Nilsson MR, Côrtes JA et al (2010) Comparative study of cattle Antirabies vaccines. Zentralblatt für Veterinärmedizin R B 20:398–404. https://doi.org/10.1111/j.1439-0450.1973.tb01141.x
119. Yakobson B, Taylor N, Dveres N et al (2015) Cattle rabies vaccination—a longitudinal study of rabies antibody titres in an Israeli dairy herd. Prev Vet Med 121:170–175. https://doi.org/10.1016/j.prevetmed.2015.05.004
120. Moya MI, Pacheco LF, Aguirre LF (2015) Relación de los ataques de Desmodus rotundus con el manejo del ganado caprino y algunas características del hábitat en la prepuna de Bolivia. Mastozool Neotrop 22:73–84
121. Constantine DG (2014) Sound and fury: non-biological sound for the selective capture of rabies-infected bats. A WDA Report from the Field. Wildlife Disease Association
122. Delpietro HA (1989) Case reports on defensive behaviour in equine and bovine subjects in response to vocalization of the common vampire bat (Desmodus rotundus). Appl Anim Behav Sci 22:377–380. https://doi.org/10.1016/0168-1591(89)90032-4
123. Lanzagorta-Valencia K, Fernández-Méndez JI, Medellín RA et al (2019) Landscape and cattle management attributes associated with the incidence of Desmodus rotundus attacks on cattle. Ecosistemas y Recur Agropecu 7:10.19136/era.a7n1.2164
124. Botto Nuñez G, Becker DJ, Plowright RK (2019) The emergence of vampire bat rabies in Uruguay within a historical context. Epidemiol Infect 147:e180. https://doi.org/10.1017/S0950268819000682
125. Lee DN, Papeş M, van Den Bussche RA (2012) Present and potential future distribution of common vampire bats in the Americas and the associated risk to cattle. PLoS One 7:1–9. https://doi.org/10.1371/journal.pone.0042466
126. Schneider MC, Aron J, Santos-Burgoa C et al (2001) Common vampire bat attacks on humans in a village of the Amazon region of Brazil. Cad saúde pública/Ministério da Saúde, Fundação Oswaldo Cruz, Esc Nac Saúde Pública 17:1531–1536. https://doi.org/10.1590/s0102-311x2001000600025
127. Gonçalves F, Fischer E, Dirzo R (2017) Forest conversion to cattle ranching differentially affects taxonomic and functional groups of Neotropical bats. Biol Conserv 210:343–348. https://doi.org/10.1016/j.biocon.2017.04.021
128. De Andrade FAG, Gomes MN, Uieda W et al (2016) Geographical analysis for detecting high-risk areas for bovine/human rabies transmitted by the common hematophagous bat in the Amazon region, Brazil. PLoS One 11:1–15. https://doi.org/10.1371/journal.pone.0157332
129. Streicker DG, Allgeier JE (2016) Foraging choices of vampire bats in diverse landscapes: potential implications for land-use change and disease transmission. J Appl Ecol 53:1280–1288. https://doi.org/10.1111/1365-2664.12690
130. McCarthy TJ (1989) Human depredation by vampire bats (Desmodus rotundus) following a hog cholera campaign. Am J Trop Med Hyg 40:320–322. https://doi.org/10.4269/ajtmh.1989.40.320
131. Gilbert AT, Petersen BW, Recuenco S et al (2012) Evidence of rabies virus exposure among humans in the Peruvian Amazon. Am J Trop Med Hyg 87:206–215. https://doi.org/10.4269/ajtmh.2012.11-0689
132. Gómez-Benavides J, Laguna-Torres VA, Recuenco S (2010) El real significado de ser mordido por murciélagos hematófagos en las comunidades indígenas Amazónicas del Perú. Rev Peru Med Exp Salud Publica 27:657–658. https://doi.org/10.1590/S1726-46342010000400030
133. Hayman DTS (2016) As the bat flies. Science 354:1099–1100. https://doi.org/10.1126/science.aaj1818

Chapter 9
Conflicting Perspectives on a Rabies Outbreak in a Venezuelan Rainforest, 2007–2008

Charles L. Briggs

Abstract Prevention and control of infectious diseases, such as rabies, involve seemingly complicated variables that require multiple frameworks of conceptualization, documentation, analysis, and response, particularly within the realm of major health disparities. Besides biomedical and technical aspects, collaborations in social science help pinpoint cultural barriers and suggest how to increase awareness of health and disease on the part of underserved populations. The description of a 2007–2008 outbreak in a rainforest in eastern Venezuela, that was clinically diagnosed as rabies, exemplifies ongoing challenges and opportunities to understand such complex multispecies relations towards resolution. Holistic focus on the perspectives and practices of scientists, clinicians, and policymakers, as well as those New World communities most affected by rabies, such as in Amazonia, may provide a fresh area for engagement among stakeholders to forge a true One Health partnership.

Keywords Inequity · Outbreak · Rainforest · Social justice · Venezuela

Introduction

Rabies presents us with a conundrum. Despite the development of effective vaccines, the World Health Organization estimates that rabies still kills some tens of thousands of people in over 150 countries annually. Recently, Rupprecht and Salahuddin [20] suggested that "…remaining barriers to widespread rabies vaccination include an interrelated set of economic, cultural, social, educational, ecological and technological factors…". As such, this is a complex equation, involving

C. L. Briggs (✉)
Berkeley Center for Social Medicine and the Department of Anthropology, University of California, Berkeley, CA, USA
e-mail: clbriggs@berkeley.edu

C. E. Rupprecht (ed.), *History of Rabies in the Americas: From the Pre-Columbian to the Present, Volume I*, Fascinating Life Sciences,
https://doi.org/10.1007/978-3-031-25052-1_9

variables that require multiple frameworks of conceptualization, documentation, and analysis.

As a zoonotic disease, rabies emerges through webs of engagement between humans, animals, and viruses. In the Americas, rabies moves between humans, bats, domesticated animals, livestock, and many other species. In assessing human rabies in the Peruvian Amazon, Gilbert et al. [8] specified factors that help shape this interaction: "…common risk factors for human RABV infection in the Amazon include poor housing conditions, small population groups in remote areas, poor access to health services, and a general lack of awareness or cultural barriers regarding the transmission of rabies by bats…". The latter is often framed as a major stumbling block to prevention and health education programs. Social scientists, like myself, are sometimes invited into investigations to help pinpoint these cultural barriers and suggest how to increase awareness of health and disease on the part of underserved populations. My experience in a 2007–2008 outbreak in a rainforest in eastern Venezuela that was clinically diagnosed as rabies would seem to position me to undertake this task. Rather than fitting into this slot, however, my goal in this chapter is to suggest that more effectively understanding these complex interrelations and developing more effective vaccination, education, and other prevention programs would benefit from a radically different approach to bringing the social sciences into research on rabies virus transmission. Rather than claiming that only anthropologists can solve these problems, however, I suggest that anthropologists have much to learn from veterinarians, clinicians, biologists, ecologists, public health specialists, and community members about complex multispecies relations. Accordingly, I propose new forms of collaboration in the conclusion.

Traditionally, as a medical anthropologist, my core task would have been "ethnomedical" in nature, documenting the contrastive health "beliefs" and practices of populations whose perspectives on health and disease differ sharply from biomedical forms of knowledge [18]. Anthropology's distinctive methodology, 'ethnography', was formerly confined to long-term research on underserved populations, such as those most affected by human cases of rabies. As a science, however, medical anthropology has moved beyond this limited focus to suggest that addressing concerns about social and cultural barriers requires in-depth fieldwork with *all* parties, that is, studying the perspectives and practices of scientists, clinicians, and policymakers as well as those of populations most affected by human rabies cases. Thus, what would happen if we turned questions of "awareness or cultural barriers" around in such a way as to render visible the obstacles to investigation and intervention encountered by all stakeholders, thereby enabling them to contribute to new, path-breaking understandings of rabies? Moreover, a recent shift, often called multispecies ethnography, has widened the scope to include nonhumans, and it is showing promise in studying rabies.[1]

This chapter moves in four stages. First, I describe an outbreak that killed 38 persons classified as indigenous in Delta Amacuro State of Venezuela in 2007–2008

[1] See Nadal [16] and Sodikoff [22] for examples.

that was clinically identified as rabies.[2] A second section explores why the disease remained undiagnosed for over a year despite efforts by epidemiologists, pediatricians, toxicologists, and other specialists. The third section uses this conundrum to look at an approach that is often used in investigating rabies outbreaks and designing programs aimed at lowering relevant risk factors, Knowledge, Attitudes, and Practices (KAP) surveys. The final section outlines my proposal for new forms of collaboration between stakeholders that are geared to creating partnerships that will be more effective in studying rabies outbreaks and designing vaccination and prevention programs. It attempts to forge a relationship between medical anthropology and multispecies ethnography on the one hand and One Health approaches on the other.

An Epidemic Puzzle

A mysterious outbreak began in July 2007 in Mukoboina, a settlement of some eighty residents and a dozen thatch-roofed houses perched on stilts above the mud and water of a midsized tributary of the Orinoco River (Fig. 9.1). Gabriel Torres developed a fever, which did not respond to antipyretics, phytotherapy, or to treatments by the local healer. The fever persisted, followed by symptoms—including

Fig. 9.1 Houses in Mukoboina, 2010

[2] I described the methodology employed in this work below. For further information on the outbreak and our investigation, see Briggs and Mantini-Briggs [1]. At the request of the parents, who "want lots of people to know what happened to our children," I use their names. I supply pseudonyms for all other persons mentioned except for members of the investigatory team on which I participated.

heavy salivation—that did not seem to fit any illness that residents had previously seen. Parents Wilmer and Zoila Torres secured passage by motorized canoe to the nearest clinic, some 20 min away, but he died en route. The following month, the couple reported, his little sister Yuri "developed an identical fever. It was the same sickness." After the local healer again failed, they took her the next morning to the nearby nursing station. The nurse placed her in a hammock to examine her. He hooked up an IV and tried antipyretics and other treatments. Nevertheless, he had no idea what was causing the symptoms. When Yuri was not responding to treatment and was getting weaker, he sent the family to the Nabasanuka Clinic, which has a resident physician and more experienced nurses (Fig. 9.2). During the trip they noticed, 'Yuri just kept getting sicker'. Crying out continually, by the time they reached the clinic, she was unable to walk. The newly graduated physician and his team of experienced, bilingual nurses were no more successful in diagnosing and treating Yuri, and she died there before the doctor could refer her upriver to the hospital in Tucupita, the state capital.

Then, 12 days later, a cousin, 11-year-old Ángel Gabriel developed "the same sickness." The disease presented as a strange, unknown fever, progressing more rapidly than anything they had previously witnessed. For parents Matilse Carrasquero and Gracian Florín, however, fever, headache, sore throat, and body aches, following on day three by unusual salivation, were all too familiar symptoms. Ángel Gabriel asked for food and water, but quickly rejected them as soon as they were presented. The parents similarly started with the local healer, but he soon realized that the disease was the same one that had foiled his diagnostic and treatment efforts twice already. The healer urged the family to paddle their canoe across the river to visit the nurse. José Pérez had not received any information from the physician after he treated Yuri, so he could only treat a similarly high fever with analgesics,

Fig. 9.2 The Nabasanuka clinic, with Dr. Clara Mantini-Briggs in the foreground, 2008

antipyretics, and an IV. When interviewed several months later, he reflected: "I thought, 'darn, what could it be? Patients come here with a fever all the time. I bring it down with an analgesic, antipyretic, or some other medication, and usually they recover in a couple of days. But with this fever, I use medicines and it's just the same, just the same—it even gets worse!'" He transported the family to the clinic himself, hoping to see how the physician would treat the child and give him advice. Pérez recalled that "Unfortunately, "the only words [the doctor] said to me were, 'When you get this kind of case, bring the patient to me immediately. We don't know what disease it is either.'"

Ángel Gabriel is the first patient that Dr. Ricardo Cáceres remembers treating in the outbreak. When Clara Mantini-Briggs and I visited him in July 2008, he reflected that Carrasquero and Florín told him that Ángel Gabriel had been ill for 3 days with a high fever, body aches, headache, convulsions, and excessive salivation. After growing delirious, he said he was fighting a monster. Cáceres continued: "I had never seen any disease like that." Before he could hospitalize the child in Tucupita, the parents visited a healer, returning the next morning. Ángel Gabriel "had a temperature of 41 degrees [Celsius], an intense headache, and great difficulty breathing. I administered rehydration solution and an antipyretic to control the fever." A stethoscope revealed "severe hoarseness and a crackling sound," so Cáceres administered an inhaler. "But when we were preparing the boat to refer the little boy [to the hospital], he died at 11:00 AM." Cáceres reported that three more Mukoboina children (6 and 8 years of age) arrived with similar symptoms, and he referred them immediately to the hospital.

By September four more Mukoboinan children had died, followed by one in January 2008, part of a second wave of cases that included deaths in surrounding communities. One family, Odilia Torres and Romer Torres, lost all three children, leaving only an infant born during the outbreak. All of the parents tried desperately to save their children's lives. Venezuelan and Cuban epidemiologists visited Mukoboina repeatedly, but they could not pinpoint the cause. When pediatricians and other specialists in the Tucupita hospital could not diagnose or treat the children, they referred them to tertiary care facilities in larger cities. Extensive lab work yielded hypotheses, including mercury poisoning, none of which yielded a confirmed diagnosis or provided physician Cáceres and nurse Pérez with additional diagnostic or treatment tools. All of the patients died in 2–7 days after the onset of symptoms. A third wave of cases that covered an even broader area began in June 2008. The sixteen deaths in 2 months included six young adults. In total, thirty-eight died in an outbreak that had continued for more than a year.

Missing Pieces and a Diagnostic Failure

What went wrong in the diagnostic process? My goal is not to criticize the clinicians or epidemiologists who attempted to diagnose the outbreak. They were conscientious and well-trained professionals who were doing their best to sort things out and save lives. No patient was denied care, and parents were never presented with a bill, despite the cost of specialized hospital care and long ambulance rides. Indeed, it is a common occurrence for rabies cases to be misdiagnosed, and it can be particularly difficult to distinguish the paralytic form from Guillain-Barré syndrome [6]. Might it accordingly be important to investigate what physicians know and what information they might be lacking as well as exploring how knowledge—such as having read a section on rabies in a medical textbook—can fail to translate into a diagnosis? Here, my interview with Cáceres was quite telling.[3] He was a *medico rural*, recently graduated from the leading school of medicine in Venezuela and conducting his obligatory year of employment with an underserved population. At the beginning of the outbreak he was dedicated, well liked by his patients and the community. He noted that his practice up until that point had yielded diseases where his knowledge base perfectly matched cases presenting in his clinic, mainly diarrheal and respiratory illnesses. "But afterwards the first cases of the children came to me, without any known cause." He noted that the parents were very nervous," because they could see that the children were going to die. The parents were clearly signaling that their experience—and that of the healer and nurse—had proved insufficient.

This moment of rupture when Cáceres realized that the experience he had obtained in medical school had fallen short is crucial: "I had never seen a disease like that." Knowledge, for this dedicated and well-trained physician, consisted of pathologies that he routinely observed in his clinic and other diseases that epidemiological patterns and medical training might have led him to include in his differential, which for this case consisted of pneumonia, malaria, dengue fever, tetanus, Guillain Barre Syndrome, meningitis, a cerebral tumor, and medication toxicity. Rabies was not on this list. What is crucial is that the physician then tried to expand his knowledge base to encompass the disease. He referred patients to urban hospitals where specialists could diagnose it. He then requested over the radio, by letters, and in visits to the state capital access to information being generated in these sites, but no information was forthcoming. Frustrated, he broke protocol to visit the tertiary facilities where his patients had died and the institute of hygiene in the national capital, seeking to expand the range of diagnostic possibilities he could pursue, but no one had answers. He returned to his professor in medical school, asking if she could come to the delta, knowing that her clinical experience was broader. She sent medical supplies and four students in their last year of medical school to help out for 6 months.

The physician's efforts to expand his knowledge base were limited in four crucial ways. First, Cáceres reproduced the hierarchically-ordered organization for the production of medical information, which projects it as created in what Bruno Latour [10] refers to as centers of calculation, such as leading hospitals, laboratories, and schools of medicine and public health, and then as being implemented by practitioners in other sites. Unless you are the expert at the center, you are a receiver of knowledge produced elsewhere. Secondly, it never occurred to the physician that

[3] This interview was conducted jointly with Dr. Clara Mantini-Briggs, as was most of the research.

the answer might lie in his own vicinity and that he could have a hand in producing it. Third, it was even more unthinkable to him that his patients, other indigenous residents, or the clinic's bilingual nurses, who could bridge medical systems in the Delta, might become collaborators in resolving the mystery. Finally, there was one person he never thought to visit in the Regional Health Office headquarters, an official who was similarly never invited on any of the epidemiological commissions during the outbreak, whose opinion was only solicited once as it unfolded. This was the Director of Zoonosis in the Regional Health Office, a veterinarian. Humans were dying, so the needed professional competence must lie somewhere in medicine or epidemiology, with experts on *human* health. Here is a crucial way that definitions of knowledge can result in delayed or failed diagnoses of rabies.

This orientation is hardly confined to Cáceres, Venezuela, or the Americas. Indeed, it is a product of medical training more broadly. Patients are sites for the extraction of information, for garnering fragmented details of names, dates of birth, symptoms, blood, tissue, etc. Only physicians can then transform this information into diagnostically-relevant facts.[4] This issue of attitude is particularly acute when the "population" in question is deemed to be poor and rural, especially indigenous. As we have documented in previous work, the indigenous/nonindigenous binary is primarily defined in medical terms: an *indígena* is a person whose purported culture makes him or her believe in spirits and healers, not doctors, and renders him or her incapable of even understanding what a physician tells them [1]. New physicians get socialized into this attitude, into viewing indigenous patients in this fashion during their orientation in the state capital. Their patients are cast as empty vessels who need to be filled with medical knowledge, even as their "cultural beliefs" purportedly render them incapable of learning it [3]. Another important dimension of the physician's attitudes was his evaluation of multilingualism, that is, of the fact that many of the parents, whose first language is Warao, had limited ability to converse in Spanish, the only language he knew. He saw this issue as minor and mechanical, one that could be solved by having nurses translate.

This attitude toward the parents as potential collaborators and toward multilingualism translated into practices that impeded physicians, nurses, and epidemiologists from expanding their knowledge bases. The bilingual nurses, such as Pérez, spoke to the parents. Nevertheless, their clinical training led them to focus primarily on standard protocols: eliciting the patient's name, age, symptoms, and date of symptom onset. Centered on details that might help them diagnose and develop a treatment plan for patients whose health was deteriorating rapidly, their questions did not make room for the parents to provide their own detailed accounts of what they had observed and the forms of treatment they tried. In the clinic, physician Cáceres utilized the same constrained mode of interacting with parents and patients. Even in the case of families who had already accompanied children to urban hospitals and were facing their second or third encounter with the disease, no space was made for them to provide their observations. Cáceres sometimes grew frustrated

[4]Among the many medical anthropologists who have explored this issue, see Good [9].

with parents when they took their child to a local healer, "against medical advice," when the doctor's diagnostics failed and treatment did not work. Even as he was dedicated to saving his patients' lives, Cáceres did not see the parents as having much to contribute to the diagnostic process. With each increase in the level of care as patients were referred to urban hospital and tertiary-care facilities, the parents' role progressively diminished: specialists relied on Cáceres's referral report. They announced their clinical decisions to the parents rather than asking for their input.

The residents of Mukoboina were elated when the Regional Epidemiologist arrived with a commission after metropolitan specialists suspected mercury poisoning in the death of a six-year-old girl. The parents began to assemble in the community representative's house, but the epidemiologists declined to participate in a meeting. They rather traveled to homes of each of the children who had died, elicited basic diagnostic information, observed the family's activities, and searched the premises for clues (such as pots of mercury). Like Cáceres, they viewed language differences as being of such little importance that they sometimes used their motorboat operator—rather than bilingual nurses—to translate their questions and Mukoboinans' responses. The epidemiologists left without sharing their findings with residents. When this pattern was repeated in the course of visits by several epidemiological commissions, the parents grew increasingly frustrated and less inclined to participate, even as the epidemiologists complained that they "had difficulty acquiring information due to the closed character of the community."[5]

Knowledge, Attitudes, and Practices (KAP) Surveys: What They Leave Out

Now I want to step back a bit from the particulars of this case to reflect on how researchers often document and analyze these sorts of problems as they arise in studying bat-transmitted rabies virus outbreaks. Given their common use in investigating rabies outbreaks, it might seem as if surveys would provide the best tool for documenting these sorts of knowledge gaps and barriers—for listening to local populations. A KAP survey can be used to measure "a general lack of awareness or cultural barriers regarding the transmission of rabies by bats," which is viewed as constituting "common risk factors for human RABV infection" [8]. Discounting the importance of investigating these issues is not my intention. As detailed below, I participated in a community-driven effort to diagnose the disease from July–August 2008 and then spent several years conducting research on what had taken place. Indeed, when we suspected bat-transmitted rabies and began to ask people about bat bites and discuss prevention measures, we did not find a single member of the local population who was aware that bats can transmit rabies virus. Some individuals

[5]This quotation comes from a report of 7 August 2007 compiled by the Office of Regional Epidemiology in Tucupita.

actively discounted the possibility that bats could be the vector. Two factors are particularly relevant here: First, most houses lack exterior walls on at least two sides (Fig. 9.3). Nocturnal *Desmodus rotundus* bites thus take place periodically. Nevertheless, health education campaigns—which focus mainly on diarrheal diseases—have not, as far as we know, ever included information on bites, including proper techniques for washing wounds and prevention of bat attacks. Second, neither clinical visits nor public health efforts prompt prophylactic rabies vaccination, even for children and young adults who have been bitten repeatedly. It thus would seem far from clear that cultural barriers or a lack of awareness are limited to indigenous laypersons in the Delta.

In a paper that lays out KAP, Médecins du Monde [13] defines knowledge broadly as "a set of understandings." Nevertheless, the discussion moves on to suggest that "the degree of knowledge assessed by the survey helps to locate areas where information and education efforts remain to be exerted." Herein lies the rationale for focusing on knowledge, its centrality for improving public health:

> KAP questions tend to reveal not only characteristic traits in knowledge, attitude and behaviours about health related to religious, social, traditional factors, but also the idea that each person has of the body or of disease. These factors are often the source of misconceptions or misunderstandings that may represent obstacles to the activities that we would like to implement and potential barriers to behaviour change.

Fig. 9.3 House in Santa Rosa de Guayo, 2009

Two things bear pointing out. First, "knowledge" is actually a misnomer here. When articulated in this fashion, a KAP does not locate and map knowledge. Rather, it starts from a position of authority, an implicit claim to know what constitutes knowledge in relation to a disease or public health issue before an investigation begins. The K dimension thus focuses more on *nonknowledge* [7]. Thus, K is a negative definition that prompts a search for "misconceptions or misunderstandings" rather than what people actually know and how their knowledge might be useful.

Second, "knowledge" applies not to everyone but to "a population or target group," certainly not to the "we" that is projected by the health professionals who conduct a project or those who read their reports and articles. "We" are simply presumed to have knowledge, thereby empowering KAP implementers to assess the nonknowledge of "the population." If nonknowledge is part of the problem with rabies diagnosis and public health measures, the gaps in the knowledge possessed by physicians, nurses, and epidemiologists would also seem crucial. Rabies involves very complex problems of the interaction of viruses, humans, and other animals in shifting, entangled ecological, demographic, and disease patterns. If KAP directs the vision of researchers, clinicians, and planners away from assessing the knowledge of professionals, it will thereby exacerbate gaps that make solving such puzzles even harder. Hence, a KAP creates social boxes that place "communities" and the "we" of health professionals in distinct sectors. Not only does it locate deficits in only one of these seemingly opposed sectors, but it projects transfers of information as moving only in one direction, from professionals to members of underserved communities. It also creates a static view of knowledge, construed as fragmented components that can be measured at one or more particular points in time rather than holistic, ongoing processes of discovery. During the outbreak, however, all parties were thrown off guard by the insufficiency of their knowledge. Delta parents, who regularly move between a variety of healing systems, attempted to build algorithms that would enable them to combine their strengths as they actively assessed what each nurse, doctor, hospital, or healer knew and how they could help. A major source of their frustration was that this process failed to yield the knowledge that would save their children.

The same contradiction is apparent in the definition of attitude, defined as "a way of being, a position."[6] Again, we would seem to all have attitudes. Nevertheless, the discussion similarly redefines attitudes as located in the subject population. Professionals either do not seem to have attitudes or their attitudes are not be related to the creation or persistence of health problems. The term practices—which, unlike knowledge or attitudes is more located in individuals than populations—is positioned in similar terms, dividing a target population that exhibits actions and a "we" who does the tracking. Like knowledge, practices are thus defined negatively, focusing either on the absence or insufficiency of actions defined as desirable for good health outcomes or on practices that are detrimental.

[6] The term "attitude" is sometimes used in the singular and sometimes plural in the literature.

How are these three terms related in the model? Changing knowledge and attitudes should induce the practices that "we" deem required to overcome the health problem addressed by the project. The relationship should be linear, transparent, and automatic. The KAP is thus designed to measure what are projected as deviations from this trajectory. Nevertheless, the KAP relationship is deemed to be fundamentally precarious or perhaps impossible: "It is interesting to note that numerous studies have often shown a low and sometimes no connection between attitude and practices."[7] By defining knowledge in advance and measuring what seems to be absent in the population's knowledge base, KAP projects affected communities as having nothing positive to offer. Beyond precluding the possibility that generations of observations of interactions between humans and other species might help scientists sort out complex and shifting multispecies relations, KAP can exacerbate chasms between the "we" of health professionals and the "them" of target populations, potentially complicating efforts to foster the sorts of collaborations that are needed to tackle acute health problems in underserved areas. I return to this point below.[8]

A Community-Driven Alternative

A return to the specifics of the Delta outbreak can suggest alternative ways of approaching these thorny problems of knowledge gaps and barriers. By July 2008, the deaths, the lack of a diagnosis, and the parents' exclusion from attempts to achieve one, were beginning to undermine health care in the Lower Delta. Many parents no longer trusted physicians, and some even stopped taking their children to the clinic for other illnesses or preventive care visits.

The parents repeatedly expressed their anxiety and frustration to Conrado Moraleda. Conrado lives next door to the clinic, his son is employed there as a nurse, and Conrado served at the time as the head of the Neighborhood Health Committee that supports the clinic staff. Conrado visited the clinic when each patient arrived, and he listened extensively to the parents. Working in collaboration with the physician, he made several visits to health officials in the state capital to request concerted action as the second wave of cases began. In February 2008, he traveled to the capital with some of the parents. In a public meeting of the state's

[7] See Farmer [4] for a critique of assumptions regarding knowledge deficits as causing health problems and infusions of information as needed to overcome them, which he sees as directing attention away from underlying structural factors.

[8] A number of anthropologists have critiqued KAP. To cite just two examples, Nichter [17] argued that much health knowledge is embodied rather than explicit and objectified and thus difficult to access through KAP surveys. Launiala [11] suggests that KAP is geared to measuring public health concepts and thus misses "other types of knowledge, such as culture-specific knowledge of illness notions and explanatory models," and that KAP data are affected by the contexts in which questionnaires are used.

legislative assembly, he presented information on the cases, thereby placing additional pressure on health authorities, who retaliated by making media statements that blamed the parents and healers for the deaths. When the third wave of cases began in June 2008 and a diagnosis seemed as elusive as ever, Conrado decided to organize an "investigatory team" with his brother Enrique.[9] Venezuelan public health physician, Dr. Clara Mantini-Briggs, and I stumbled onto the outbreak in July 2008 while working with residents to establish a new model for health care. Old friends, the Moraledas recruited us to participate as physician and anthropologist, along with nurse Norbelys Gómez and healer Tirso Gómez.

We visited 14 communities in which the strange deaths had occurred. In each, Conrado, Enrique, Norbelys, and Tirso collaborated with local leaders in organizing a meeting that included the parents, other relatives, healers, and neighbors. Individuals who came forward could speak as long as they liked, uninterrupted, relating their experience in the outbreak and presenting their observations. Their narratives were accordingly structured by their own perceptions of what was relevant, not ours. Although they generally praised nurses and physicians, they detailed how their input had been silenced and how they painfully discovered that autopsies been performed without notification, consent, and, crucially, without providing the parents with the results. Only after each witness had finished could Clara and I ask for elaboration of details that might be clinically relevant.

As the parents told their stories, a common pattern emerged. The symptoms reported were fever, headache, body aches, a feeling of itching in the feet that turned to numbness and then paralysis that ascended, difficulty swallowing, hydrophobia, seizures, strange dreams or hallucinations, and excessive salivation, resulting invariably in death in 2–7 days. Clara's presumptive clinical diagnosis of rabies was supported by examining a patient, Elbia Torres Rivas, for several days until the time of her death, even as we provided palliative care. As soon as the parents' narratives revealed these symptoms, we began asking about animal bites. Most patients had been bitten nocturnally by vampire bats, 1–1.5 months before the onset of symptoms. Residents reported that the frequency of these attacks had increased dramatically, just before the beginning of the outbreak. Once we suspected rabies, we spoke with community members about cleansing bites with soap and water and using mosquito netting or other means of preventing bat attacks. We noted that vaccination is recommended whenever a human is bitten by an animal that might be rabid. Our team took a report to national health officials, requesting a thorough investigation and effective countermeasures.

The methodology crafted by the team differed from previous attempts at diagnosis in important ways. Elicitation by clinicians and epidemiologists cast health professionals and participants in opposing roles: one party could determine the questions, ask them, and assess when responses constituted appropriate answers, versus other individuals who were confined to answering the questions. Surveys

[9] I use first names for members of the team, in part, because it would otherwise be hard to distinguish the Moraledas and the Gómezes.

generally provide even less opportunity to elicit information that does not fit the parameters of the instrument. As Aaron Cicourel [2] suggests, persons who fill out surveys and their respondents actively collaborate in finding ways to fit knowledge (beliefs, opinions, etc.) into preconceived boxes. Unless they are designed collaboratively and ethnographically, in ways I specify below, interviews and surveys are like epistemological steamrollers that flatten the ways that those participating in the research ordinarily collaborate in producing, circulating, interpreting, and evaluating knowledge. Moreover, grossly unequal power relations that accord professionals, who are generally outsiders, the right to decide what counts as knowledge, who will elicit it, and how it will be used—thwarts collaboration, thereby creating the very "potential barriers to behaviour change" that KAP attributes to the "population's" knowledge and attitudes. Epidemiologists' experience in Mukoboina suggests what happens when research participants demand that their own knowledge-making practices be included: the visitor's refusal to hold meetings and listen to the parents' narratives led to inconclusive investigations and strong feelings of frustration and anger on both sides.

As a community-inspired and community-driven project, the investigation inaugurated by Conrado and Enrique was collaborative from start to finish. A crucial goal of the democratization of knowledge production was to meet the parents' long-standing requests to provide witness to the extent of the parents' efforts to save their children, the value of their children's lives, and the importance of their deaths. At the same time, the major focus for everyone was diagnosing the disease and ending the outbreak. Space was made for multiple medical logics and forms of authority. Rather than relying exclusively on one genre, the events made space for multiple forms, including narratives, community problem-solving, clinical interviewing, epidemiological-oriented queries, and, in two cases, the laments that relatives were singing over the children's bodies. All participants spoke the language of their choice. Most of the investigation thus unfolded in Warao, and translation (when required) was undertaken later to avoid interrupting speakers and disrupting knowledge-making practices.

One additional discursive genre emerged. As I noted, the identification of *Desmodus rotundus* as the vector that seemed to be transmitting rabies met with some opposition, including from one team member, Tirso Gómez. As we were compiling the report that we would take to national health authorities, Tirso suddenly stood up and performed the Myth of Bat. The story takes place in a primordial time in which animals had human form and could converse with humans, even as signs of difference were apparent. A young man "who seemed just like everyone else" had a strange characteristic. Rather than going out to hunt and collect forest products during the day, he worked alone at night and slept during the day. When asked about his strange behavior, he replied "I can't see during the day." After he married, his wife tried to convince him to go out and work during the day with her parents, but he refused. In time, people discovered that nocturnally he crept into women's hammocks and sucked their blood, "and so a number of people were dying like this." The men killed him, cut up his flesh, threw it into the fire, and then threw the charred remains in the cardinal directions. The myth concludes "that was the transformation of the bat, the burned pieces transformed into bats.... And that's why bats like people's blood."

Tirso's performance challenged the process by which the preparation of a scientific report was displacing indigenous perspectives on human-nonhuman relations. By presenting vampire bats as a vector possibly responsible for rabies virus transmission, we were adopting what Eduardo Viveiros de Castro [23] refers to as a Western "naturalist" ontology in presenting bats as an element of nature that only gained human significance by becoming a rabies virus vector. The myth, on the other hand, pointed to a long history of intimate bat-human relations—never particularly friendly and always involving human blood. *Desmodus rotundus* fly into a darkened house, land silently, climb ropes holding hammocks, slide down to the edge, puncture a face, head, finger, or toe, and lap up the blood, generally without waking the victim. Bats read human breathing sounds like we read human voices, performing "a rather sophisticated analysis of the sounds based on multiple parameters" despite their quite low volume [5]. Often returning nightly to attack the same person, they develop an intimate relationship of exchange with a human as their saliva and its contents enter the human's body and human blood enters the bat's digestive track. Rabies viruses enter into this complex exchange if they move between the bat's saliva and human nerve cells.

Being challenged to see the outbreak through the lens of these ongoing multispecies relations revealed other relevant details. When we returned to the Delta in July 2008, many families had adopted cats for the first time. Our queries suggested that residents were responding to a notable increase in *Desmodus rotundus* attacks. We asked for other changes in the ecology of human-nonhuman relations that had taken place during the previous year. Several residents reported that after exhibiting flu-like signs, massive numbers of chickens died just before the beginning of the outbreak. They also reported strange deaths of carnivores, sloths, and other animals in the forest and invasions of nearby islands by cattle and pig ranchers. In short, when community members became partners in knowledge production, their distinct attitude toward human-nonhuman relations yielded a host of clues, not only as to the source of the deaths but the broader patterns of ecological relations in which they were located.

Conclusion: Rethinking Trans-disciplinary Research Design

I have explored the parameters of the Delta outbreak to draw attention to the broader ecologies of knowledge-making in which clinical diagnosis, epidemiological investigation, and broader One Health research grapples with the complexities of vampire bat-transmitted rabies virus. I have tried to demonstrate that all stakeholders had important potential contributions to make in figuring out what was taking place, even as I pointed to ways that health professionals constrained the sort of collaboration that ultimately proved productive. I have suggested that the use of only a narrow range of social science methods, specifically KAP surveys, can amplify the effects of standard clinical and epidemiological practices in not only failing to draw on these broader ecologies of evidence but actually rendering health professionals unable to see that they exist. Given that my goal is to offer a different starting point and suggest a broader range of tools, now it is my turn to make some recommendations, which I have organized under the acronym CESS.

C stands for *collaboration*. My experience working on and in one outbreak suggests to me that bat-transmitted rabies offers a best-case scenario for collaboration. At the beginning of the Delta outbreak, everyone—parents, physicians, nurses, epidemiologists, and healers—were all on the same page: everyone wanted to know what was killing the kids and to prevent more deaths. Adopting a One Health perspective would have expanded possibilities for collaboration. In the Delta case, bringing specialists in zoonosis, veterinarians, cattle owners, and ecologists into dialogue with physicians, nurses, epidemiologists, and public health officials would have been crucial. My chapter would suggest that bringing social scientists—rather than just using specific methodologies based in the social sciences—on board in investigations could be extremely valuable. Rather than relegating them to a separate slot devoted to isolating social and cultural factors, however, social scientists should be involved in all facets of the outbreak investigation, in part so that they can learn from their veterinarian, physician, nurse, epidemiologist, biologist, and ecologist colleagues. One Health can, however, signal collaboration between "experts". Leaving out patients, relatives, other residents, healers, and community representatives in knowledge-making is perhaps the quickest route to squandering this initial sense of common cause, as the epidemiologists' experience in Mukoboina would suggest. Our team's experience suggests that collaboration means much more than inviting local participation in work that is inaugurated and controlled by professionals but requires a community-based foundation in which nonprofessionals play key roles from the start.[10]

E is for *ethnography*. Putting aside debates and complexities, it can be defined as generally in situ (but now also remote) research on a particular setting or problem that is open-ended. It uses such tools as observation, interviewing, focus groups, and informal conversations to discover information not previously known to the researcher and to learn the terms in which local participants conceptualize and use it. At the same time that it is anthropology's signature contribution, it is now widely used in the social sciences and humanities and in applied research on health care and much more. Long limited to human perspectives, relations, and actions, received perspectives are increasingly giving way to multispecies (or more-than-human) ethnography. The focus is not just on how people think about nonhumans but the active roles of nonhumans and environments in shaping contemporary worlds.

An initial ethnographic focus for exploring rabies outbreaks (or areas in which cases sometimes appear) would be to identify the sorts of ecologies of evidence that I have discussed. What are the different forms of knowledge that can illuminate

[10] A key approach in public health is Community-Based Participatory Research [14]; others now use the term "community-driven" [15].

multispecies relations and health problems? Who produces and exchanges them? What gaps exist? What barriers are apparent to collaboration in knowledge-making? Keeping questions of language, genre, and context in mind are crucial to document modalities that become invisible or peripheral when only dominant languages and professional methodologies are taken for granted. As I have argued, restricting such questions to scrutiny of "a population or target group" and thus failing to study health professionals would doom the results of ethnographic inquiry from the start. Ethnographic research cannot be adequately conducted *on* humans and nonhumans by outsiders but involves collaborative knowledge production and openness to ways that nonhumans offer valuable insights. It involves learning *with* and *from* participants, not just learning about them. True, having conducted ethnography in the Delta for over two decades and speaking Warao was a great asset when I stumbled onto the outbreak. Nevertheless, medical anthropologists have devised techniques of "rapid ethnographic assessment" for use precisely when problems are pressing and time is precious.[11]

The first S stands for *surveys*. I certainly do not wish to disparage the collection of quantitative data or the use of surveys, as I have conducted them myself. If surveys or other formal instruments are needed to complement qualitative ethnographic research, the time to design them is after the ethnography phase has produced significant results. My discussion of KAP points to the importance of assessing knowledge, not nonknowledge—documenting the different perspectives, attitudes, and practices that can contribute to understanding and solving health problems rather than stigmatizing nonprofessionals for approaching health issues differently. Taking the results of the ethnographic inquiry as a starting point and collaborating with nonprofessionals in designing them can yield questions that draw on rather than flatten local knowledge-making practices and can phrase questions in terms that are relevant to respondents, not just those who design questionnaires. Again, it is important that surveys include health professionals as respondents.

The final S is for *sharing*. Even as I discussed how problems with diagnosing the outbreak pointed to gaps in clinical and epidemiological knowledge production, my goal has not been to present a romantic portrait of Amazonian indigenous peoples as possessing "traditional ecological knowledge" that can unlock all secrets. The COVID-19 pandemic and its differential effects on racial minorities in the United States and elsewhere suggests that research needs to be translated into ways of making changes designed to improve health conditions, including health policies and practices that challenge the racial and other inequities and forms of structural violence that engender health problems and thwart health communication and education programs [4]. Nevertheless, clinical and epidemiological practices that exclude or demean community voices and methodologies that stigmatize people for their nonknowledge of professional perspectives jeopardize collaboration when it is vitally needed to turn things around. Health education "campaigns," whether

[11] See Manderson and Aaby [12] for a foundational essay and Sangaramoorthy and Kroeger [21] for a recent, comprehensive guide aimed at a broad audience.

undertaken in clinical settings or media-based, all too often take the form of the unidirectional, top-down infusions of biomedical knowledge envisioned by KAP-based projects [19]. Such attempts to clear away “misconceptions or misunderstandings that may represent obstacles to the activities that we would like to implement and potential barriers to behaviour change” are framed as a silver bullet, even when their purveyors deem them unlikely to succeed.

The example that I have analyzed suggests that in the complex settings in which rabies cases and outbreaks emerge, no one has all the knowledge that is needed or the practices that can easily prevent, control, or eliminate rabies. Thus, sharing points to the need for multi-directional exchanges of knowledge. If health communication is seen as a collaborative and dynamic process that continually builds new knowledge rather than simply imparts what is already known, it will be much more likely to earn the active participation of all parties—and to prepare them to join one another in facing what whatever new health challenges lie right around the corner.

References

1. Briggs CL, Mantini-Briggs C (2016) Tell me why my children died: rabies, indigenous knowledge, and communicative justice. Duke University Press, Durham
2. Cicourel A (1974) Theory and method in a study of Argentine fertility. Wiley, New York
3. Das V (2015) Affliction: health, disease, poverty. Fordham University Press, New York
4. Farmer P (1999) Infections and inequalities: the modern plagues. University of California Press, Berkeley
5. Gröger U, Wiegrebe L (2006) Classification of human breathing sounds by the common vampire bat, *Desmodus rotundus*. BMC Biol 4:18. https://doi.org/10.1186/1741-7007-4-18
6. Gadre G, Satishchandra P, Mahadevan A, Suja MS, Madhusudana SN, Sundaram C, Shankar SK (2010) Rabies viral encephalitis: clinical determinants in diagnosis with special reference to paralytic form. J Neurol Neurosurg Psychiatry 81:812–820. https://doi.org/10.1136/jnnp.2009.185504
7. Geissler PW (2013) Public secrets in public health: knowing not to know while making scientific knowledge. Am Ethnol 40(1):13–34
8. Gilbert AT, Petersen BW, Recuenco S, Niezgoda M, Gómez J, Laguna-Torres VA, Rupprecht C (2012) Evidence of rabies virus exposure among humans in the Peruvian Amazon. Am J Trop Med Hyg 87(2):206–215
9. Good BJ (2008) Medicine, rationality, and experience: an anthropological perspective. Cambridge University Press, Cambridge
10. Latour B (1987) Science in action. Harvard University Press, Cambridge, MA
11. Launiala A (2009) How much can a KAP survey tell us about people's knowledge, attitudes and practices? Some observations from medical anthropology research on malaria in pregnancy in Malawi. Anthropol Matters 11(1):1–13
12. Manderson L, Aaby P (1992) An epidemic in the field? Rapid assessment procedures and health research. Soc Sci Med 35(7):839–850
13. Médecins du Monde (2011) The KAP survey model (Knowledge, attitude and practices). https://www.medecinsdumonde.org/en/actualites/publications/2012/02/20/kap-survey-model-knowledge-attitude-and-practices
14. Minkler M, Wallerstein N (eds) (2008) Community-based participatory research for health: from process to outcomes, 2nd edn. San Francisco, Jossey-Bass

15. Montoya MJ, Kent EE (2011) Dialogical action: moving from community-based to community-driven participatory research. Qual Health Res 21(7):1000–1011
16. Nadal D (2020) Rabies in the streets: interspecies camaraderie in urban India. Penn State University Press, University Park
17. Nichter M (1993) Social science lessons from diarrhea research and their application to ARI. Human Organization 52(1):53–67
18. Nichter M (2008) Global health: why cultural perceptions, social representations, and biopolitics matter. University of Arizona Press, Tucson
19. Pigg SL, Pike L (2001) Knowledge, attitudes, beliefs, and practices: the social shadow of AIDS and STD prevention in Nepal, South Asia. J S Asian Stud 24(sup001):177–195
20. Rupprecht CE, Salahuddin N (2019) Current status of human rabies prevention: remaining barriers to global biologics accessibility and disease elimination. Expert Rev Vaccines 18(6):629–640
21. Sangaramoorthy T, Kroeger KA (2020) Rapid ethnographic assesments: a practical approach and toolkit for collaborative community research. Routledge, New York
22. Sodikoff GM (2017) Multispecies epidemiology and the viral subject. In: Heise UK, Christense J, Nieman M (eds) The Routledge companion to the environmental humanities. Routledge, London, pp 112–119
23. Viveiros de Castro E (1998) Cosmological deixis and Amerindian perspectivism. J R Anthropol Inst 4(3):469–488

Chapter 10
Rabies, Medicine, and Culture: Dogs, Disease, and Urban Life in the United States, 1840–1920

Jessica Wang

Abstract The specter of rabies has cast long shadows throughout human history. This essay surveys some key medical and cultural challenges associated with dogs and disease in U.S. urban settings in the nineteenth and early twentieth centuries. Physicians puzzled over the constitutional and contagionist dimensions of *rabies canina*, and struggled to identify therapeutic strategies to counter the unremittingly deadly disease of hydrophobia in symptomatic human patients. Meanwhile, competing claims of the dramatic curative properties of madstones and other vernacular remedies tested urbanites' abilities to distinguish truth from bunk in a nineteenth-century society preoccupied with the problems of reliable knowledge in a democratic society. In an urban world that encompassed both humans and domesticated animals, dogs themselves and their unruly presence on city streets also fomented 'mad dog' panics, as well as generalized anxieties about urban insalubrity and the tenuous status of civilization in the modern metropolis. Rabies was more than a disease: it fueled imaginings about the city itself and its myriad dangers.

Keywords Communications · Control · Dogs · Health department · New York · Reporting · Urbanization

Rabies has exerted a tenacious hold on the human imagination and elicited a sense of horrified fascination for millennia. As the historian of medicine Lise Wilkinson once noted, written accounts of rabies date "further back in time than any other infectious or contagious disease."[1] The earliest possible known references appeared

---

[1] Lise Wilkinson, "Understanding the Nature of Rabies: An Historical Perspective," in *Rabies*, ed. James B. Campbell and K. M. Charlton (Boston and Dordrecht: Kluwer Academic Publishers, 1988), 1.

J. Wang (✉)
University of British Columbia, Vancouver, British Columbia, Canada
e-mail: jessica.wang@ubc.ca

C. E. Rupprecht (ed.), *History of Rabies in the Americas: From the Pre-Columbian to the Present, Volume I*, Fascinating Life Sciences,
https://doi.org/10.1007/978-3-031-25052-1_10

about four thousand years ago, on Mesopotamian tablets that referred to enraged or frenzied dogs in terms that suggested behavior out of the ordinary and illness rather than plain viciousness. Texts in Sanskrit, classical Chinese, and ancient Greek bore witness to mad dogs' incursions in the Mediterranean, the Indian subcontinent, and China some 2000–2500 years ago, while Byzantine sources, medieval Islamic texts, and European writings in the centuries prior to the early modern period spoke to the durability and historical continuity of human experiences with rabies.[2] By the nineteenth-century, when physicians and laypersons in the United States compiled natural histories of the disease and pondered over its conundrums, their ruminations rested upon thousands of years of human experience with a scourge associated most often with dogs or other canine species, characterized by uncontrollable spasms and other neurological signs, and invariably fatal.

To contemplate the history of rabies from the mid-nineteenth to the early twentieth century in the United States is to encounter a world of medical understanding and strangeness both particular to its own historical moment and yet familiar from a contemporary perspective. This essay explores some of the cultural dimensions of rabies from the perspective of both medical knowledge and urban society. Both physicians and the public at large found in rabies opportunities to test their powers of discernment and ability to make sense of a perplexing and horrifically deadly affliction. Rabies and the dogs that transmitted the virus also offered a cultural terrain upon which American urbanites projected all of their multiple anxieties about the city itself, its unruly human inhabitants, and its alien and alienating built spaces.

The differences between past and present start with the disease itself. At one level, the malady in the past was not quite the same in a philosophical sense as rabies today. Every era and society possess its own conceptual apparatuses for defining and thinking about illness and disease. In the early nineteenth century, American physicians viewed hydrophobia and rabies as distinct conditions. Rabies, or *rabies canina*, referred to the canine madness that afflicted dogs and that could be transmitted to other animals via dog bite. When medical texts referred to a "virus" of rabies, they meant a poison or toxin at work, and not a self-replicating biological agent. Hydrophobia connoted the human variant of *rabies canina*, and some physicians argued that it should be considered a separate disease altogether. The distinction was not absolute, however. Although physicians rarely invoked rabies to reference human cases, they not infrequently described animals as suffering from hydrophobia. In lay discussions, such as newspaper stories about rabies outbreaks, the slippage in terminology appeared regularly in descriptions of rabid animals—particularly horses and livestock, but also dogs as well—as afflicted with hydrophobia. Sometimes the word choice hinged on whether the animal was a raging rabid perpetrator that had run on the loose and threatened infection with

[2] P. B. Adamson, "The Spread of Rabies into Europe and the Probable Origin of This Disease in Antiquity," *Journal of the Royal Asiatic Society of Great Britain and Ireland* 109 (July 1977): 140–44; Jean Théodoridès, *Histoire de la Rage: Cave Canem* (Paris and New York: Masson, 1986), chs. 1 and 2; James H. Steele, "History of Rabies," in *The Natural History of Rabies*, ed. George M. Baer, vol. 1 (New York: Academic Press, 1975), 1–29, on 1–5.

every vicious snap and bite, or the helpless hydrophobic victim of a rampaging mad dog. At other times, linguistic usages were simply arbitrary. Not until the 1880s did hydrophobia begin to lose favor among physicians, who increasingly found the term inaccurate, imprecise, and too redolent of folk knowledge or popular belief. The shift in terminology occurred only gradually, however, and rabies did not become the standard label until the 1920s.

The constitutionalism at the center of nineteenth-century American medicine's conceptual apparatus also defined the gap between past and present. Prior to the rise of germ theory, medical authority hinged on constitutionalism, that is, the sum total of an individual's specific bodily condition, habits, state of mind, and relationship to local conditions, which required a knowledgeable and observant physician to diagnose illness properly and develop a course of treatment suited to the physiology, temperament, and environmental setting of a given patient. From this perspective, disease states emerged almost exclusively from disturbances in bodily equilibrium, with only rare exceptions explained by a specific condition or an infectious disease agent.[3] When the prominent Massachusetts physician and Revolutionary War veteran James Thacher scorned "ridiculous specifics" as remedies in his 1812 treatise on hydrophobia, for example, his opprobrium signaled not just derision for measures considered more or less as dubious in the early nineteenth century as they are today, but the defense of an entire system of medical knowledge.[4] Thacher and other physicians over two successive generations disdained specific remedies because true medical practice did not rely on a set formula out of a bottle to effect a sure-fire cure for a given condition. Hence, the American physician J. G. Norwood, in his 1838 edition of William Buchan's *Domestic Medicine*, urged his readers to reject "pompous prescriptions" and promises of "great cures."[5] Patent nostrums such as an "anti-spasmodic tincture" advertised in the *Brooklyn Daily Eagle* in 1846, therapeutic mixtures based on salt, vinegar, and other common household items, secret family recipes for curative poultices or elixirs, or the application of madstones and other folk remedies all smacked of superstition and quackery, not simply because allopaths thought they relied on homespun nonsense or the humbuggery of mystery. Rather, skilled and artful treatment rested upon the physician's discerning eye and ability to determine the bodily disturbances at work in a particular patient, and to correct those imbalances by drawing judiciously from the established pharmacopeia of the materia medica.[6]

---

[3] John Harley Warner, *The Therapeutic Perspective: Medical Practice, Knowledge, and Identity in America, 1820–1885* (Cambridge, MA and London: Harvard University Press, 1986), 64–65.

[4] James Thacher, *Observations on Hydrophobia, Produced by the Bite of a Mad Dog, or Other Rabid Animal* (Plymouth, MA: Joseph Avery, 1812), 17.

[5] William Buchan, *Domestic Medicine: A Treatise on the Prevention and Cure of Diseases…Revised, Enlarged, and Adapted to the Diseases of the United States…by J. G. Norwood, M.D.* (Cincinnati: U. P. James, 1838), xi.

[6] Advertisement for an "Anti-Spasmodic Tincture," *Brooklyn Daily Eagle*, 15 June 1846, 2. For a more extensive account of rabies remedies and allopaths' medical authority, see Jessica Wang, *Mad Dogs and Other New Yorkers: Rabies, Medicine, and Society in an American Metropolis, 1840–1920* (Baltimore: Johns Hopkins University Press, 2019), 90–105.

Thus, when it came to treating an agonized hydrophobia patient, the allopathic perspective held that only a proper physician could determine whether unremitting spasms required mercury-based compounds or other depletive measures to calm a nervous system in overdrive, or stimulative strategies to prop up bodily systems in disarray and in need of tonicity to restore proper functioning.[7] When John H. Griscom administered opium to a hydrophobia patient at New York Hospital in 1855, his choice reflected not merely an attempt to sedate the patient, but an effort at therapeutic stimulation, since medical professionals understood opium as one of the "medicines which excite the whole system into action."[8] Griscom also employed ammonia carbonate, brandy, and beef tea, all known for their different stimulative effects. In addition, some nineteenth-century reports suggested that ammonia carbonate provided an antidote to venomous snake and insect bites, so Griscom likely hoped the chemical compound might have some efficacy against the toxic bite of a mad dog.[9]

That we today find such measures unimpressive is not the point. Nineteenth-century American physicians also knew full well the uniform fatality of rabies, but they attempted to mitigate the disease processes at work as best they could, according to the principles, scientific concepts, and discretionary art of medical practice as understood in their place and time. Moreover, Griscom himself was no ordinary physician. As a pioneer in public health and a skilled medical professional who combined a sense of religious conscience with an ardent commitment to sanitary reform, he held prominent positions within the city's elite medical institutions and philanthropic circles. His approach to medicine and public health reflected, as the historian James H. Cassedy once observed, a "preoccupation with physiology, vital statistics, ventilation, local politics, and precision in medical terminology," and his command in these areas represented cutting-edge professionalism in the mid-nineteenth century.[10] The stimulative measures that he tried on his suffering rabies patient rested upon established principles of judicious medical perception circa 1855.

---

[7] Wang, *Mad Dogs and Other New Yorkers*, 96–98.

[8] For the quotation on opium's stimulative effects, see [John C. Gunn], *Gunn's Domestic Medicine, or Poor Man's Friend, in the Hours of Affliction, Pain, and Sickness*, 4th ed. (Springfield, OH: John M. Gallagher, 1835), 619. Griscom's full case record of the patient he treated can be found in "Rabies Canina (Dr. Griscom)," 14 May [1855], from Patient Medical Records, [Case Book], Medical Division, New York Hospital, vol. 22, pp. 51–55, Medical Center Archives, Cornell University Weill Medical College, New York, NY. In accordance with the regulations of the Weill Medical School, I have not identified the patient, even though he was named in contemporaneous press coverage of the case.

[9] Alfred Stillé, *Therapeutics and Material Medica. A Systematic Treatise on the Action and Uses of Medicinal Agents, Including Their Description and History*, vol. 1 (Philadelphia: Blanchard and Lea, 1860), 675–76.

[10] Charles E. and Carroll S. Rosenberg, "Pietism and the Origins of the American Public Health Movement: A Note on John H. Griscom and Robert M. Hartley," *Journal of the History of Medicine and Allied Sciences* 23:1 (January 1968): 16–35; James H. Cassedy, "The Roots of American Sanitary Reform 1843–47: Seven Letters from John H. Griscom to Lemuel Shattuck," *Journal of the History of Medicine and Allied Sciences* 30:2 (April 1975): 136–47, quotation from Cassedy, 139.

Although constitutionalism ruled the day in Griscom's time, physicians also understood rabies as a rare instance of transmissible disease. As Thacher noted, "Whether this disease arises spontaneously in the body, or results from a specific contagion, rabid animals are capable of communicating the infection, through the medium of their teeth and saliva, to others, and also to the human species."[11] Early nineteenth-century physicians still debated the question of whether rabies erupted spontaneously, due to some kind of bodily derangement, or if it resulted from a contagious process, but either way, the rabid animal developed a transmissible condition. Within a few decades, medical consensus shifted decisively towards transmission as nearly the sole route by which the disease propagated itself, although the potential for long incubation periods suggested constitutional factors determined whether human exposure to the rabies "virus" resulted in actual disease. In 1856, for example, Job Lewis Smith, then a young physician making his career in New York City, published a remarkable review of 131 case reports of hydrophobia. Based on his investigation, he thought spontaneous emergence of rabies in dogs and hydrophobia in humans occurred in rare instances, and he also cited constitutional factors in whether or not victims of bites from rabid dogs developed active cases of hydrophobia.[12] Smith would later make his mark as a founding figure, alongside his contemporary Abraham Jacobi, in pediatric medicine. That same year, Thomas W. Blatchford, a physician in upstate New York who maintained a lively presence in state- and national-level medical associations, issued a special report on hydrophobia at the behest of the American Medical Association. Blatchford weighed in on the question of constitutionalism versus contagion by stressing that "genuine rabies… *never proceeds from any other cause*" than transmission from the bite of a rabid animal, but he also cited the "*constitutional irascibility*" of the dog as a source of canine susceptibility to rabies.[13] As late as the 1870s, medical treatises left room for the possibility of spontaneous rabies in extremely rare cases. The gradual rise of germ theory, however, meant that over the next twenty years, they increasingly embraced disease specificity over constitutional disturbance and accepted that microscopic disease agents caused specific medical conditions, including rabies.

Allopathic medical conceptions of rabies co-existed in an uneasy dynamic with popular ideas about preventatives and cures. Sometimes, a shared conventional wisdom crossed medical and lay understandings. For example, prior to Pasteurian rabies vaccination, which first became available in Paris in late 1885, cauterization or excision of the wound offered an imperfect but useful preventative measure for bites from suspected mad dogs. Early medical statistics suggested that cauterization of the wound reduced considerably a person's chances of developing rabies from

[11] Thacher, *Observations on Hydrophobia*, 15.

[12] J. Lewis Smith, *Report of a Case of Hydrophobia; with Statistical Observations* (New York: Holman & Gray, Steam Printers, 1856), 13–15, 33–35, 43–44.

[13] Thomas W. Blatchford, *Hydrophobia: Its Origins and Development, as Influenced by Climate, Season, and Other Circumstances. Being the Report of the Special Committee Appointed by the American Medical Association, and Read at the Meeting in Detroit, MI, May, 1856* (Philadelphia: T. K. and P. G. Collins, Printers, 1856), 9.

the bite of a rabid dog. The French veterinary expert Henri M. Bouley, in an extensive lecture on hydrophobia that the esteemed French émigré veterinarian Alexandre Liautard translated and published in English in 1874, recommended cauterization with a hot iron and cited French statistics that showed a decline in hydrophobia mortality from 84% in cases of untreated wounds versus 31% when bite victims likely exposed to rabies sought immediate cauterization.[14] News accounts also regularly stressed that cauterization or excision constituted the only effective means of preventing dangerous animal bites from progressing to hydrophobia. Newspapers advised dog bite victims to have their wounds cauterized by either chemical means or a hot iron as soon as possible, and city dwellers generally knew to make their way to a pharmacist or physician as needed. Even years post-Pasteur, cauterization remained an important line of defense, especially since rabies vaccination was not easily accessible until the development of injection kits with a long shelf life that could be mailed to doctors' offices in the early 1900s. In 1899, experiments on guinea pigs suggested the dramatic utility of cauterization, if properly performed with the strongest chemical agents. Under laboratory conditions, cauterization with nitric acid within 24 h of injection with a solution containing rabies virus produced survival rates of 91%. Consequently, the study's authors stressed the importance of cauterization even for bite victims who had the means to seek vaccination.[15]

Where medical and lay opinion converged on the benefits of cauterization, in other respects, popular culture generated an unwieldy mixture of ideas and practices at odds with professional understandings. The notion of the "dog days" of summer and heightened canine susceptibility to animal madness during the hottest months of the year continued to circulate, despite Blatchford's AMA report and other mid-nineteenth century studies that documented the year-round threat of rabies. Newspapers perpetuated the seasonal myth by accelerating their rabies coverage during the summer months, even as they also solemnly reminded audiences that modern medical data demonstrated rabies' year-round occurrence. As one reader commented acerbically to the *New York Times* in June 1903, "Hydrophobia, Mr. Editor, is the Summer newspaper's standby until the sea serpent commences to be reported."[16] Stories about mad dogs on the run, gripping accounts of suffering and dying patients, and reports of purported preventive and therapeutic regimens ranging from the mildly plausible to the completely outrageous all provided the allure of amusement and spectacle in the era of the penny press and its nineteenth- and early-twentieth century version of infotainment (Fig. 10.1).

Over the decades, American newspapers, magazines, and medical journals reported on a dizzying array of folk treatments, family recipes, secret formulas, and patent remedies that claimed prophylactic or therapeutic effects on rabies. Physicians' favorite choices from established chemical arsenals, new botanical

---

[14] H. Bouley, *Hydrophobia: Means of Avoiding Its Perils and Preventing Its Spread*, trans. A. Liautard (New York: Harper & Brothers, Publishers, 1874), 44.

[15] "The Effect of Cauterization or Cleansing of Wounds Infected with Rabies, after an Interval of 24 h," *Medical Record* 55 (21 Jan. 1899): 107–08.

[16] Frederic Griffith, "Merely Tetanus" [Letter to the editor], *New York Times*, 15 June 1903, 6.

Fig. 10.1 A summer scene in New York City, 1874 illustration from Harper's magazine

sources of drugs from around the world, and medical and lay theories alike about the therapeutic benefits of vapor baths as methods for preventing or treating hydrophobia also made the rounds. All such remedies had their respective partisans and earnest defenders, even if news stories frequently adopted a tongue-in-cheek attitude towards curative claims beyond the allopathic materia medica. At the same time, the uniform fatality of symptomatic rabies also made every proposed treatment both equally plausible and equally outlandish. Moreover, the cultural meaning and significance of nineteenth-century hydrophobia therapeutics derives from more than simply whether or not they worked. Beneath the entertainment value of strange and mysterious remedies lay more profound tests of truth and discernment in a democratic society.

Reports of madstones and their virtues testify to the premium that nineteenth-century urban culture placed on challenging Americans' ability to distinguish between truth and bunk. Madstones were the nineteenth-century American version of bezoars, snake stones, or tabasheer—objects derived either from particular mineral formations or from animals' bodies that were purported to have spectacular healing properties (Fig. 10.2). In the American south, parts of the Midwest, and as far west as Oklahoma and Texas, local madstones earned powerful reputations for their ability to protect or cure by filtering poisons from the body. In these regions, a fearful dog bite victim might go to the possessor of a madstone, who would soak the stone in warm milk and apply it to the wound. If the madstone stuck, that meant it was removing toxins. After several hours or even a couple of days, the madstone would fall off and undergo another soaking. At this point, the appearance of green scum in the milk signified the poisons removed successfully from the body, and the bite victim could depart with a confident feeling of freedom from the danger of hydrophobia.

Some madstone owners provided their services free of charge, while others supplemented their income by charging perhaps \$5 or \$10 per treatment. Named stones, such as the Mascot in Kansas City or the Fauquier madstone of Fauquier County, Virginia, boasted hundreds of successes, and failure only if patients arrived already symptomatic with rabies. Even after Pasteurian rabies vaccination first became readily available in the United States, starting with New York City in 1890, the madstone remained a more viable choice for many Americans. By 1899, only New York City, Chicago, and Baltimore offered preventive, postexposure rabies vaccination, and the procedure required an extensive series of injections with attenuated virus of graduated strengths in order to build immunity over a period of 2 to 3 weeks.[17] The

Fig. 10.2 Examples of various Madstones or Bezoars used historically in folk medicine for treatment against poisons or infections, such as rabies. (Courtesy Gerhard Elsner, German Pharmacy Museum in Heidelberg Castle, https://commons.wikimedia.org/wiki/File:Bezoare.jpg)

[17] Anti-Rabic Institutions," *Journal of the Massachusetts Association of Boards of Health* 9 (July 1899): 46–47.

prohibitive costs of a long journey and protracted absence from home meant that despite provisions for free vaccination for the indigent, madstones constituted a far more accessible option in many parts of the country, especially for those Americans already attuned culturally to believe in their wondrous capacities.

In the urban northeast, madstones occupied a distinctly different cultural space, as objects that provided city dwellers a chance to test their powers of discerning judgment in a society nearly obsessed with the specter of deceit, fraud, and humbug in an era of scammers and men on the make. The rise of large cities and the world of strangers that they represented deeply unsettled Americans, particularly the specter of enterprising chiselers ready to fleece trusting rubes fresh from the countryside with all manner of schemes and swindles. The New York police force in the 1860s, for example, identified a seemingly endless array of scams carried out by con artists—the "Spanish Prisoner" racket, the work of "damper sneaks," the machinations of the "watch-stuffer"—in what the historian Karen Halttunen once described as a world of "swindles, frauds, forgeries, counterfeiting activities, and other confidence games."[18] The ever-looming threat of the confidence man represented just one example of the jarring mixture of opportunity and fraud that accompanied the dizzying new possibilities of the market in an expanding national economy. Indeed, the entire market economy itself, with its anonymous relationships in which fortunes were so easily won and lost, all too easily resembled little more than a gamble and a confidence game.

In response, nineteenth-century urban culture generated a host of institutions and sensational episodes that confronted popular audiences with all manner of opportunities to cultivate keen powers of observation and judgment. Through such forums, narratives and displays of natural marvels allowed Americans to test their ability to ward off dangerous credulity in a world rife with confidence men and their deceptions. P. T. Barnum's American Museum, which operated in New York City from 1841 to 1865, constituted one such examination site. In 1842, for example, his famed display of the Feejee Mermaid, a mammalian-piscine hybrid purportedly discovered on an expedition to Fiji by Dr. J. Griffin of the British Lyceum of Natural History, entertained viewers with the chance to come to their own conclusions about a fantastic specimen. Although the actual object was rather more grotesque-looking than the mermaid appellation suggested, viewers could choose either to appreciate this bizarre creature from a distant and exotic sea as a representation of nature's strange and ever-diverse qualities, or they could condemn it as total bunk and hokum.[19]

Of course, the Feejee Mermaid was a hoax. Barnum hired the supposed "Dr. Griffin" to help sell the story of a strange natural wonder on display, and the object itself likely came from Japan, where craftsmen made these odd taxidermical

[18] Karen Halttunen, *Confidence Men and Painted Women: A Study of Middle-Class Culture in America, 1830–1870* (New Haven and London: Yale University Press, 1982), 7–8.

[19] James W. Cook, *The Arts of Deception: Playing with Fraud in the Age of Barnum* (Cambridge, MA and London: Harvard University Press, 2001), ch. 2.

constructions from the bodies of monkeys and fish.[20] But visitors were also in on the act, and they patronized the American Museum for the fun of determining for themselves whether the things they saw were truth or bunk in a natural world that contained all manner of bizarre phenomena. Nor was the line between fact and fiction necessarily so clear-cut. From a mid-nineteenth century perspective, was the Feejee Mermaid really all that more unlikely than Chang and Eng Bunker, the conjoined twins from Siam (now Thailand), joined at the sternum, who made their fortunes by exhibiting themselves and astounding audiences in the 1830s and early 1840s? Notably, the Bunkers also toured Europe at Barnum's behest in the late 1860s. Or in the professional world of American medicine, the same journal that railed against quackery in 1845 also reported in a tone of perfect seriousness about a case of spontaneous combustion of a living human being 6 years later, which suggests the fine line between truth and bunk in the mid-nineteenth century.[21] Nor did the Feejee Mermaid compare all that unfavorably with the *Medical Record*'s description in 1890 of a "parasitic foetus," a sensationally freakish oddity that would have raised eyebrows if brandished by Barnum, but that we would recognize today as a case of fetal resorption or *fetus in fetu*.[22]

In a society full of patent medicines and hyped-up therapeutic claims, dramatic medical advancements also raised questions about discerning judgment and how to distinguish between innovation and humbug. Thus, Louis Pasteur's development of preventive rabies vaccination through the application of new experimental techniques in the bacteriological laboratory appeared dubious to some reputable physicians precisely because the French savant's methods seemed overly reminiscent of the same forms of mystery and promises of wondrous cures that surrounded secret formulas and patent remedies. In November 1885, George F. Shrady, editor of the *Medical Record*, viewed Pasteurian rabies vaccination with a high degree of skepticism: "Pasteur's process, by which he prepares his virus and his subjects, is a very complicated one, and reads more like the performance of a conjurer than the methods of a scientist." Shrady became convinced of the validity of Pasteur's results within a few years, but his initial reaction vividly illustrates a central conundrum of the self-consciously modern age of the late nineteenth century, namely, the question of how one was to distinguish between the truth of seemingly miraculous new medical developments, versus the bunk of the quack or the con man.

In a culture defined by constant tests of credulity, urban newspaper stories about madstones challenged readers to decide for themselves the truth or falsehood of fantastic medical possibilities. Madstones themselves hardly ever appeared as actual prophylactic or therapeutic options in large cities. In the 80-year period that I

[20] On the likely Japanese origins of the Feejee Mermaid, consult the account by Harvard University's Peabody Museum: https://www.peabody.harvard.edu/feejee-mermaid (accessed March 2021).

[21] Charles E. Lee, editor, "To the Patrons of the Journal," *New York Journal of Medicine* 5 (July 1845): i (for a statement against quackery); W. J. Byrne, "Case of Spontaneous Combustion," *New York Journal of Medicine*, n.s. 7 (November 1851): 399, reprinted from the *Nashville Journal of Medicine and Surgery*.

[22] "A Parasitic Foetus," *Medical Record* 37 (15 Mar. 1890), 304.

studied from 1840 to 1920, I did not come across a single reference to treatment by madstone in New York City, although I did find mentions of madstones applied by a Chicago bookseller in 1888 and an unspecified bearer in Cleveland, Ohio in 1893.[23] Madstones did attract attention in press accounts, which ranged across straightforward reports about the objects' natural histories and treatment records, capsule stories about hydrophobia cases that testified to successful and failed usages, tongue-in-cheek coverage of madstones' status as curiosities, and uncompromising debunkings. Taken together, this jumble of narratives testified to a society that yearned for mystery and wonder, yet cast a skeptical eye upon extraordinary claims.

For every story that described madstones as a sham, another documented their strengths. In 1892, a commentary in the *Brooklyn Daily Eagle* typified madstone denunciations when it described "stories of mysterious stones that have come from the far east or that were bought at enormous prices from voodoos and hoodoos and wizards" as "all humbug."[24] Other writers attempted to maintain the allure of wonder, such as a report from Kentucky in 1872, also published by the *Eagle*, that described a girl whose suffering had reached such an advanced stage that "her eyes…were of a dark green, and of that strange expression which is peculiar to animals [in] a state of hydrophobia" until the "victory" of the madstone.[25] Some medical practitioners wrote in to scoff, while others recounted cases in which humans treated by madstone lived, while livestock animals bitten by the same rabid dog died.[26] Occasionally, an individual author had difficulty deciding whether to validate or debunk. When the *New York Times* published a detailed story on the Fauquier madstone in 1871, the lede centered on the ubiquity of humbug: "There may be limits to human credulity, but we do not often get a glimpse of them…belief in things absurd, and, as matters of physical science, impossible, was never in the world's history more prevalent than now." The Fauquier madstone, according to the article, constituted just such an exemplar of "widespread credulity." After describing the madstone's provenance and seemingly reliable testimonials about the hydrophobia cures it had effected, however, the *Times* concluded rather more agnostically that "the story of the 'Fauquier Mad-stone' certainly opens the door to a fresh field of conjecture, wherein the part played by the imagination may be newly analyzed, and, perhaps, more fully explained."[27] Ultimately, for this hard-boiled reporter, the madstone stretched plausibility but did not break it. In an era of constant and seemingly marvelous discoveries, including germ theory and its revelations about a living world invisible to the naked eye, or x-rays and their validation of strange, invisible natural forces, finding the proper balance between gullibility and nihilism

[23] *Brooklyn Daily Eagle*, 24 June 1888, p. 4; "Death from Hydrophobia," *Brooklyn Daily Eagle*, 9 October 1893, p. 5. Tellingly, the Chicago madstone was originally from South Carolina.

[24] "The Madstone Humbug," *Brooklyn Daily Eagle*, 10 July 1892, p. 6.

[25] "The Kentucky Girl and the Madstone," *Brooklyn Daily Eagle*, 30 July 1872, 2.

[26] "Madstones and Scientists," *Brooklyn Daily Eagle*, 26 July 1884, 4.

[27] "The 'Fauquier Mad-Stone,'" *New York Times*, 1 February 1871, 4.

posed a fundamental challenge, particularly in an urban world seemingly filled with cheats and scoundrels ready to pounce on every passing mark. Madstones and multiple other rabies preventatives and cures constituted just one part of an urban cultural environment replete with tests of credulity.

Dogs themselves and the dangers they represented also shaped the perception and experience of urban life. The robust and chaotic public spaces of the nineteenth-century American city, with their teeming human throngs, horse-drawn vehicles, cattle drives, and roaming livestock, threw people and animals together on a daily basis. Urban dairies, stables for cows and horses, slaughterhouses, and other businesses kept large, domesticated animals within city limits. Poor and working-class residents not infrequently lived side-by-side with cows, pigs, goats, various poultry species, and other animals to supplement their meager subsistence. Although New York City tried to ban loose hogs starting in the 1860s, swine continued to mix on city streets until the end of the nineteenth century, while goats and chickens remained until well into the 1910s. The fact that San Francisco still had a "two-cow limit" as of 1912 says volumes about the necessity of animal-keeping as a means of urban survival.[28]

Within this urban world of domesticated animals, cats and dogs roamed freely, and in untold numbers. In 1912, New York City estimated its dog population at 400,000 to 500,000, but the figure seems little more than arbitrary.[29] City residents certainly viewed their public spaces as filled with dogs of all sizes and colors. As the *Brooklyn Daily Eagle* put it in 1870, "Let an observer visit the least frequented as well as the most public thoroughfare, and he is sure to be confronted more frequently than he desires with roaming specimens of the canine race, from the huge mastiff to the diminutive poodle."[30] Owned dogs, strays, and animals with an in-between community status—known, fed, and played with by locals, but without established homes—mingled freely on city streets, where they amused or harassed human passers-by. Not surprisingly, ordinary moments could quickly ramp up to sheer terror, when suspicious-looking animals acted strangely or aggressively, and the cry of "mad dog!" sent crowds scrambling (Fig. 10.1).

Whenever dogs threatened, canine-human interactions enacted small dramas of urban social life and its demographically defined rhythms of age, gender, and class. In New York City on a spring day in 1889, for example, "a large black dog" bit a little girl at play in front of her house at 77 Hudson Street, about half a mile northwest of City Hall. The dog immediately went on the run and attacked another nine people along a mile-long path, including two 15-year-old boys from Brooklyn and Harlem and two men from Jersey City, in addition to five men and boys from the lower reaches of Manhattan. At Cortlandt Street, the animal's final victim grabbed

---

[28] Wang, *Mad Dogs and Other New Yorkers*, 19–20; Andrew A. Robichaud, *Animal City: The Domestication of America* (Cambridge, MA and London: Harvard University Press, 2019), 122.

[29] "Extreme Rigor Needed in Fighting the Ever Increasing Hydrophobia Malady," *New York Tribune*, 9 June 1912, p. A4; Wang, *Mad Dogs and Other New Yorkers*, 24.

[30] "Dogs at Large," *Brooklyn Daily Eagle*, 8 June 1870, p. 2.

the dog by the collar and held on until a policeman dispatched it by gunshot.[31] Similar episodes took place on a regular basis in late nineteenth-century New York City, and such momentary snapshots from daily life encapsulated larger patterns of rabies risk, as well as the general mobility of people and animals in, out, and about the city.

As this brief vignette indicates, area residents converged on Lower Manhattan, where City Hall, Wall Street, and commercial office space made the locale a center of municipal governance, finance, and business. Ferry traffic defined commuter networks that tightly integrated Brooklyn and nearby parts of New Jersey into New York City's economic life already by the mid-nineteenth century. Such connections might explain how a young Brooklynite and two men from Jersey City came to cross paths with the same aggressive dog on a random spring day in the late 1880s. Exactly what business brought an adolescent male all the way from Harlem remains unknown, but by the 1880s, elevated railroads reached all the way to the then predominantly German immigrant community of Harlem some seven or eight miles to the north, which tied uptown and downtown across the distant reaches of Manhattan.

Age and gender strongly conditioned vulnerability to dog bites and their attendant danger of rabies virus exposure, which made the little girl, adolescent boys, and men bitten by the same black dog absolutely typical. At the turn of the century, a child's chances of suffering a dog bite probably ran more than double the risk that an adult faced. In part, the heightened danger simply reflected the sheer extent to which children and dogs shared the streets. At a time of cramped and frequently unsanitary living conditions, particularly among the working-class immigrant communities that fueled the new industrial economy of the late nineteenth and early twentieth centuries, urban children spent an inordinate amount of time outdoors, where they worked and played. Dogs, whether owned or stray, also wandered about on their own, in an era that considered such animal peregrinations to be perfectly normal. The dog's collar in the aforementioned incident, for example, indicates the animal's status as a pet rather than a stray. Children at play on their own front stoops or in small yards, as in the case of the girl cited above, or hawking newspapers, people-watching, scavenging for goods, playing in schoolyards and city streets, enjoying outings to local candy stores, or otherwise participating in the mundane rituals of childhood inevitably crossed paths with unruly canines engaged in their own routines of urban street life.[32]

It is also no coincidence that men and boys predominated among the victims of the black dog that menaced pedestrians in lower Manhattan in the spring of 1889. Boys counted disproportionately among dogs' victims, out of a combination of both dogs' aggression and the tendency of male children to tease and bait animals at a time when Americans expected relatively little in the way of self-discipline and good behavior from either population. With the founding of the American Society for the Prevention of Cruelty to Animals in New York City in 1866, along with the

[31] "Two Dogs Run Amuck," *New York Tribune*, 22 April 1889, p. 3.

[32] Wang, *Mad Dogs and Other New Yorkers*, 38–40.

establishment of SPCAs in Philadelphia and Boston shortly thereafter, animal welfare advocacy and its ethic of kindness to animals established a toehold on the American cultural landscape. Decades would pass, however, before free-range dogs disappeared from U.S. city streets, and dogs trained to resist provocations from children became a standard expectation. In the nineteenth and early twentieth centuries, encounters between children and strange dogs ended badly all too often.[33]

Among adults, men outnumbered women as dog bite and rabies victims by a wide margin. Bouley identified statistics from France that suggested a 3:1 ratio among male and female hydrophobia victims. Reliable figures are difficult to come by, but data culled from newspaper reports suggests comparable ratios at work among dog bite victims in New York City in the 1860s, 1880s, and 1900s, and even more highly skewed numbers in deaths from rabies. Such conditions reflected the powerful role of gender in molding everyday urban life, in which women did not congregate on city streets to nearly the same extent as men. When unescorted women did spend time out of doors, they could count on the intercession of men and boys to chase off unwanted canine attention, albeit at the cost of the catcalls and other forms of male harassment that discouraged women from venturing out on their own in the first place. Among hydrophobia deaths in the greater New York City area reported in the *New York Times* during the 1900s, four out of the five women who fell victim to rabies died following bites from their own pet dogs, whereas only eleven out of twenty-three male victims had been bitten by animals known to them. As this simple statistic indicates, profoundly gendered differences in everyday social interactions extended to human relationships with domesticated animals, and men far more than women experienced the city as a world of strangers.[34]

These realities of animal-borne disease melded with all manner of fears about urban life. The discordant effects of urbanization itself, with its blend of strangers from all over the world, the roiling upheaval of the new industrial economy, an ever-quickening pace of life, and an environment laden with filth, stagnant indoor spaces, and challenged waste-disposal systems raised profound questions about the basic viability of the city. In an era ridden with apprehensions about urban existence and its threat to health and morals, some observers in New York City wondered whether dogs' nuisance value and susceptibility to canine madness meant they should be banned altogether. As one parent whose children barely escaped being bitten on city streets the previous summer plaintively asked early in 1870, "Why are dogs, any more than venomous serpents, tolerated? Why should a civilized community be kept in constant dread of a creature that is liable at any moment to inflict upon one or more of its members the most dire catastrophe possible?"[35] A decade later, sanitary inspector Charles P. Russel also objected to "vagrant curs" as an affront to civilized values and declared, "No such dangerous beasts should ever be permitted to

[33] *Ibid.*

[34] *Ibid*, 40–43.

[35] Letter to the Editor, "Slaughter the Dogs," *New York Herald*, 6 January 1870, p. 5.

infest the thoroughfares of a civilized town."[36] In 1887, New York mayor Abram Hewitt also created a stir when he cited the threat of rabies and suggested that "right minded persons should refrain from keeping dogs in the city."[37] *Tribune* reader "F.S.D." followed up in June 1890 by citing dogs' general uselessness and the dwindling numbers of pigs, cows, and other domesticated animals in the city: "Is it not time to take care of the dog as we have taken care of all other animals except the cat, the twin relic of barbarism?"[38] Questions about dogs' utility or lack thereof, their contribution to urban disorder, and the danger of rabies blended together to target dogs as an obstacle to civilization itself (Fig. 10.3).

The debate reflected deep-seated ambivalence about the city and its unhealthy environment. Urbanites wondered if city life itself might render some dogs more susceptible to canine madness than others. The rural-urban divide, with its anxieties about the insalubrity of the city versus the healthfulness of country living, meshed with misgivings about urban dogs. The city simultaneously symbolized not just civilization, but also ill health, devolution, and moral corruption, in contrast to the countryside, which well-established cultural tropes portrayed as superior both physically and morally. Agrarian life cultivated health and virtue, in contrast to the muck, grime, vice, and degradation of the city.[39]

Ideas about dogs' proper environment merged with discourses about place, health, and disease. Opponents of urban dogs regularly insisted that such animals properly belonged in the countryside, which suited their innate nature, and not in the city. As one Brooklynite observed in 1870, the old adage, "Let dogs delight to bark and bite/For God hath made them so" might "do very well for the country, but I am sure it was never intended for the latitude of this city....Of what use is a dog in the city?"[40] Ongoing efforts to expel livestock from urban spaces also raised expectations of a canine-less future safe from rabies. The *Brooklyn Daily Eagle*, for example, explained in June 1876 that the dog "is as much out of place in our crowded thoroughfares as the cow, the goat or the pig would be....In the country the dog is surrounded by conditions that fit his nature, and as a consequence *he rarely if ever*

[36] "Unmuzzled Curs," *New York Herald*, 2 March 1881, p. 8.

[37] Quoted in [untitled], *Brooklyn Daily Eagle*, 24 June 1887, p. 4. For reactions to Hewitt's pronouncement, see "Many Mayors on Dogs," *New York Herald*, 25 June 1887, p. 3; Letter to the Editor from "A Woman," "The Dogs' Indignant Friends," *New York Times*, 26 June 1887, p. 16; "The Case of the Dog," *New York Tribune*, 27 June 1887, p. 4; "The Fourth of July," *New York Times*, 1 July 1887, p. 4.

[38] Letter to the Editor from F. S. D., "To Prevent Hydrophobia," *New York Tribune*, 15 June 1890, p. 7.

[39] For a classic account of the agrarian myth in American history, consult Richard Hofstadter, *The Age of Reform: From Bryan to F.D.R.* (New York: Vintage Books, 1955), 23–36. By the late nineteenth century, overcivilization seemed to be encroaching upon country living as well, in the form of ragweed and other allergens that led American hay fever sufferers to begin to think about agriculture as a problem of civilization. Gregg Mitman, *Breathing Space: How Allergies Shape Our Lives and Landscapes* (New Haven and London: Yale University Press, 2007), 43.

[40] Letter to the Editor from A. W. Adsto, "A Growl at the Dogs," *Brooklyn Daily Eagle*, 21 June 1870, p. 3.

Fig. 10.3 Friend or menace – Competing images of dog control in nineteenth century New York City. (Illustration by Paul Frenzeny, Harper's Weekly, #1334, p. 444, July 15, 1882)

goes mad there [italics mine]. The city is no more fitted for him than it is for the propagation of swine."[41] Dogs' persistence, however, meant that the clarion call bore repeating nearly 15 years later: "There is no necessity for an untethered dog in the streets more than for an unharnessed horse, or the ancient and honorable municipal pig, or the contemplative can chewing goat; and there is, aside from hydrophobia, a much greater aggregate of nervousness attending the breach of good manners and good citizenship involved in obtruding the dog's presence everywhere in public that his master or mistress is sure to go."[42]

According to this line of thought, the healthful environment of rural life helped to keep canine madness at bay in the countryside, whereas both hydrophobia and generalized nervousness plagued dogs' and humans' lives in the city. Critics of urban dogs contended that the animals' habits, temperaments, and fundamental nature meant they could only remain healthy if kept in open spaces. Colonel George E. Waring, the well-known reformer who overhauled New York City's street-cleaning system in the mid-1890s with his famed "White Wings," a corps of disciplined street cleaners clad in white uniforms, succinctly phrased the sanitarian position on dogs. "A dog in the city," he declared, "is a dog out of place."[43] As the

---

[41] "Dogs and Dog Days in the City," *Brooklyn Daily Eagle*, 12 June 1876, p. 2.

[42] "Rambler," "Walks About the City," *Brooklyn Daily Eagle*, 27 July 1890, p. 6.

[43] "The Street-Cleaning Commissioner and the Dogs," *Our Animal Friends* 23 (Apr. 1896): 171. On Waring, see Martin V. Melosi, *Garbage in the Cities: Refuse, Reform, and the Environment*, rev.

Tribune observed in the summer of 1899, "A dog's habitat is the open country, where he can get the exercise and liberty which nature designed him to have. A dog that is shut up in a city house or cellar and restricted to the range of a tiny back-yard—and in a flat denied even that meagre privilege—is not in his natural habitat....[Dogs] belong in the suburbs and the open country, and not in the overcrowded and strictly policed town."[44]

Consequently, some urban observers associated indulgent and sedentary indoor pethood with the creation of a canine menace. As early as 1856, the *Brooklyn Daily Eagle* proclaimed, "There cannot be a doubt that confinement predisposes dogs to rabies, however well they may be tended and pampered; witness the many instances that occur among house pets."[45] More than a decade later, the *Eagle* added a gendered twist on the connection between coddled pets and the threat of canine nervousness when it contended that "the most dangerous class of dogs kept are the little pets of the ladies, the shivering, snarling little poodles, attenuated grayhounds [sic], and terriers."[46] This association of pampered animals with women in the fashionable elite tied feminized luxury to the moral dissipation of wealth and social refinement.[47] Other commentaries in the 1870s and 1880s also continued to debate the question of city dogs' debilitating indoor lifestyles as a source of rabies and other diseases. As the *New York Tribune* warned in 1876, "experience shows that it is not alone the vagrant, homeless, ill-treated dog that is likely to go mad, for the pets of the parlor and the kennel are equally exposed to the disease."[48]

The emergence of the vogue surrounding neurasthenia in the 1890s further reinforced the perceived connections between the enervated condition of spoiled pet dogs and their susceptibility to rabies. Medical theories about how the hustle and bustle of modern life left behind jangled and frayed nervous systems in its wake produced a veritable epidemic of urban nervousness that affected humans and animals alike. The noisy, chaotic rhythms of city life, according to the *Eagle* in March 1897, made animals "fretful and nervous," and therefore "more liable to do an injury to their owners and others," in contrast to "the brutes that live a normal life in the country."[49] *Our Animal Friends*, the lead publication of the American Society for the

ed. (Pittsburgh, PA: University of Pittsburgh Press, 2005), ch. 2.

[44] "Dogs and Their Habitat," *New York Tribune*, 6 July 1899, p. 6.

[45] "Mad Dogs in Prussia," *Brooklyn Daily Eagle*, 21 July 1856, p. 2.

[46] "The Hydrophobia Scare," *Brooklyn Daily Eagle*, 25 Feb. 1869, p. 2.

[47] See also "Women and Poodles," *Brooklyn Daily Eagle*, 17 Sept. 1880, p. 2; "Dogs and Muzzles," *New York Times*, 19 Apr. 1903, p. 6; Edward Marshall, "Pasteur Expert Sounds Warning Against Pet Dogs," *New York Times*, 27 Aug. 1911, p. SM11.

[48] "Hydrophobia," *New York Tribune*, 22 May 1876, p. 4. See also, for example, Letter to the Editor from "E.S.C.," "Dogs," *Brooklyn Daily Eagle*, 13 Apr. 1872, p. 1; "Mad Dogs," *New York Times*, 28 June 1874, p. 4; and "Scientific Gossip," *New York Times*, 26 Feb. 1882, p. 4. Some medical opinion even suggested that cancer attacked animals deprived of natural vigor by domestication: "they are made more subject to the disease when in a state of domestication than in their natural wild condition." "Cancer and the Doctors," *New York Herald*, 5 Dec. 1887, p. 9.

[49] "Ugly Dogs to Be Watched," *Brooklyn Daily Eagle*, 29 Mar. 1897, p. 6.

Prevention of Cruelty to Animals, put the issue bluntly when it alleged that dogs suffered from the same "nervous atmosphere of human society....In short, many dogs at the present time are affected with neurasthenia."[50] Consequently, as the *Brooklyn Daily Eagle* warned its readers, since "some dogs, as well as some people, have nerves," as a result "the pampered, overfed" dog also "has fits or rabies, and does harm."[51] From a turn-of-the-century perspective, city life seemed fit for neither man nor beast, and it was all too easy to imagine that the rattled nervous systems of overindulged pets devoid of canine vigor might become all too prone to hydrophobia.

A competing cultural discourse, however, blamed the stray cur—untamed, mired in filth, and already diseased—as the more likely candidate for hydrophobia. Nineteenth-century anxieties about the "disorderly classes" and the class conflict that accompanied an age of immigration, urban poverty, and industrialization melded with perceptions of dogs and disease. For example, the *Brooklyn Daily Eagle* postulated in May 1863 that the "ill-fed, ill-favored curs" running loose on city streets "are the animals most liable to rabies, and form par excellence the 'dangerous classes' of the canine race."[52] Racialized ideas about blood and lineage pervaded such notions. As the *New York Tribune* speculated in 1877, if hydrophobia ever emerged spontaneously in dogs, "it is probably so oftenest in the curs 'of low degree'—the street-pariahs of the race. Good blood tells in this as well as in the other matters of discretion, intelligence, and affection."[53] Nor were these ideas merely a matter of popular perception. Physician L. L. Dorr, writing in the *New York Medical Journal* in 1881, outlined the degeneration that he claimed came of racial mixing, whether in humans or in animals. In humans, he argued, "the crossing of different races only produces inferiors, natural criminals, and idiots." Cross-breeding in animals, Dorr alleged, produced similarly dire outcomes. Rabies, he speculated, originated from "the violation of this seeming law of Nature," namely, a natural order that supposedly kept breeds "pure and of the original race." He concluded, "If so, the prevention would be easy—that is, to allow none but dogs of a pure and original breed to live."[54]

The juxtaposition of dogs, health, and hygiene dovetailed perfectly with elites' cultural anxieties about the dangers of urban life, whether in the form of their fixation on the uncontrollable urban masses who threatened social upheaval, or the overcivilization that they feared portended their own declining health and fortunes. Stray dogs' filthiness spoke to the conditions that kept cities mired in barbarity and mocked pretensions of modernity and progress. For a white social elite nearly obsessed with the specters of urban disorder, unruly working-class populations, and

---

[50] "Dr. Elmer Lee on Rabies, Hydrohpobia, and the Pasteur Treatment," *Our Animal Friends* 25 (November 1897): 49–50, quotation on p. 50.

[51] "Care of Dogs in Summer," *Brooklyn Daily Eagle*, 15 July 1901, p. 8, excerpted from the *Babylon Signal*.

[52] "The Dog Ordinance," *Brooklyn Daily Eagle*, 11 May 1863, p. 2.

[53] "Dog Days in May," *New York Tribune*, 9 May 1877, p. 4.

[54] L. L. Dorr, "Rabies—A Possible Cause and a Probable Preventive," *New York Medical Journal* 34 (Nov. 1881), 470–76, on 473.

racial degeneration, dirty and dangerous mutts on the loose symbolized the tenuous hold of political stability and civilized values in the life of the American city.

The impossibility of health in the built environment's confined, unnatural spaces, however, also suggested the debilitating effects of overcivilized, neurasthenic lives. In accordance with Americans' more generalized anxieties about the city, when New Yorkers declared that dogs could not flourish in urban environments, the silent implication was that humans also fared best in the salubrious countryside, far away from the city's heightened mortality rates. If dogs did not belong in the city, then perhaps human beings could not thrive there either.

Defenders of dogs countered these grim portents of canine-generated decline with their own civilizational arguments. Animal welfare advocates contended that kindness towards animals encouraged the development of gentler, more refined values, particularly in children. In that sense, domestic animals promised to make urban life more civilized. Former New York mayor and governor of the state John T. Hoffman, who was also an ASPCA partisan, used such arguments in December 1885, when he defended dogs' urban presence by praising the society's mission of "not only humanizing animals, but of civilizing boys and men."[55] Pro-canine New Yorkers also reminded their fellow citizens of dogs' virtues. The *New York Tribune*, for example, responded to Mayor Hewitt's proposal in 1887 to make New York a non-canine city by declaring the dog "one of the most useful, amiable, faithful, and companionable of animals, and also one of the most intelligent." In case that encomium left any room for doubt, the *Tribune* added that "the natural tendency of the dog is toward beneficence."[56] Where opponents of dogs saw disease and dissolution, dogs' partisans identified the most admirable of character traits. As the *New York Herald* observed in 1883, "Next to religion and the love of wife and children comes the love of a good dog."[57] No doubt some readers would have ranked the dog's love first.

New York never launched serious efforts to ban dogs entirely. Policy conflicts revolved around the regulation of owned dogs and the elimination of strays, and calls for wholesale expulsion, however impassioned, never gained significant political traction.[58] Within the nineteenth-century ideology of domesticity, dogs' uselessness actually favored their hold on the city, as part of the sentimental family. Animals valued only for their utility became marginalized geographically when their absence better suited urban sanitation, as in the gradual removal of slaughterhouses and abolishment of livestock from city centers. Canine and feline family members,

[55] Letter to the Editor from John T. Hoffman, "Ex-Governor Hoffman on Dogs," *New York Herald*, 24 December 1885, p. 2.

[56] "The Case of the Dog," *New York Tribune*, 27 June 1887, p. 4.

[57] "Chat By the Way," *New York Herald*, 27 May 1883, p. 19.

[58] On the history of canine animal control in New York City, see Wang, *Mad Dogs and Other New Yorkers*, ch. 6.

however, were another matter entirely.[59] To that extent, the dog lovers easily won the day in a city in which most people, whether they liked dogs or not, assumed their presence as an inevitable part of human society. As a cultural phenomenon, however, New Yorkers' ruminations about dogs underscored the fears of civilizational decline that tailed urban expansion during the long turn of the century. Yet the ascendance of civilized values also threatened, through the proliferation of shattered nerves and constitutions left in tatters by an excess of refinement. The debate over whether dogs suffered from too little civilization or too much intertwined concerns about modernity, social development, and racial fitness together with fears of rabies.

Although rabies was a relatively rare disease in the United States during the nineteenth and early twentieth centuries and has blessedly remained so ever since, it nonetheless loomed large in medicine and culture. For physicians, hydrophobia confounded and challenged both basic medical understanding and hopes to treat and cure illness. The disease's resistance to all therapeutic measures once symptoms set in also generated all manner of creative possibilities for treatment, which left ordinary Americans to their own devices in deciding what to believe or dismiss in a world of modern progress that generated extraordinary novelties and complete flim-flam on a regular basis. The turn-of-the-century city itself emerged from a self-consciously modern age that transformed fears about dogs and rabies into more deeply seated anxieties about urban life as a source of dizzying and dangerous forms of change. From a potent brew of dogs, disease, and culture emerged nightmarish, canine-inflected images of a society in turmoil, in which the disruptive consequences of progress and regress fueled a rabid American imagination.

Acknowledgements This essay includes material adapted from Mad Dogs and Other New Yorkers: Rabies, Medicine, and Society in an American Metropolis, 1840–1920 and is published with the kind permission of Johns Hopkins University Press.

[59] On pets' familial position and role in the upbringing of children within the middle-class American family, see Susan J. Pearson, *The Rights of the Defenseless: Protecting Animals and Children in Gilded Age America* (Chicago and London: University of Chicago Press, 2011), 29–33 and 37; Katherine C. Grier, *Pets in America: A History* (Chapel Hill: University of North Carolina Press, 2006), 127–31; and Jennifer Mason, *Civilized Creatures: Urban Animals, Sentimental Culture, and American Literature, 1850–1900* (Baltimore and London: Johns Hopkins University Press, 2005), 13–18. On livestock and the spatial segregation of the American city, consult Robichaud, *Animal City*, chs. 2 and 3.

Chapter 11
Historical Disparities in Health: Rabies Surveillance, Risk Factors and Prevention

Amira Roess, Kis Robertson, and Sergio Recuenco

Abstract In the Americas, as throughout much of the world, rabies is a neglected disease because of who it affects, primarily poorer, indigenous, black, minority, rural inhabitants, and those who come into close contact with suspect wildlife and domestic animals. There is a broader context that should be considered when reviewing current disparities and proposing solutions for this zoonosis. For example, the colonial past of the Americas is manifested in structural racism in every country and has led to persistent disparities in every facet of life, particularly in applicable disease surveillance, basic health education, overt national policies, and tangible health care access and delivery. This is a nearly universal reality that non-white populations face lower access and outcomes in education, financial, and health outcomes. In this chapter we review the history of rabies in the Americas and recent developments within this broader societal context.

Keywords Rabies · Health · Colonialism · Structural racism · Disparities · New world · Old word · Indigenous populations

Introduction

The untold story of rabies in the New World, as with many health conditions, is entwined with that of colonialism and exploitation [1]. First, in current times, the disparities related to rabies, as with many infectious diseases, fall

A. Roess (✉)
Department of Global and Community Health, George Mason University, Fairfax, VA, USA
e-mail: Aroess@gmu.edu

K. Robertson
DVM, College Park, MD, USA

S. Recuenco
Facultad de Medicina San Fernando, Universidad Nacional Mayor de San Marcos, Lima, Peru

C. E. Rupprecht (ed.), *History of Rabies in the Americas: From the Pre-Columbian to the Present, Volume I*, Fascinating Life Sciences,
https://doi.org/10.1007/978-3-031-25052-1_11

disproportionately on (1) those sub-populations in the Western Hemisphere that inherited the legacy of the Colonial era, including the original populations of the Americas (i.e., the original indigenous inhabitants of the Americas, the Native Americans as they are referred in North America) and populations of color (i.e., primarily those who are descendants of slaves, rural dwellers, and those with the lowest income); and (2) those countries that more globally bear the brunt of their colonial past, those that are the most politically insecure, have the largest proportion of nonwhite populations, and the lowest GDPs. For example, Haiti and the Dominican Republic (i.e., the island of Hispaniola, as the first permanent settlement during European colonization) continue to be the worst performers in terms of economic development and health outcomes of any country in the Western Hemisphere, including in terms of rabies prevention and control [2–4].

Second, a deeper impact of colonialism on the overall lack of recorded history of infectious diseases, including rabies, is a product of the fact that the cultures of the indigenous populations were largely decimated through forced assimilation, other lethal infectious diseases, broad geographic incursions, and structural genocide [5, 6]. In the case of rabies, its history, and that of how zoonoses and other diseases first impacted indigenous populations as told from their perspective, is largely lost. This difference is made starker when compared to the rich body of knowledge of the subject from the Old World, going back to antiquity. Europeans were well acquainted with classical rabies (at least in dogs, but not in bats), which begs the question that if canine rabies was present in the Americas during the fifteenth century (especially in major population centers within Mesoamerica and the Andes), would it have been mentioned in colonial records? [7, 8] As regards the potential utility of paleoanthropology in the health of pre-colonial populations, little can be gleaned beyond suggestions of microbial involvement, as most infectious diseases provide evidence only via bones and teeth [9].

Lastly, the modern implementation of public health disease surveillance, of rabies and everything else, is parallel to this history, with higher income areas having better representation in surveillance data, and thus resources allocated to address identified health problems. This reality was highlighted most recently during the COVID-19 pandemic at both the global and local levels, with the highest income populations overrepresented in basic case counts at local and country levels. At the country level, entire low-income countries were barely represented in global case counts due to lack of testing resources and public health surveillance infrastructure upon which to build upon [10].

Missing Historic Records Set the Stage for Current Disparities on Rabies

Technology brought from the Old World to the New led to greater opportunity for indigenous people's information to vanish and for their history to be lost (i.e., because written, colonial tradition outpaced, and eventually replaced, the largely oral tradition of the original indigenous populations). The ancient Inca method of recording events and history in knots, the Kipu, was sophisticated and in use by the time Spanish colonizer arrived, but its interpretation was lost and efforts to decipher them are ongoing [11]. In contrast, the history of the Old World was retained because the emergence and spread of writing occurred at a gradual enough pace to coexist with the oral transfer of intergenerational knowledge. The slower replacement of the oral tradition there provided opportunities for documentation of ancient mythology, folklore, and history that were not possible in the New World. This is in contrast with the rapidity of which whole populations, their languages and cultures were decimated in the New World. Few ancient mythologies from the New World remain in the first place and the fact that no hints of hydrophobia/rabies have been found in this context do not serve as absolute proof that rabies did not impact the pre-colonial New World. Nevertheless, it is known that indigenous populations had interactions with dogs and bats and one review proposes that enough "...circumstantial evidence suggests that bat rabies was already present pre- Columbian colonization..." [12]. For example, fatal outcomes from bat bites were recorded to have been experienced by man and livestock alike by the colonists in the early 1500s, but no detailed descriptions of the symptoms nor direct observations by the colonists or by the original inhabitants of the Americans exist [12]. Whether canine rabies was simply introduced by Colonists to a naïve population or if it was circulating at a low level in sylvatic populations is a matter that will likely be settled only by enhanced surveillance and deeper molecular science. To date, such work so far suggests that it was likely that rabies was maintained by New World bats prior to dog-transmitted rabies [13–17]. If rabies was indeed endemic in the pre-Columbian Americas, it seems unlikely that generations of indigenous people would have been complete strangers to the disease, either in animals or humans. Whether their experiences would have been distinctive and non-sporadic enough to merit a place in their oral histories can only be speculated.

Canine rabies was not recorded by the colonists shortly after their arrival during the late 1400s, but reports emerged centuries later. Canine rabies would become the most common cause of human rabies and phylogenetic data support the notion that these viruses were introduced via incubating dogs from the Old World likely by accident, as were other pathogens [18]. The earliest and most well-known reports of large rabies outbreaks in the Western Hemisphere come from the 1700s and are written accounts primarily about the rabies exposures and cases observed and experienced by colonists [19–22]. In Texas, anecdotal accounts by cowboys were recorded in the mid-1800s and a detailed history of rabies there is largely told through the colonists perspective, with no prior mention by the local Native Americans [21]. This is largely the case throughout the Americas. Large rabies outbreaks among coyotes and other wildlife species were recorded during this time period in the Western part of the US, among Canadian settlers, and among other colonists. The size of some of the outbreaks suggests that, particularly in North America during the 1700–1800s, they would certainly have resulted in rabies virus exposures and cases, not only among the colonists, but also among exploited Native Americans, enslaved peoples, and their animals. This may be anticipated, given that certain dogs were bred specifically and used widely during the later Colonial Period to direct their aggression and to restrain, control, track, capture, or kill selected minorities, such as recalcitrant Indians or runaway slaves, a practice in part that evolved into the twentieth century [23]. Nevertheless, such case reports of hydrophobia among indigenous and enslaved peoples are notably absent from historical record. The colonial medical profession may have neglected such populations, in large part due to the development of a 'racial inferiority mythology' [24].

Canine Rabies in the Americas

By the twentieth century, rabies cases in humans have been dramatically reduced throughout the Americas. The laudable goal of elimination of dog-mediated human cases of rabies by 2030 globally [25] and PAHO's earlier target data by which to eliminate rabies from the Hemisphere resulted in numerous investments in the hemisphere [26]. Before appreciating the success of the elimination program in the Americas it is necessary to acknowledge that the complex colonial history of the Americas has implications on present day determinants of health, including structural barriers to care and investments in preventive services This directly traces back to colonial and post-colonial policies by the white establishment that systematically subjugated brown and black populations. Colonial policies physically relocated or exterminated indigenous populations, forcibly moved millions of Africans across treacherous seas to serve as slaves for centuries throughout the Americas, and eventually relegated power to the urban elite primarily through their blood ties. These historic realities have led to continuing discrimination in every aspect of society for these populations in every country of the Americas. In health, this is manifested by disproportionate mortality rates among indigenous Americans, populations of color, lower infant survival rates, and poorer incomes [27–29]. In addition, the lowest income populations – Indigenous Americans, populations of color, etc., have no political agency and similarly are faced with disparities in every aspect of health and development [30–33]. While some countries created government structures to approach indigenous peoples' health, a significant impact is still to be seen [30, 34].

In both the popular media and the limited historic (and not so historic) medical accounts in the Americas, descriptions of rabies among non-Caucasians are complex, if included. For example, in one recent book about the topic, rabies cases among minoritized populations in New York are not mentioned at all, while detailed accounts of cases among Caucasians are widespread [7]. Older peer-reviewed articles support the idea that populations of color faced disrespect at the health care interface [35]. The latter is a theme that is prevalent in the Americas with recent attention highlighting the disrespect that women of color face during childbirth in medical facilities [36, 37]. In numerous symbolic scenes from American literature, we find rabies representing evil, racism and domestic violence (for example, in the novels To Kill a Mockingbird and in Their Eyes Were Watching God) [38, 39]. In the former, the white local lawyer kills the rabid dog, saving the Alabama town from a “real” disease, while the town continues to suffer prejudicially from the social “disease” of racism [40]. The black protagonist in the latter piece does not receive medical care after getting bitten by her rabid husband, Tea Cake, who is also black, regardless of the availability of prophylaxis and despite the doctor telling the wife that her husband has rabies [38]. In the 1966 film, ‘Rage’, an unvaccinated Mexican construction worker, Fortunato, dies of rabies, after being drug into town and restrained, but the exposed alcoholic Caucasian doctor, Dr. Reuben, who is bitten by his own dog (which was bitten by Fortunato’s cat), is able against all odds to journey into town in time for vaccination [41]. Similarly, socioecomic differences in the US ‘Jim Crow Deep South’ are showcased symbolically in the 1991 film version of *Paris Trout*, based upon the novel by Peter Dexter, that won a National Book Award. Set during a rabies outbreak within a 1949 Georgia county, Rosie Sayers is a 12-year-old African American child, bitten by a fox. At the local clinic, her wound is barely wiped by an uncompassionate white nurse, who then washes thoroughly her own hands with soap and water. The older condescending white doctor offers Rosie the choice of a ‘shot in the stomach’ (brandishing a huge needle!) or a ride to her rural home. The traumatized child chooses the latter. Once reunited with her family, her mother examines Rosie’s wound and reassures her that she has not been ‘poisoned’ from the bite, which is left untreated, as a metaphor of poverty, race, and ignorance [42]. The comparison of rabies to a toxin by Rosie’s mother is a reflection of the generic term *virus* to have meant historically a ‘slimy poison’, before the finer etiological understanding of transmission. Moreover, such a misunderstanding harkens to an earlier time in the racially-divided southern US during the 1800s, when bites from ‘blue gum Negroes’ were believed erroneously to result in acute poisoning and an horrific death akin to hydrophobia [43].

Disproportionate Animal Bite Injuries – Native American and Rural Populations

Despite a lack of in-depth meaningful surveillance, several studies have reported a significantly higher rate of dog and other animal bites among Native American populations [44–46] with one study reporting up to 10 times more observed cases versus non-Native American populations [46, 47]. Compared to their urban counterparts, rural populations are more likely to experience dog bites [47] whereas access to PEP is more readily available in urban areas, particularly in low-income countries [48]. Young children (<5 years old) disproportionately experience severe injury from bites [46, 49], and young adult males are more likely to experience bites as a result of interaction with wildlife. For example, the average number of days between exposure and postexposure prophylaxis (PEP) is 4 days among Native populations in Canada and the number of animal bites has been increasing over time with no explanation, in part due to limited resources to explore this trend [48]. Few studies from lower-income countries in the Americas have produced similarly detailed information but the information available highlights the disparity in animal bites between native Americans and non-Native Americans with similar trends in demographics of severe animal bite injury throughout the Americas [48]. For example, in Brazil, more bites from wildlife are observed in rural areas and with the control of dog rabies the risk of rabies falls disproportionately among rural and indigenous populations [50], yet access to PEP and human health care remains limited in those areas [51].

Urban/Rural Divides

Regarding the stark urban/rural divide, it is worthwhile to note that in 1983, the Veterinary Public Health unit of the Pan American Health Organization (PAHO)/WHO Regional Office for the Americas began their official program for elimination of dog-mediated human rabies, with an initial objective to eliminate rabies from the principal cities of Latin America (urban). While it is logistically easier to focus on urban areas, it does further widen the gap between urban and rural health, with urban populations having better access to health and social services, economic and educational opportunities. After 1992, the program expanded to non-urban areas. Overall, the program has been a remarkable success with a decrease by more than 90% in the numbers of human and canine rabies cases in the Americas, particularly in urban areas [52]. Based upon PAHO surveillance data, human cases transmitted by dogs during 2021–22 were only reported in Bolivia, Cuba, and Haiti [53].

Distinctions in Strategies Based Upon Health Economics

In South America, canine rabies was eliminated from Uruguay, in 1968, and Chile in 1990. Argentina still has sporadic animal cases, but no more human rabies since 2008 [54]. Not unexpectedly, many of those successful countries are the ones with the highest GDP in the Hemisphere. Even among the wealthiest countries, the ability to avoid disparities in access to rabies prevention is tenuous in light of the global rise in nationalism (manifested in vaccine or public health nationalism) and xenophobia. In North American countries, for example, canine rabies was successfully eliminated through resource-intense canine vaccination efforts, anti-stray animal programs, and timely response when suspected cases and exposures were reported [55]. The multi-pronged approach taken regionally resulted in a significant decrease in human rabies cases and provided a model of what is possible in a high-resourced context. Nevertheless, outside of North America, limitations persist including incomplete data, and any removal of resources will drastically reverse progress.

The expense of these systems cannot be underestimated. Human PEP is expensive, outside of the Ministry of Health's budget in most countries. For example, in the US, approximately \$300 million (\$245 million – \$510 million) is spent annually for rabies prevention, vaccination, and control and the bulk of this spending comes from the medical care related to PEP. Approxiamtely30,000–60,000 courses of vaccination are given annually, costing about $228 million [55].

In many parts of the US, there is a stark disparity between the costs of animal and human vaccination. For animals, the price of rabies vaccination can vary between $10 to $20 for a single dose that lasts 1–3 years. In some jurisdictions, owners are guaranteed an affordable threshold above which the government will subsidize. In many cases, the government will cover the cost of testing if an animal is suspected of being rabid. In contrast, the cost of human PEP is out of reach for many low-income populations, many of whom are also from minority populations. The lack of meaningful access to health care in the US continues to disproportionately affect low-income and minority populations [56–59].

Trinational cooperation (US-Mexico and US-Canada), which began at least in the mid-1900s, remains critical to prevent and maintain elimination of canine rabies, and to reduce the incidence of other forms of rabies [60]. Cooperation between border states has been jeopardized in recent years with anti-immigration rhetoric and operations to forcibly detain and expel populations. These have led to fear among animal production workers who are most at risk of rabies virus exposure. Many are wary of approaching health departments for any health care, including vaccination [61, 62].

In Latin America, the dramatic reduction in dog-mediated human rabies cases was achieved by a complex program that included mass vaccination of over 50 million dogs annually, vaccination of people at risk of rabies exposure via PrEP or PEP, and public health surveillance.

The success of these campaigns in Latin America was due in no small part to the coordinating role of the public health sector and the involvement of many stakeholders in rabies management [63].

In contrast, canine rabies is difficult to control in the poorest New World countries, including Haiti, the Dominican Republic, and Guatemala, and in countries with persistent health and development disparities among indigenous and rural populations (e.g., Bolivia, Honduras, Peru, and Brazil), and significantly in the Amazon Basin (i.e., which includes territories in nine South American countries and more than 350 ethnic groups). Between 2013 and 2016, dog-mediated human rabies was reported only in Bolivia, Brazil, the Dominican Republic, Guatemala, Haiti, Honduras, Peru and Venezuela. In 2016, 10 deaths due to dog-mediated rabies were officially reported in the Americas – 8 in Haiti and 2 in Guatemala, and 23 deaths were reported due to rabies in species other than dogs, with 3 in Brazil, 2 in Colombia, 1 in Guatemala, 2 in Mexico and 15 in Peru [63, 64].

Historical Realities: The Example of Hispaniola

Elimination of dog-mediated rabies on the island of Hispaniola should have been achievable earlier in this century. Many islands have been able to successfully eliminate dog-mediated rabies and/or prevent the introduction of canine rabies through strict implementation of quarantine. The methods to do so have been in use for centuries. For example, several areas in Europe were free of canine rabies, long before the advent of vaccination. In the twentieth century, the UK reported its last infection in 1922 and last death from indigenous rabies in 1902. Similarly, most island nations in the Caribbean have no human or dog rabies cases. The standard public health approach includes initial investments to identify infected animals and vaccinate all dogs. From there, the key is to quarantine dogs that are imported.

Comparably, Haiti has the highest rate of human rabies deaths in the Western Hemisphere with estimates of ~two human infections per week. About seven deaths are officially reported and this is an underestimate largely due to less than ideal surveillance, testing infrastructure, and limited awareness about the disease [2]. The economic and social realities of life in Haiti are directly linked to poor uptake in the vaccination program with low-income and rural populations particularly less likely to participate in interventions, canine vaccination, and general human and animal health care, leaving them vulnerable to exposures [65, 66].

At a national level, the historic realities, which include often failing binational cooperation between Haiti and the Dominican Republic, political instability in both, and a weak public health care system have led to the stagnation in rabies elimination progress. Numerous regional missions to Haiti have been undertaken, some to establish collaborations between Haiti, Brazil, and the US CDC [2], but notably absent is their sovereign neighbor to the east. Like many global policy situations, neighboring countries that should have a shared goal that is only achievable in collaboration may refuse that collaboration due to historic and political reasons. In the case of the countries in Hispaniola, their political strife has its origins in how the Spanish and French Colonists divided the island, and how each took different approaches to their exploitation of the original peoples and the land. The eventual occupation of the Dominican Republic by Haiti sowed seeds of resentment that are seen in the present-day treatment of people of Haitian origins who are born in the Dominican Republic. All of this historical strife makes binational cooperation difficult, even in public health matters that can only be resolved by such collaboration [67–72].

The Conundrum of Bat Rabies

Globally, bat rabies accounts for a small proportion of human cases, but in the higher income countries in the Americas, it now accounts for the majority of deaths [52, 64]. In North America, this may in part be due to the fact that people are much more likely to seek PEP following bites from carnivores compared to interactions with bats [73]. In the USA, during 2021, four of the five reported cases were due to bat exposures [61]. Within Latin America, vampire bat rabies is an important cause of livestock morbidity and mortality and affects both subsistence and commercial farmers, who tend to be lower-income and disproportionately brown. Because the range of the vampire bat spans from Mexico to northern Argentina, a significant proportion of the human population is at risk, especially in rural and remote areas.

The Disproportionate Burden of Vampire Bats in Amazonia

One of the major challenges to dealing with vampire bat-mediated rabies is that it affects primarily rural populations in the subtropical and tropical areas of the Americas [74]. The incidence of human infection with rabies virus increased significantly in Latin America, especially in more remote and rural areas of the Amazon rainforest, where this obligate hematophagus bat feeds on humans [75] and where public health access to care and surveillance are extremely limited [76].

Historically, those living in remote portions of Amazonia and in rural areas in Latin American have been marginalized from political power. Structural racism in many countries is demonstrated by the fact that the right to vote by indigenous peoples was only introduced recently. State offices remain largely absent in rural areas and in several countries national identification registration is still a challenge. The legacy of such policies that lead to the 'invisibility' of indigenous populations continues to be felt acutely in all facets of life, including in the undercounting of indigenous peoples and the uncontested timber, mining, and oil exploitation of Amazonia. Without a political voice or influence, investments needed to combat bat rabies for those in remote areas, especially among indigenous peoples, generally go towards other competing priorities of those who are historically (and currently) prioritized. Considering the historic, and closely related current, realities of the Western Hemisphere, it is unlikely that these investments will be made by countries without regional pressure.

Bat Rabies in Colombia

In 2004, the largest human rabies outbreak up to that point was reported in Colombia in Bajo Baudó. It was caused by vampire bats. Six months later another outbreak was reported [77, 78]. An in-depth analysis was undertaken to identify populations with human rabies virus exposure due to dogs, cats, bats, or farm animals between 2007–2016 (n = 666,411 exposures). In urban areas exposure was primarily due to interaction with cats and dogs, while in rural areas human exposure was due to contact with infected farm animals. Urban areas with the highest poverty rates in the country and high human rabies exposure due to farm animals and populations of color were the most impacted. Regarding exposures due to bat bites, the most were observed among indigenous populations, particularly in the Amazon region [79]. Similar trends have been reported in Brazil, Peru and Ecuador [76, 80, 81]. Investments to public health and health care systems in such historically marginalized populations need to be prioritized, particularly in the light of the looming complexities of rabies prevention and control moving forward [82].

Other Vulnerable Populations at Risk

In the US, two human rabies cases linked to bat exposure have inadvertently highlighted disparities in health care access of incarcerated individuals and of those detained by immigration officials, both of whom are disproportionately from vulnerable populations (i.e., people of color, and low-income) [83, 84]. In the first case a 29-year-old individual who was incarcerated was infected by a rabies virus variant associated with insectivorous bats. There was no documentation of contact with, or scratches from, a bat. However, the individual worked outdoors around the prison on a farm and multiple duties that involved contact with livestock and trash and debris. He was the first case of rabies in Virginia since 1953 and the only case in the US in 1998. Following his diagnosis, family members who came to visit him, incarcerated individuals, prison staff and health care and other providers received PEP [83]. In a second case, a 28-year-old man from Central America who was detained in a facility in Texas, died from rabies in 2013. No bite marks or known rabies exposures were reported. The man had owned a dog in Guatemala, which died from unknown causes in 2011, and family members reported that they were unaware of any history of animal bites. This was the first reported case of a person in federal custody in the US, while infectious. Detainees and detention facility staff members received PEP. A higher proportion of detainees compared to facility staff members (41% versus 2% respectively) were given PEP following public health recommendations based on public health recommendations [84]. Both cases highlight that although human-to-human transmission of rabies is rare, individuals detained are particularly vulnerable since prisons, detention centers and similar settings provide extensive opportunities for exposure. The latter case also highlights the uneven response of countries to the International Health Regulations notification that went out, with lower income countries not conducting contact tracing [84]. In both cases, inadequate monitoring of health led to a death in each facility and exposures to multiple individuals. If not motivated by a fundamental desire to provide basic health care because it is a human right, then countries should consider the importance of doing so to protect their own and other populations.

Recent trends to tighten immigration control by inspecting health care facilities for illegal migrants and by pressuring or requiring health care workers, hospitals and other health care providers to report undocumented individuals can result in outbreaks – and there should be no link between these two. In one case, an undocumented 16 year old victim of rabies from Mexico did not seek health care in the US and his parents did not fully cooperate with health authorities out of fear of deportation [60]. In the US and other countries in the Americas, there has been a long and well-documented legacy of unethical research practices and deliberate withholding of effective treatments from incarcerated individuals and among populations of color which has led to the persistent mistrust in the medical system [85–88]. There are undoubtedly more unrecognized vulnerable populations related to rabies. For example, a recent report uncovered populations of fishermen exposed to vampire bites in rabies endemic areas of Brazil, which had not been previously recognized [80].

Disparities in Basic Health Knowledge

A few studies have examined knowledge about bats and rabies. In areas at risk of bat rabies, investigators found low knowledge [89, 90]. Overall, populations of color in the US have significantly lower knowledge [90]. In addition, disparities in knowledge of low-cost rabies vaccination clinics was found among populations of color and low-income populations [91]. Compliance with vaccination was low and most individuals were willing to report suspect rabid animals, but did not know whom to call. White respondents preferred online information delivery, whereas Latinos and African Americans preferred postal mail, suggesting the need for tailored communication messaging and dissemination [91]. While such investments have not been made, support in other, more strict enforcement were implemented.

Inequalities in Animal Control Enforcement

It has long been documented that minority and low-income populations are disproportionately affected by policing and enforcement. Indigenous American populations in nearly each country in the hemisphere, black and brown populations, rural populations, and low-income populations are acknowledged to fall prey to enforcement of nearly every law [92, 93]. This evidence goes back centuries. The Spanish colonists brutalized indigenous populations with dogs. Centuries later, so too did the police during the civil rights movement in the US during the 1960s (Fig. 11.1).

It is therefore not surprising that this continues to be the case for animal control enforcement that targets black and brown populations. In one recent review, significantly higher rates of animal confiscation, relinquishment, and euthanasia, lower return to owner, and extended lengths of stay in animal shelters were documented among low-income populations in the US, particularly those who were from minoritized populations [94]. Instead, interdisciplinary and culturally appropriate and collaborative approaches should be utilized to improve knowledge of compliance with local animal health regulations [95].

Fig. 11.1 Sculpture dedicated to the Foot Soldiers of the Birmingham Civil Rights Movement. Kelly Ingram Park, Birmingham, Alabama. United States Alabama Birmingham, 2010. (Photograph by Carol M. Highsmith. https://www.loc.gov/item/2010636972/)

Moving Forward in the Americas?

Steps to prevent and control rabies are well known, with rapid evaluation and correct indication of PEP being the cornerstone to avoid human rabies, and dog rabies vaccination the key to eliminate canine rabies [96]. Heterogeneous access to life-saving biologics for humans and for dogs is an unsolved issue for decades, where the most at risk have the least access, including populations living in the remote villages such as the Andes Mountains, Amazon Basin, and rural areas of Central America and the Caribbean [97, 98]. Trained and updated healthcare personnel are lacking where they are most needed. Surveillance of bites in humans is neglected and the practice is largely absent, due to local logistics and political difficulties but most importantly due to the prevailing design of health care systems that do not address zoonoses properly and largely neglect indigenous and rural populations [99]. This has only been compounded by the COVID-19 pandemic [100]. Unrecognized risk groups for bat bites and dog exposures are emerging related to occupation and residence in risk areas where attention was lacking due to be less economical important for society [80].

Structural racism, where indigenous peoples receive less attention from health care and other government structures and who the media portray as being "far" from their target demographics, in "dangerous areas", and being "less developed" is tied to the systemic realities that have grown from a lack of understanding of cultural traditions and beliefs. That is especially problematic for rabies, a disease where the incubation periods are long, making it difficult to relate bite exposures with disease onset, and also because neurological manifestations are confounded with paranormal activity or "magic" which is then dismissed because of its link to traditional beliefs [101]. The health systems usually lack translational approaches to improve outreach to diverse groups. A usual practice is to blame the rabies victims and their communities for not seeking medical attention in time, deviating attention away from the structural barriers that perpetuate the risk for rabies in a discriminatory way for those populations at the highest risk [102–105].

From all rabies virus exposures scenarios in the world, the rural Amazon Basin is the area with highest frequency of animal bites, where vampire bats feed not only on cattle or other domestic animals, but also on humans, especially on children [76, 106]. Effective interventions to avoid those bites have not been developed. Poverty and environmental factors combine to make housing conditions in that part of the world accessible to bats and despite known house modifications to address this issue, no government has launched a program that utilizes this approach [107]. While vampire bat bites occur with high frequency, they remain mostly ignored as an exposure source by victims and governments, despite a plethora of research studies that highlight the range of potential pathogens carried by them [108–114].

Brazil has been promoting essential improvements in health indicators by implementing free-access health programs, which successfully reduced the prevalence of neglected zoonosis in urban areas, such as rabies. Despite constant efforts from the authorities to monitor and control the disease, sylvatic rabies remains a current issue in Amazonian communities. The inequalities among Amazonian areas challenge the expansion of high-tech services and limit the implementation of enhanced laboratory surveillance to effectively avoid outbreaks in human and non-human hosts, which also reproduces a panorama of vulnerability in risk communities. Because rabies is a preventable disease, the prevalence in the particular context of the Amazon basin highlights the failure of surveillance strategies to predict transmission events and indicates the need to adapt the public policies to a "One Health" approach. A much-needed strengthening of the sylvatic and urban surveillance networks to achieve the "Zero by 30" goal (which is inextricable from multilateral efforts to combat the progressive biome's degradation) should be supported [108–114].

Animal bites receive sporadic and temporary attention when deadly outbreaks kill native children in clusters, but no long-term meaningful intervention investments follow. One example shows the use of research evidence to support addressing this widespread problem. In Peru, the highest animal bite frequency ever reported was shared with local communities. This information exchange prompted a change to the traditional vaccination paradigm – from vaccinating only those exposed – to universal PrEP vaccination of high-risk populations. The outcome was the first massive program for rabies PrEP in high-risk populations, endorsed by the WHO, and funded by the government [76, 115]. Such PrEP did successfully prevent human rabies due to vampire bat bites where it was implemented. However, bites continued and more than a decade later, the program still does not cover all targeted areas, due to the same structural barriers that originally kept those populations at the bottom of the list of national health priorities [115, 116]. Inexcusably, rabies biologics are still inaccessible to the most high-risk indigenous populations.

Conclusions

In the shadow of the COVID-19 pandemic, the global rise in public health nationalism, the infodemic, and increased xenophobia, all public health systems are at risk, especially those that have previously been under resourced, as is the case of rabies in Amazonia. Historically, the Americas has under resourced public health, medical and development resources for indigenous, black, and brown populations and prior progress has been stalled or reversed due to the COVID-19 pandemic. Even before the pandemic, dog-mediated rabies was difficult to control within the poorest New World countries (e.g., Haiti, the Dominican Republic, Bolivia) because resources were not allocated to problems of the poor. The same is often true in countries with persistent health and development disparities among indigenous and rural populations. Clearly, this includes not only the Neotropics but also those in the Arctic. Society has the tools to control rabies, and even selectively eliminate viral variants, but what is missing is a prioritization of effective programs. While the highest income locations have systems that deliver biologics at the first sign of viral exposure, a passive approach is used in indigenous areas where the decision-makers have deemed it acceptable to wait for the next human rabies outbreak to occur before a response is initiated. This is an unacceptable public health paradigm that accepts, and even institutionalizes, rabies deaths in poor or indigenous peoples and continues the colonial tradition of blaming individual behaviors, absent of a direct and honest assessment of the structural institutions that play a larger role in continued disparities [85, 86]. A change is urgent and there is no excuse to delay immediate expansions of massive vaccination PrEP programs to all high-risk areas with limited access to rabies biologics.

References

1. Byrd WM (1990) Race, biology, and health care: reassessing a relationship. J Health Care Poor Underserved 1(3):278–296
2. Millien MF, Pierre-Louis JB, Wallace R et al (2015) Control of dog mediated human rabies in Haiti: no time to spare. PLoS Negl Trop Dis 9(6):e0003806. https://doi.org/10.1371/journal.pntd.0003806
3. UNICEF (2022) Haiti. Published 2022. https://data.unicef.org/country/hti/. Accessed 2 Oct 2022
4. World Food Programme (2022) Haiti Country Brief. Published 2022. https://www.wfp.org/countries/haiti. Accessed 2 Oct 2022
5. McMillen CW (2016) Beyond Germs: Native Depopulation in North America. Amerind Studies in Anthropology. Edited by Catherine M. Cameron, Paul Kelton, and Alan C. Swedlund. Foreword by John Ware. Published online 2016
6. Orr R, Sharratt K, Iqbal M (2019) American Indian erasure and the logic of elimination: an experimental study of depiction and support for resources and rights for tribes. J Ethn Migr Stud 45(11):2078–2099. https://doi.org/10.1080/1369183X.2017.1421061
7. Wang J (2019) Mad dogs and other new Yorkers: rabies, medicine, and Society in an American Metropolis, 1840–1920. Johns Hopkins University Press

8. Wasik B, Murphy M (2013) Rabid: a cultural history of the world's most diabolical virus. Penguin
9. Martin DL, Goodman AH (2002) Health conditions before Columbus: paleopathology of native North Americans. West J Med 176(1):65–68
10. Whittaker C, Walker PG, Alhaffar M et al (2021) Under-reporting of deaths limits our understanding of true burden of covid-19. BMJ 375:n2239
11. Urton G (2017) Inka history in knots: reading khipus as primary sources. University of Texas Press
12. Vos A, Nunan C, Bolles D et al (2011) The occurrence of rabies in pre-Columbian Central America: an historical search. Epidemiol Infect 139(10):1445–1452
13. Hughes GJ, Orciari LA, Rupprecht CE (2005) Evolutionary timescale of rabies virus adaptation to North American bats inferred from the substitution rate of the nucleoprotein gene. J Gen Virol 86(5):1467–1474
14. Kuzmin I, Hughes G, Botvinkin A, Gribencha S, Rupprecht C (2008) Arctic and Arctic-like rabies viruses: distribution, phylogeny and evolutionary history. Epidemiol Infect 136(4):509–519
15. Kuzmina NA, Kuzmin IV, Ellison JA et al (2013) A reassessment of the evolutionary timescale of bat rabies viruses based upon glycoprotein gene sequences. Virus Genes 47(2):305–310
16. Smith JS, Seidel HD (1993) Rabies: a new look at an old disease. Prog Med Virol 40:82–106
17. Troupin C, Dacheux L, Tanguy M et al (2016) Large-scale phylogenomic analysis reveals the complex evolutionary history of rabies virus in multiple carnivore hosts. PLoS Pathog 12(12):e1006041
18. Guerra F (1993) The European-American exchange. Hist Philos Life Sci 15(3):313–327
19. Nair R (2019) B.C. rabies death revives interest and fear in age-old disease. CBC. Published July 21, 2019. https://www.cbc.ca/news/canada/british-columbia/rabies-history-1.5217363. Accessed 6 Aug 2021
20. Notes From the Frontier (2019) Rabies on the Frontier: terrifying scourge of the wild. Notes From The Frontier Published December 6, 2019. https://www.notesfromthefrontier.com/post/rabies-on-the-frontier. Accessed 6 Aug 2021
21. Rabies: A Historical Perspective. Published 2022. https://dshs.texas.gov/idcu/disease/rabies/history/historyInTexas.pdf. Accessed 2 Oct 2022
22. Unanue JH (2022) Observaciones sobre el clima de Lima, y sus influencias en los seres organizados, en especial el hombre – Digital Collections – National Library of Medicine. https://collections.nlm.nih.gov/catalog/nlm:nlmuid-9111245-bk. Accessed 3 April 2022
23. Boisseron B (2015) Afro-Dog. Transition, i can be lightning. Indiana University Press, on behalf of the Hutchins Center for African and African American Research at Harvard University, pp 15–31
24. Sanders R (1992) Lost tribes and promised lands: the origins of American racism. HarperPerennial, New York
25. Minghui R, Stone M, Semedo MH, Nel L (2018) New global strategic plan to eliminate dog-mediated rabies by 2030. Lancet Glob Health 6(8):e828–e829
26. Rationale for investing in the global elimination of dog-mediated human rabies. Published online 2015. https://rabiesalliance.org/resource/rationale-investing-global-elimination-dog-mediated-human-rabies. Accessed 2 Oct 2022
27. Cosby AG, McDoom-Echebiri MM, James W, Khandekar H, Brown W, Hanna HL (2019) Growth and persistence of place-based mortality in the United States: the rural mortality penalty. Am J Public Health 109(1):155–162
28. Small MJ, Allen TK, Brown HL (2017) Global disparities in maternal morbidity and mortality, vol 41. Elsevier, pp 318–322
29. Wehby GL, Gili JA, Pawluk M, Castilla EE, López-Camelo JS (2015) Disparities in birth weight and gestational age by ethnic ancestry in south American countries. Int J Public Health 60(3):343–351

30. Axelsson P, Kukutai T, Kippen R (2016) The field of indigenous health and the role of colonisation and history. J Popul Res 33(1):1–7
31. Ko D, Bal A, Bird Bear A, Sannino A, Engeström Y (2021) Transformative agency for justice: addressing racial disparity of school discipline with the indigenous learning lab. Race Ethn Educ 25:1–24. Published online
32. Marrone S (2007) Understanding barriers to health care: a review of disparities in health care services among indigenous populations. Int J Circumpolar Health 66(3):188–198
33. Michiel Oosterveer T, Kue YT (2015) Primary health care accessibility challenges in remote indigenous communities in Canada's north. Int J Circumpolar Health 74(1):29576
34. Ramírez NA, Ruiz JP, Romero RV, Labonté R (2011) Comprehensive primary health care in South America: contexts, achievements and policy implications. Cad Saúde Pública 27:1875–1890
35. Denison GA, McAlpine JG, Gill D (1937) Rabies deaths in Alabama: analysis of case histories with regard to treatment. Am J Public Health Nations Health 27(9):869–874
36. Davis DA (2019) Obstetric racism: the racial politics of pregnancy, labor, and birthing. Med Anthropol 38(7):560–573. https://doi.org/10.1080/01459740.2018.1549389
37. Grilo Diniz CS, Rattner D, Lucas d'Oliveira AFP, de Aguiar JM, Niy DY (2018) Disrespect and abuse in childbirth in Brazil: social activism, public policies and providers' training. Reprod Health Matters 26(53):19–35. https://doi.org/10.1080/09688080.2018.1502019
38. Hurston ZN. Their eyes were watching god. Virago; 1937
39. Lee H (2010) To kill a mockingbird. Random House
40. Jankowski C (2021) The symbolism of the dog being shot in "To Kill a Mockingbird" | Synonym. https://classroom.synonym.com/symbolism-dog-being-shot-to-kill-mockingbird-21914.html. Accessed 11 Aug 2021
41. Rage (1996) Columbia Pictures. https://www.imdb.com/title/tt0060882/?ref_=tt_urv
42. Dickson F (2019) Deep southern, diversified, & re-imagined. Southern movie 35: "Paris Trout". Published 2019. https://fosterdickson.com/2019/05/01/southern-movie-35-paris-trout-1991/. Accessed 3 Oct 2022
43. Prahlad A (2006) The greenwood encyclopedia of African American folklore. Greenwood Press
44. Bjork A, Holman RC, Callinan LS, Hennessy TW, Cheek JE, McQuiston JH (2013) Dog bite injuries among American Indian and Alaska native children. J Pediatr 162(6):1270–1275
45. Castrodale LJ (2007) Hospitalizations resulting from dog bite injuries – Alaska, 1991–2002. Int J Circumpolar Health 66(4):320–327. https://doi.org/10.3402/ijch.v66i4.18273
46. Mediouni S, Brisson M, Ravel A (2020) Epidemiology of human exposure to rabies in Nunavik: incidence, the role of dog bites and their context, and victim profiles. BMC Public Health 20(1):584. https://doi.org/10.1186/s12889-020-08606-8
47. Raghavan M, Martens P, Burchill C (2014) Exploring the relationship between socioeconomic status and dog-bite injuries through spatial analysis. Rural Remote Health 14(3):2846. https://doi.org/10.22605/RRH2846
48. Eng TR, Fishbein DB, Talamante HE et al (1993) Urban epizootic of rabies in Mexico: epidemiology and impact of animal bite injuries. Bull World Health Organ 71(5):615–624
49. Aenishaenslin C, Simon A, Forde T et al (2014) Characterizing rabies epidemiology in remote Inuit communities in Québec, Canada: a "one health" approach. EcoHealth 11(3):343–355. https://doi.org/10.1007/s10393-014-0923-1
50. Benavides JA, Megid J, Campos A, Hampson K (2020) Using surveillance of animal bite patients to decipher potential risks of rabies exposure from domestic animals and wildlife in Brazil. Front Public Health 8:318. https://doi.org/10.3389/fpubh.2020.00318
51. Benavides JA, Megid J, Campos A, Rocha S, Vigilato MAN, Hampson K (2019) An evaluation of Brazil's surveillance and prophylaxis of canine rabies between 2008 and 2017. PLoS Negl Trop Dis 13(8):e0007564. https://doi.org/10.1371/journal.pntd.0007564
52. Vigilato MAN, Clavijo A, Knobl T et al (2013) Progress towards eliminating canine rabies: policies and perspectives from Latin America and the Caribbean. Philos Trans R Soc Lond B Biol Sci 368(1623):1–8

53. PAHO (2022) Rabies. Published 2022. https://www.paho.org/en/topics/rabies#:~:text=Coordinated%20actions%20for%20the%20regional%20elimination%20of%20human,of%20the%20regional%20epidemiological%20surveillance%20system%20of%20rabies
54. Rupprecht CE, Mani RS, Mshelbwala PP, Recuenco SE, Ward MP (2022) Rabies in the tropics. Curr Trop Med Rep 9:1–12. Published online
55. Pieracci EG, Pearson CM, Wallace RM et al (2019) Vital signs: trends in human rabies deaths and exposures — United States, 1938–2018. MMWR Morb Mortal Wkly Rep 68:524. https://doi.org/10.15585/mmwr.mm6823e1
56. CDC – Cost of Rabies Prevention – Rabies (2019) Published June 11, 2019. https://www.cdc.gov/rabies/location/usa/cost.html. Accessed 4 April 2022
57. Goodman B, Miller A (2022) The high cost of surviving rabies. https://www.webmd.com/a--to-z-guides/news/20180220/the-high-cost-of-surviving-rabies. Accessed 4 April 2022
58. Kliff S (2018) Why a simple, lifesaving rabies shot can cost $10,000 in America. Vox Published February 7, 2018. https://www.vox.com/policy-and-politics/2018/2/7/16851134/rabies-treament-expensive-emergency-room. Accessed 11 Aug 2021
59. Recuenco S, Cherry B, Eidson M (2007) Potential cost savings with terrestrial rabies control. BMC Public Health 7(1):47. https://doi.org/10.1186/1471-2458-7-47
60. CDC (2009) Imported human rabies—California, 2008. JAMA 302(10):1051–1052
61. Ma X, Bonaparte S, Toro M et al (2022) Rabies surveillance in the United States during 2020. J Am Vet Med Assoc 1(aop):1–9
62. Malerczyk C, DeTora L, Gniel D (2011) Imported human rabies cases in Europe, the United States, and Japan, 1990 to 2010. J Travel Med 18(6):402–407
63. World Health Organization (2018) WHO Expert Consultation on Rabies. World Health Organization. https://apps.who.int/iris/bitstream/handle/10665/272364/9789241210218-eng.pdf?ua=1
64. Human rabies: 2016 updates and call for data. Releve Epidemiol Hebd. 2017;92(7):77–86
65. Barbosa Costa G, Ludder F, Monroe B et al (2020) Barriers to attendance of canine rabies vaccination campaigns in Haiti, 2017. Transbound Emerg Dis 67(6):2679–2691. https://doi.org/10.1111/tbed.13622
66. Schildecker S, Millien M, Blanton JD et al (2017) Dog ecology and barriers to canine rabies control in the Republic of Haiti, 2014-2015. Transbound Emerg Dis 64(5):1433–1442. https://doi.org/10.1111/tbed.12531
67. Amaya AB, De Lombaerde P (2021) Regional cooperation is essential to combatting health emergencies in the global south. Glob Health 17(1):9. https://doi.org/10.1186/s12992-021-00659-7
68. Homedes N, Ugalde A (2003) Globalization and health at the United States–Mexico border. Am J Public Health 93(12):2016–2022. https://doi.org/10.2105/AJPH.93.12.2016
69. Javed S, Chattu VK (2020) Strengthening the COVID-19 pandemic response, global leadership, and international cooperation through global health diplomacy. Health Promot Perspect 10(4):300–305. https://doi.org/10.34172/hpp.2020.48
70. Ridding Haiti of Rabies (2022) Merck animal health. https://www.merck-animal-health.com/about-us/featured-stories/ridding-haiti-of-rabies/. Accessed 4 April 2022
71. Seetahal JFR, Vokaty A, Vigilato MAN et al (2018) Rabies in the Caribbean: a situational analysis and historic review. Trop Med Infect Dis 3(3):89. https://doi.org/10.3390/tropicalmed3030089
72. Wallace RM, Reses H, Franka R et al (2015) Establishment of a canine rabies burden in Haiti through the implementation of a novel surveillance program. PLoS Negl Trop Dis 9(11):e0004245. https://doi.org/10.1371/journal.pntd.0004245
73. Birhane MG, Cleaton JM, Monroe BP et al (2017) Rabies surveillance in the United States during 2015. J Am Vet Med Assoc 250(10):1117–1130. https://doi.org/10.2460/javma.250.10.1117

74. Streicker DG, Recuenco S, Valderrama W et al (2012) Ecological and anthropogenic drivers of rabies exposure in vampire bats: implications for transmission and control. Proc Biol Sci 279(1742):3384–3392
75. Schneider MC, Romijn PC, Uieda W et al (2009) Rabies transmitted by vampire bats to humans: an emerging zoonotic disease in Latin America?: La rabia transmitida por murciélagos vampiros a humanos: ¿una zoonosis emergente en América Latina? Rev Panam Salud Pública 25(3):260–269. https://doi.org/10.1590/S1020-49892009000300010
76. Gilbert AT, Petersen BW, Recuenco S et al (2012) Evidence of rabies virus exposure among humans in the Peruvian Amazon. Am J Trop Med Hyg 87(2):206–215. https://doi.org/10.4269/ajtmh.2012.11-0689
77. Romero-Sandoval N, Escobar N, Utzet M, Feijoo-Cid M, Martin M (2014) Sylvatic rabies and the perception of vampire bat activity in communities in the Ecuadorian Amazon. Cad Saúde Pública 30:669–674. https://doi.org/10.1590/0102-311X00070413
78. Valderrama J, García I, Figueroa G et al (2006) Outbreaks of human rabies transmitted by vampire bats in Alto Baudó and Bajo Baudó municipalities, department of Chocó, Colombia, 2004-2005. Biomedica 26(3):387–396. https://doi.org/10.7705/biomedica.v26i3.357
79. Arias Caicedo MR, Xavier D d A, Arias Caicedo CA, Andrade E, Abel I (2019) Epidemiological scenarios for human rabies exposure notified in Colombia during ten years: A challenge to implement surveillance actions with a differential approach on vulnerable populations. PloS One 14(12):e0213120. https://doi.org/10.1371/journal.pone.0213120
80. de Paula Silva N, de Andrade E d A, Cardoso D et al (2021) Assessment of crab fishermen's exposure to rabies virus in a typical Amazonian community. Zoonoses Public Health 68(8):973–981. https://doi.org/10.1111/zph.12869
81. Romero-Sandoval N, Parra C, Gallegos G et al (2013) Haematophagous bat bites in Ecuadorian Amazon: characterisation and implications for sylvatic rabies prevention. Public Health Action 3(1):85–89
82. León B, González SF, Solís LM et al (2021) Rabies in Costa Rica – next steps towards controlling bat-borne rabies after its elimination in dogs. Yale J Biol Med 94(2):311–329
83. CDC Human Rabies – Virginia, 1998; 1999. https://www.cdc.gov/mmwr/preview/mmwrhtml/00056456.htm. Accessed 11 Aug 2021
84. CDC (2014) Rabies death attributed to exposure in Central America with symptom onset in a U.S. detention facility – Texas, 2013, vol 63. U.S. Center for Disease Control, pp 446–449. http://www.proquest.com/docview/1534528072/abstract/58CC9EB818944C12PQ/1. Accessed 4 April 2022
85. Brave Heart MYH, Chase J, Elkins J, Altschul DB (2011) Historical trauma among indigenous peoples of the Americas: concepts, research, and clinical considerations. J Psychoactive Drugs 43(4):282–290
86. Paradies Y (2016) Colonisation, racism and indigenous health. J Popul Res 33(1):83–96
87. Robinson TA (1955) A report on research at Mississippi State penitentiary. Miss Doct 32(11):315–316
88. Washington HA (2006) Medical apartheid: the dark history of medical experimentation on black Americans from colonial times to the present. Doubleday Books
89. Lankau EW, Cox SW, Ferguson SC et al (2015) Community survey of rabies knowledge and exposure to bats in homes – Sumter County, South Carolina, USA. Zoonoses Public Health 62(3):190–198. https://doi.org/10.1111/zph.12135
90. Palamar MB, Peterson MN, Deperno CS, Correa MT (2013) Assessing rabies knowledge and perceptions among ethnic minorities in Greensboro, North Carolina. J Wildl Manag 77(7):1321–1326
91. Palamar MB, Correa MT, Peterson NM, DePerno CS (2017) Public preference for pet-rabies prophylaxis: opportunities and information dissemination. Trop Med Infect Dis 2(3):46. https://doi.org/10.3390/tropicalmed2030046
92. Kahn KB, Martin KD (2016) Policing and race: disparate treatment, perceptions, and policy responses. Soc Issues Policy Rev 10(1):82–121

93. Perry B (2009) Impacts of disparate policing in Indian country. Polic Soc 19(3):263–281. https://doi.org/10.1080/10439460902871348
94. Hawes SM, Hupe T, Morris KN (2020) Punishment to support: the need to align animal control enforcement with the human social justice movement. Animals 10(10):1902. https://doi.org/10.3390/ani10101902
95. Kunkel A, Jeon S, Joseph HC et al (2021) The urgency of resuming disrupted dog rabies vaccination campaigns: a modeling and cost-effectiveness analysis. Sci Rep 11(1):12476. https://doi.org/10.1038/s41598-021-92067-5
96. Lembo T, Partners for Rabies Prevention (2012) The blueprint for rabies prevention and control: a novel operational toolkit for rabies elimination. PLoS Negl Trop Dis 6(2):e1388
97. Carvalho-Costa FA, Tedesqui VL, de Jesus Nascimento Monteiro M, Bóia MN (2012) Outbreaks of attacks by hematophagous bats in isolated riverine communities in the Brazilian Amazon: a challenge to rabies control. Zoonoses Public Health 59(4):272–277. https://doi.org/10.1111/j.1863-2378.2011.01444.x
98. Guzman-Holst A, De Antonio R, Prado-Cohrs D, Juliao P (2020) Barriers to vaccination in Latin America: A systematic literature review. Vaccine 38(3):470–481. https://doi.org/10.1016/j.vaccine.2019.10.088
99. Bastos V, Mota R, Guimarães M et al (2021) Challenges of rabies surveillance in the eastern Amazon: the need of a one health approach to predict rabies spillover. Front Public Health 9. https://www.frontiersin.org/article/10.3389/fpubh.2021.624574. Accessed 4 April 2022
100. Tidman R, Thumbi S, Wallace R et al (2022) United against rabies forum: the one health concept at work. Front Public Health 10
101. Nadal D, Hampson K, Lembo T, Rodrigues R, Vanak AT, Cleaveland S (2022) Where rabies is not a disease. Bridging healthworlds to improve mutual understanding and prevention of rabies. Front Vet Sci 9:867266
102. Arias-Orozco P, Bástida-González F, Cruz L et al (2018) Spatiotemporal analysis of canine rabies in El Salvador: Violence and poverty as social factors of canine rabies. Plos One 13(8):e0201305. https://doi.org/10.1371/journal.pone.0201305
103. Briggs CL (2017) Uncovering a tragic flaw in revolutionary health policies: From health and communicative inequities to communicative justice in health. Salud Colect 13(3):411. https://doi.org/10.18294/sc.2017.1152
104. Clavijo GC, Arevalo JD, Pretel VLO, Gutiérrez JP (2016) Percepciones sobre la vacunación de la rabia silvestre en población Awajún de la provincia de Condorcanqui, Amazonas, Perú, 2013. Physis Rev Saúde Coletiva 26(1):201–224. https://doi.org/10.1590/S0103-73312016000100012
105. da Costa LJC, Fernandes MEB (2016) Rabies: knowledge and practices regarding rabies in rural communities of the brazilian amazon basin. PLoS Negl Trop Dis 10(2):e0004474. https://doi.org/10.1371/journal.pntd.0004474
106. Schneider MC, Aron J, Santos-Burgoa C, Uieda W, Ruiz-Velazco S (2001) Common vampire bat attacks on humans in a village of the Amazon region of Brazil. Cad Saúde Pública 17(6):1531–1536. https://doi.org/10.1590/S0102-311X2001000600025
107. Ormaeche MM, Gómez-Benavides J (2007) Factores De Riesgo Para Mordeduras Por Murciélagos Hematófagos En El Valle Del Río Apurímac: Risk factors for vampire bat bites in the apurimac river valley. Rev Peru Med Exp Salud Pública 24(1):89–92
108. Alves RS, Canto Olegário J, Weber MN et al (2022) Detection of coronavirus in vampire bats (*Desmodus rotundus*) in southern Brazil. Transbound Emerg Dis 69(4):2384–2389. https://doi.org/10.1111/tbed.14150. Published online May 28, 2021
109. Bai Y, Recuenco S, Gilbert AT et al (2012) Prevalence and diversity of Bartonella spp. in bats in Peru. Am J Trop Med Hyg 87(3):518–523. https://doi.org/10.4269/ajtmh.2012.12-0097
110. Gómez-Benavides J, Laguna-Torres VA, Recuenco S (2010) El real significado de ser mordido por murciélagos hematófagos en las comunidades indígenas amazónicas del Perú. Rev Peru Med Exp Salud Pública 27(4):657–658

111. Mateus J, Gómez N, Herrera-Sepúlveda MT, Hidalgo M, Pérez-Torres J, Cuervo C (2019) Bats are a potential reservoir of pathogenic Leptospira species in Colombia. J Infect Dev Ctries 13(04):278–283. https://doi.org/10.3855/jidc.10642
112. Tao Y, Shi M, Conrardy C et al (2013) Discovery of diverse polyomaviruses in bats and the evolutionary history of the Polyomaviridae. J Gen Virol 94(4):738–748. https://doi.org/10.1099/vir.0.047928-0
113. Tong S, Zhu X, Li Y et al (2013) New world bats harbor diverse influenza A viruses. PLoS Pathog 9(10):e1003657. https://doi.org/10.1371/journal.ppat.1003657
114. Villena FE, Gomez-Puerta LA, Jhonston EJ et al (2018) First report of Trypanosoma cruzi infection in salivary gland of bats from the Peruvian Amazon. Am J Trop Med Hyg 99(3):723–728. https://doi.org/10.4269/ajtmh.17-0816
115. Kessels JA, Recuenco S, Navarro-Vela AM et al (2017) Pre-exposure rabies prophylaxis: a systematic review. Bull World Health Organ 95(3):210–219C. https://doi.org/10.2471/BLT.16.173039
116. Chauca DPB (2021) Experiencia de control de rabia silvestre en amazonas. Oficina de Epidemiología, Prevención y Control de Enfermedades DIRESA AMAZONAS:20pp. Experiencia-control-de-rabia_Amazonas.pdf (dge.gob.pe)

Chapter 12
The Health Economics of Rabies in the Americas: An Historical Summary and a Synthesis of the Literature

S. A. Shwiff, K. H. Ernst, S. S. Shwiff, and V. R. Brown

Abstract Rabies is a unique zoonosis from a health economics perspective. This chapter describes the health economic impacts of the disease in the Americas, the species impacted, and examines the efforts made to understand the trade-offs between disease prevention and human prophylaxis. In almost all cases, vaccinating humans exposed to rabies is the more costly alternative to rabies prevention efforts in the host species. In the New World, significant efforts to vaccinate dogs and wildlife have reduced the number of rabies-related human deaths and other associated costs. Examining the health economics of rabies management in the Americas provides compelling evidence that rabies mass vaccination campaigns are not only economically efficient, and save lives, but they are also the foundation for rabies elimination worldwide.

Keywords Cost effectiveness · Economics · Health impact · Optimization · Public support · Statistics

S. A. Shwiff (✉) · K. H. Ernst
National Wildlife Research Center, United States Department of Agriculture (USDA)/Animal and Plant Health Inspection Service (APHIS)/Wildlife Services, Fort Collins, CO, USA
e-mail: stephanie.a.shwiff@usda.gov

S. S. Shwiff
Texas A&M University – Commerce, Commerce, TX, USA

V. R. Brown
National Feral Swine Damage Management Program, United States Department of Agriculture (USDA)/Animal and Plant Health Inspection Service (APHIS), Fort Collins, CO, USA

C. E. Rupprecht (ed.), *History of Rabies in the Americas: From the Pre-Columbian to the Present, Volume I*, Fascinating Life Sciences,
https://doi.org/10.1007/978-3-031-25052-1_12

Introduction

Economics is a social science concerned with how individuals make decisions and the impact of those decisions on the consumption of goods and services, as well as the efficiency, effectiveness, distribution, and value in the production of those goods and services. The field of health economics is a subset of general economics and focuses on the production and consumption of health care. Health economists examine issues like why individuals make certain health care choices, what is the efficient and effective allocation of limited health care resources, and what treatment options offer the greatest return on investment in terms of positive health outcomes. With respect to infectious diseases, health economists tend to focus on comparing the costs and consequences of prevention and treatment strategies of diseases that can impact humans. Disease transmission at the wildlife-livestock or wildlife-companion animal interface has the ability to significantly impact human health and cause significant economic losses. Evidence of this impact can be seen through zoonotic examples, such as rabies. The interfaces between wildlife, livestock, companion animals, and humans provide many avenues to create economic losses resulting from viral transmission.

From a health economics perspective, rabies is a unique disease for several reasons. First, the case fatality rate in humans is nearly 100% and rabid patients die quickly. Secondly, while there is no effective therapy for rabies, biologics exist for humans, companion animals, livestock, and wildlife and are quite efficacious at preventing the disease. The majority of the human global rabies burden is rabies virus (RABV) transmitted by dogs. Thus, the relationship between humans and dogs presents an opportunity for effectively mitigating the impact. In the absence of vaccination, post-exposure prophylaxis (PEP) is relied upon to protect humans from developing clinical rabies. The synergy of these factors in an economic sense is that most of the costs associated with RABV do not result from illness (i.e., morbidity), but are the consequence of human deaths and efforts to prevent the disease in humans.

The vast majority of global human rabies cases are transmitted by domestic dogs and, as such, canine rabies causes the greatest economic impact of all the RABV variants and is the most well studied [1, 2]. Globally between 59,000 and 69,000 people are estimated to die annually from canine RABV infection. In the Americas, rabies exists primarily in dogs and wildlife and the human mortality rate comprises 0.05% of the global death toll [2, 3]. This is a result of the high ratio of human PEP and dog vaccinations, relative to other rabies endemic areas.

Human health impacts in the Americas from RABV transmission can result from infectious contacts with dogs, wildlife (e.g., raccoons, bats, foxes, coyotes, mongoose, skunks, etc.), or livestock, which may also exchange pathogens with one another, prior to spill over into humans [4]. However, interaction with dogs is still the main pathway for human exposure to RABV either by humans directly interacting with a rabid dog or through dogs interacting with rabid wildlife and exposing their human companions in the process [5].

The main purpose of RABV management is to reduce or eliminate the risk of viral transmission from dogs to humans. Success can be measured as the damages avoided (i.e., reduced PEP, reduced animal diagnostic tests, reduced livestock losses, etc.) or in terms of the number of protected individuals [6–8]. Due to the nature of how rabies costs manifest in an economic sense, management strategies focus primarily on dog populations and include large scale vaccination programs. Estimates of the health economic impacts of management strategies are crucial to control efforts and to gain public support.

Direct prevention of rabies in human populations via vaccination is not cost effective and only occurs in very specific small regional scales and for certain occupations, like veterinarians [9]. Because humans become exposed to RABV mainly through exposure to rabid dogs, the main route to prevention of rabies in humans is through large scale vaccination of dogs. Other prevention measures for rabies include quarantine, wildlife vaccination, companion animal vaccination, livestock vaccination, culling of impacted animals and in a limited application pre-exposure prophylaxis (PrEP) [7, 8, 10–12]. The direct and indirect costs from rabies include medical treatment for animal bites; direct and indirect costs for PEP; cost of animal vaccination efforts; cost of livestock losses; and DALYs (disability adjusted life year). These are all common metrics used to measure and compare preventative practices. Typical results provided by economic analyses of rabies include the overall impact of the disease burden, cost per animal vaccinated, cost per human life saved, cost of PEP, cost of animal vaccination programs, and DALYs [13, 14]. In most cases, costs associated with human PEP make up the largest component of these costs. Unfortunately, most studies do not have sufficient data to make these studies replicable in other locations, which makes the results regionally specific.

Human rabies PEP is expensive and comprises a significant portion of the overall economic impact of rabies. Unfortunately, PEP is often in administered in situations where the exposure doesn't warrant administration and potential cost savings could result from careful risk assessments and more judicious application of PEP [8, 15–19]. Currently, PEP usually consists of a series of four to five vaccinations over a span of several weeks, accompanied by a dose of rabies immune globulin (RIG) at the anatomical site of exposure. In the U.S., medical costs for each suspected human RABV exposure, including PEP, are estimated to total almost US $3300 (adjusted to 2020) and account for 70% of the cost of each exposure event [20]. Access to PEP is often limited due to the cost as well as by geography. However, many people seeking PEP are unable to receive the full course, so these costs reflect only partial courses. Human or equine RIG is not available in many of these areas. In poorer, more rural areas, access to PEP is often restricted due to the costs regarding transportation, storage, and administration. Hence, the cost of PEP is out of reach for many people in developing countries, where daily wages are often only US$ 1 or US$ 2 [21].

In this chapter, we provide a systematic review of the health economics literature that focuses on the Americas. We examine the information to date and the contribution of this literature to the field at large. We then project future needs for the field and conclude by examining the gaps in knowledge and potential paths forward.

Health Economics Literature Review for the Americas

We used a systematic scoping review to identify literature based on the methodological guidelines outlined by the PRISMA methods [22–24]. Search terms, specific text evaluated, and exclusion criteria, based on each PRISMA step, are outlined for this review in Fig. 12.1.

To capture the breadth of studies investigating the potential economic impact of rabies, literature evaluating a myriad of costs associated with the disease were considered eligible and are reflected in the use of broad search terms. Scientific peer-reviewed literature from journals, edited book volumes, government reports, or other similar documents were considered eligible. Primary search terms were looked for within the title, abstract, and index terms. The term "rabies" was paired with terms associated with economic impacts: economic, economic impact, cost, and economics, as well as terms identifying location, including: United States, US, Americas, North America, Latin America, South America, Mexico, and Canada. Publications in English published in any year through 2020 were considered eligible, and the initial search yielded 52 articles. Once all relevant sources were identified and retrieved, each source was reviewed to ensure relevance.

Publications were then screened (stage two of PRISMA) for relevancy and were excluded if they assessed regions outside of the western hemisphere, were a duplicate version of an already encountered publication, did not pertain to rabies, did not report an economic cost of the disease, were incomplete text, or were not locatable. The same primary search terms in the title, abstract, and index terms were considered as in the first stage. Based on these criteria, 23 articles were excluded.

The remaining 29 articles were reviewed in-depth in the "eligibility" stage to ensure bearing. Secondary search terms were exmined in the full text to help determine if the publication focused on deriving economic estimates. These included the

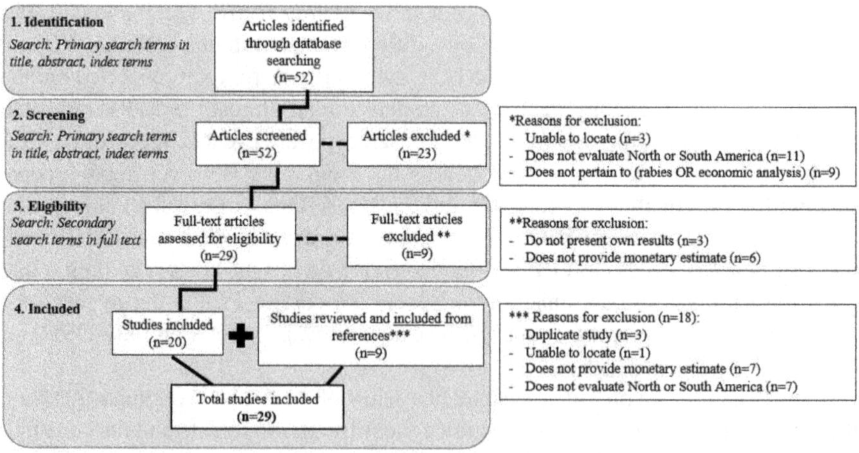

Fig. 12.1 PRISMA review steps

symbol, "$" and text, "dollar", "USD", and "value". Sources were excluded if they were a duplicate of an already reviewed study, reported economic estimates from another study, or did not report a monetary estimate.

In the final stage, 20 publications were included in the final group of relevant and acceptable publications. In addition to a full review of the entire text, 27 studies from the references of the 20 included publications were also reviewed using PRISMA steps 1–3. This yielded 9 more eligible studies, bringing the total number of selected publications to 29. Reasons for exclusion in the final stage included all previously provided exclusion criteria in stages 1–3.

Results of Related Health Economic Studies

Results from this literature review suggest that across the Americas, there has been considerable success at reducing rabies impacts through significant efforts to mass vaccinate dogs or wildlife. In the most developed parts of the Americas, canine RABV has been eliminated and while dogs are still the route to exposing humans to rabies, it is typically not canine RABV since rabies primarily exists in wildlife. The health economics of rabies examined in this review consisted of measuring the overall impact of a rabies event or vaccination program, benefit-cost analysis of an event or vaccination program, and in rare cases, examining the broader regional economic impact of a rabies outbreak or bioeconomic modeling of a potential outbreak that was halted due to rabies intervention efforts. Canada and the U.S. led the region with the most studies conducted on the health economics of rabies, followed by Mexico, which provided analyses of canine and bat RABV impacts and programs. Studies conducted in Haiti provided insight into the impact of rabies on the larger health crises occurring within that nation.

Canada and the U.S.

Rabies in the U.S. and Canada exists in wildlife, specifically, raccoons, foxes, skunks, mongoose, and bats [25]. Because of the border shared by these countries and their cooperative response to combatting rabies, they are included together in this study. This review indicated that most of the economic impacts related to the disease in Canada and the U.S. resulted from raccoon RABV [26, 27]. Historically, raccoon RABV was enzootic in parts of south-eastern Canada and in the north-eastern, mid-Atlantic and south-eastern regions of the U.S [28, 31]. Although there are very few human deaths attributable to the raccoon RABV variant in North America, rabid raccoons pose a significant threat to humans and domestic animals, as the species is well adapted to life in urban and suburban areas [28, 29]. Humans are at risk of raccoon RABV through direct exposure, and exposure to pets that have been exposed to a rabid raccoon. Though there may not be a large number of human fatalities related to raccoon rabies, this review highlighted that the presence of rabid raccoons does increase the number of livestock and pets annually tested for rabies, as well as the amount of PEP administered in that area [30]. This review also indicated that regardless of the species, oral rabies vaccine (ORV) baiting for most of the scenarios examined provided a benefit-cost ratio of greater than one, indicating economic efficiency.

To combat rabies in raccoons, ORV programs have been utilized to prevent and suppress the impacts of this RABV variant. In the U.S. and Canada, studies have indicated that by forming ORV zones along the Appalachian Ridge to strategically prevent the westward spread of raccoon rabies is economically efficient. These zones prevent the spread of RABV into other U.S. states and Canadian provinces by distributing ORV baits and thereby vaccinating raccoons that consume the baits. Studies indicated that the cost of establishing and maintaining the ORV zones is estimated to be between US $58 million and US $148 million [30]. One study estimated that the cost estimates of ORV in a single state (Ohio) between 1997 and 2000 were US $102,261 per km^2, and the mean baseline cost of blanket rabies prevention techniques for the state was estimated to be US $397,728 per year [32]. Studies have also examined the costs associated with ground vs. aerial distribution of ORV baits and concluded that ground distribution of ORV costs less than air distribution, and the cost of the baits themselves accounts for 85% of total costs [32].

Several case studies exist for raccoon rabies in Canada and the U.S. Some of these studies count the costs associated with a localized rabies outbreak. For example, a small case study from West Virginia in 2008, highlighted that at certain regional levels, raccoon RABV can have significant impacts to livestock. A total of 85 cows were euthanized to prevent the spread of RABV after one of three dead cows on the property tested positive for rabies [33]. In addition, ten people were evaluated for possible exposure following detection of these cases, and all ten received PEP, resulting in an estimated total cost of US $103,985 for this single event [33]. Another case study in Ohio measured the costs associated with euthanizing 64 calves following a confirmed death from raccoon rabies in the herd, and six humans received PEP following the incident. The total cost of this event was US $44,974 [33]. In a case study of a raccoon RABV epizootic on Long Island, New York in 2004, a benefit-cost analysis indicated that for every dollar spent on the program $1.71 was saved in terms of damages avoided [34]. Additionally, a regional economic model estimated that because of the damages avoided, the region saw employment grow by over 100 jobs and a Gross Domestic Product (GDP) increase of $9.2 million through 2019 [34].

Bioeconomic modeling has also been employed to examine the potential benefits and costs associated with stopping the spread of RABV in a host before it can become an epizootic. In Quebec, Canada, in 2006, a rabies management program stopped the spread of raccoon rabies through the province. A disease spread simulation model estimated the potential spread of raccoon rabies from the 2006 index case, and incidence rates of human PEP, animal testing, and human exposure investigations were calculated [35]. Benefits of the program, in terms of damages avoided, ranged from US $47 million to US $53 million and costs ranged from US $33 million to US $49 million. Economic efficiency was indicated for approximately half of the modelled scenarios, with the greatest benefit-cost ratios resulting from reduced future program costs [35].

While raccoon RABV makes up the majority of rabies cases and costs, skunk RABV cases comprise about 26% of all wildlife rabies cases in North America and bat RABV cases make up about 26% of cases in the US. Interestingly, the three skunk RABV variants collectively have the broadest geographical distribution in the U.S., occurring in 22 north and southcentral states as well as California [36]. In California where rabies is enzootic, striped skunks and bats act as the main wildlife reservoirs, with 70% of California rabies cases being linked to mesocarnivore species, that most often include striped skunks, spotted skunks, and gray foxes, with 30% of cases involving bats [36]. Currently, there is no ORV bait available for skunks (or bats) and vaccination of bat populations through ORV and available vaccine strategies is not feasible [37]. The most common methodology for reducing vampire bat rabies impacts is through bat control and livestock vaccination. One study that examined the economic efficiency of livestock vaccination and bat control found that livestock vaccination could be economically efficient, but bat control was not [37, 38]. In a study that utilized information from vampire bat rabies in Mexico, it was estimated that if vampire bat rabies were to spread to southern Texas, the total economic impact was estimated to be between US $seven million and US $9.2 million [38]. In another study, based on a canine rabies outbreak that began in south Texas in 1988, it was estimated that during the six years of the epizootic, 216 domestic dogs and 270 coyotes were confirmed rabid. Over those six years, aerial baiting was used to dispense vaccine-laden baits throughout the infected area, costing a total of over US$ 26 million [39]. The integration of ORV was a major contributing factor that led to elimination in the U.S. of canine rabies [40].

In 1989, Ontario initiated a red fox ORV program. In the decade prior to the initiation of the vaccination program, the average annual human PEP, positive rabies diagnostic tests and indemnity payments for livestock lost to rabies were $2,248, $1,861 and $246,809, respectively [41]. In the decade following the ORV program, a 35%, 66%, and 41%, decrease in costs were observed, respectively [41]. Benefits were measured as the damages avoided from the program and costs were those associated with ORV baiting. Multiple techniques were used to estimate four different benefit streams and in general indicated overall program efficiency in three of the four conservative scenarios [41].

Mexico

Rabies in Mexico was predominantly canine RABV and, to a much lesser extent, due to bat rabies. Vampire bats transmit RABV to humans, livestock and other animals through their hematophagous behavior and are often the species most responsible for the transmission of RABV to livestock. [3] In cattle, bat rabies can result in hide damage, weight loss, decreased milk production, and death. The common vampire bat can be found in Mexico as far north as the states of Sonora and Tamaulipas and could potentially extend into south Texas in the next few decades as a result of climate change [38]. Although most of the PEP administered in Mexico has resulted from canine RABV exposure, most of the human deaths have been associated with vampire bats [37, 42]. For example, in Hermosillo, Mexico, in 1987, a canine rabies outbreak resulted in 2.5% of city residents being bitten by a dog, 60% of whom were administered PEP. The PEP cost of this single outbreak was an estimated US $682,500 per 100,000 citizens [31]. Since that time, Mexico has embarked on a comprehensive and ambitious national dog rabies vaccination program. From 1990 through 2015, the damages avoided from this program equated to roughly 13,000 human rabies deaths, at a cost of US $300 million [43]. It was estimated that an average cost of US $23,000 per human rabies death averted, US $410 per additional year-of-life, and US $190 per dog rabies death averted [43]. Demonstrating the overall utility of mass dog vaccination programs, Mexico is now reportedly free of canine rabies.

Haiti

Haiti is a unique example in the Americas because it remains one of the few rabies-endemic countries in the Western Hemisphere and is responsible for over 70% of the rabies related deaths in the Americas [44]. This is a direct result of the inaccessibility of PEP such that less than one-third of exposed people seek medical care, which results in approximately 130 deaths annually [45]. A survey conducted in Haiti revealed that 3.2% of respondents had experienced a dog bite in the previous year, which is problematic in a country where most of the people do not seek PEP [45]. Haiti is an exception to the general experience in the Americas, in that it does not have a comprehensive rabies prevention plan for canine rabies and does not seek to actively for prophylaxis of those that have been exposed. Despite being a New World nation, rabies management in this Caribbean country is more typical of what is seen in Africa and Asia, in that it lacks a blueprint for realistic large scale dog vaccination programs and effective and efficient PEP for individuals that have been exposed.

Conclusions

Health economics provides a metric by which rabies impacts can be quantified and management success can be determined by the benefits it provides. The health economic benefits of rabies management are clear and consistent: vaccination of reservoir species (especially dogs) reduces human impacts (in terms of human death) and causes a decrease in the number of PEP courses administered. Those countries or regions most impacted by rabies are those that choose to invest less in dog vaccination efforts and more in PEP [44]. The role of dogs is crucial in all rabies prevention efforts. Even if the agent is not the canine RABV variant, dogs are still the conduit to most human exposures further emphasizing the importance of dog vaccination efforts.

This review indirectly highlights some crucial facts regarding rabies in the Americas. First, the more developed the country (e.g., Canada and the U.S.), the less likely that country is to have costs resulting from canine RABV because they have either eliminated the variant or are actively managing viral variants in the wildlife population. Additionally, in more developed countries, dogs typically receive better general health care and likely obtain routine rabies vaccinations through annual checkups which translates to lower rabies impacts. This review also demonstrated that the more developed the country, the more likely they are to have a more sophisticated calculation of benefits resulting from a rabies management program. In this case, benefits may be calculated using a bioeconomic approach that relies on an epidemiological disease spread model coupled with an economic impact model or a regional economic model. The value of explaining benefits calculated on a broader economic scale is that more individuals can recognize the economic gains resulting from suppressing or eliminating rabies in a particular region. This can garner greater public support for rabies elimination efforts which can directly impact funding for a vaccination program leading to greater overall success.

While this review highlighted the impact of rabies in different locations across the Americas, it also provided studies that had documented successful rabies elimination programs and the benefits and costs associated with those programs. Understanding the economics of successful elimination programs is the first step in crafting an effective program for a different region. Examining the global ramifications of rabies suggests that there is significant room for improvement in the fight against rabies, not only in the Americas, but especially in Asia and Africa, as well as poorer regions in other parts of the globe (44). The health economics of the fight against rabies in the Americas is part of a growing literature of rabies management efforts that clearly indicate that rabies elimination is not only feasible, but economically efficient and the components of success have been outlined in multiple locations in the Americas.

References

1. Shwiff S, Hampson K, Anderson A (2013) Potential economic benefits of eliminating canine rabies. Antivir Res 98(2):352–356. https://doi.org/10.1016/j.antiviral.2013.03.004
2. Anderson A, Shwiff SA (2015) The cost of canine rabies on four continents. Transbound Emerg Dis 62(4):446–452. https://doi.org/10.1111/tbed.12168
3. Hampson K, Coudeville L, Lembo T, Sambo M, Kieffer A, Attlan M, Barrat J, Blanton JD, Briggs DJ, Cleaveland S, Costa P, Freuling CM, Hiby E, Knopf L, Leanes F, Meslin FX, Metlin A, Miranda ME, Müller T, Nel LH, Recuenco S, Rupprecht CE, Schumacher C, Taylor L, Vigilato MA, Zinsstag J, Dushoff J, On behalf of the Global Alliance for Rabies Control

Partners for Rabies Prevention (2015) Estimating the global burden of endemic canine rabies. PLoS Negl Trop Dis 9(4):e0003709. https://doi.org/10.1371/journal.pntd.0003709

4. Jones BA, Grace D, Kock R, Alonso S, Rushton J, Said MY, McKeever D, Mutua F, Young J, McDermott J, Pfeiffer DU (2013) Zoonosis emergence linked to agricultural intensification and environmental change. PNAS USA 110:8399–8404
5. Centers for Disease Control (CDC) (1981) The cost of one rabid dog—California. MMWR Morb Mortal Wkly Rep 30(42):527
6. Kapos V, Balmford A, Aveling R, Bubb P, Carey P, Entwistle A, Hopkins J, Mulliken T, Safford R, Statters A, Walpole M, Manica A (2008) Calibrating conservation: new tools for measuring success. Conserv Lett 1:155–164
7. Mann JM, Burkhart MJ, Rollag OJ (1980) Anti-rabies treatments in New Mexico: impact of a comprehensive consultation-biologics system. Am J Public Health 70(2):128–132. https://doi.org/10.2105/ajph.70.2.128. PMID: 7352606; PMCID: PMC1619200
8. Middaugh J, Ritter D (1982) A comprehensive rabies control program in Alaska. Am J Public Health 72(4):384–386. https://doi.org/10.2105/ajph.72.4.384. PMID: 7065317; PMCID: PMC1649898
9. Blanton JD, Colwell E et al (2019) Rabies exposures and pre-exposure vaccination practices among individuals with an increased risk of rabies exposure in the United States. J Am Vet Med Assoc 252(12):1491–1150
10. Sasaki DM, Gooch JM (1983) Cost effectiveness of Hawaii's anti-rabies quarantine program. Hawaii Med J 42(7):157–160
11. Bernard KW, Fishbein DB (1991) Pre-exposure rabies prophylaxis for travelers: are the benefits worth the cost? Vaccine 9(11):833–836. https://doi.org/10.1016/0264-410x(91)90221-q
12. Slate D, Rupprecht CE, Rooney JA, Donovan D, Lein DH, Chipman RB (2005) Status of oral rabies vaccination in wild carnivores in the United States. Virus Res 111(1):68–76
13. Meltzer MI, Rupprecht CE (1998) A review of the economics of the prevention and control of rabies. Part 1: global impact and rabies in humans. PharmacoEconomics 14(4):365–383. https://doi.org/10.2165/00019053-199814040-00004
14. Meltzer MI, Rupprecht CE (1998) A review of the economics of the prevention and control of rabies. Part 2: rabies in dogs, livestock and wildlife. PharmacoEconomics 14(5):481–498. https://doi.org/10.2165/00019053-199814050-00003
15. Cantor SB, Clover RD, Thompson RF (1994) A decision-analytic approach to postexposure rabies prophylaxis. Am J Public Health 84(7):1144–1148. https://doi.org/10.2105/ajph.84.7.1144. PMID: 8017541; PMCID: PMC1614738
16. Kreindel SM, McGuill M, Meltzer M, Rupprecht C, DeMaria A Jr (1998) The cost of rabies postexposure prophylaxis: one state's experience. Public Health Rep 113(3):247–251. PMID: 9633871; PMCID: PMC1308677
17. Krebs JW, Long-Marin SC, Childs JE (1998) Causes, costs, and estimates of rabies postexposure prophylaxis treatments in the United States. J Public Health Manag Pract 4(5):56–62. https://doi.org/10.1097/00124784-199809000-00009
18. Gordon ER, Krebs JW, Rupprecht CR, Real LA, Childs JE (2005) Persistence of elevated rabies prevention costs following post-epizootic declines in rates of rabies among raccoons (*Procyon lotor*). Prevent Vet Med 68(2–4):195–222
19. Dhankhar P, Vaidya SA, Fishbien DB, Meltzer MI (2008) Cost effectiveness of rabies post exposure prophylaxis in the United States. Vaccine 26(33):4251–4255
20. Shwiff SA, Sterner RT, Jay MT, Parikh S, Bellomy A, Meltzer MI et al (2007) Direct and indirect costs of rabies exposure: a retrospective study in southern California (1998–2002). J Wildl Dis 43(2):251–257
21. Fields GS (2011) Poverty and low earnings in the developing world. Cornell University, Industrial and Labor Relations (ILR) School, Ithaca, New York, 30 pp. Available at: http://digitalcommons.ilr.cornell.edu/workingpapers/152. Accessed 22 June 2018
22. Peters MD, Godfrey CM, Khalil H, McInerney P, Parker D, Soares CB (2015) Guidance for conducting systematic scoping reviews. Int J Evid Based Healthc 13(3):141–146

23. Moher D, Liberati A, Tetzlaff J, Altman DG, Prisma Group (2009) Preferred reporting items for systematic reviews and meta-analyses: the PRISMA statement. PLoS Med 6(7):e1000097
24. Moher D, Shamseer L, Clarke M, Ghersi D, Liberati A, Petticrew M, Shekelle P, Stewart LA (2015) Preferred reporting items for systematic review and meta-analysis protocols (PRISMA-P) 2015 statement. Syst Rev 4(1):1. https://doi.org/10.1186/2046-4053-4-1
25. Sterner RT, Meltzer MI, Shwiff SA, Slate D (2009) Tactics and economics of wildlife oral rabies vaccination, Canada and the United States. Emerg Infect Dis 15(8):1176–1184
26. Uhaa IJ, Dato VM, Sorhage FE, Beckley JW, Roscoe DE, Gorsky RD, Fishbein DB (1992) Benefits and costs of using an orally absorbed vaccine to control rabies in raccoons. J Am Vet Med Assoc 201(12):1873–1882
27. Meltzer MI (1996) Assessing the costs and benefits of an oral vaccine for raccoon rabies: a possible model. Emerg Infect Dis 2(4):343–349. https://doi.org/10.3201/eid0204.960411. PMID: 8969251; PMCID: PMC2639934
28. Anderson A, Shwiff SA et al (2014) Forecasting the spread of raccoon rabies using a purpose-specific group decision-making process. Hum Wildl Interact 8(1):14. Available at: https://digitalcommons.usu.edu/hwi/ vol8/iss1/14. Accessed 22 Mar 2021
29. Pieracci EG, Pearson CM, Wallace RM, Blanton JD, Whitehouse ER, Ma X et al (2019) Vital signs: trends in human rabies deaths and exposures—United States, 1938–2018. Morb Mortal Wkly Rep 68(23):524–528
30. Kemere P, Liddel MK, Evangelou P, Slate D, Osmek S (2000) Economic analysis of a large-scale oral vaccination program to control raccoon rabies. In: Proceedings 3rd National Wildlife Research Center (NWRC) special symposium on human conflicts with wildlife: economic considerations (L. Clark, ed.), 1–3 August, Fort Collins, Colorado, United States of America. United States Department of Agriculture, Fort Collins, Colorado, United States of America, 109–116. Available at: www.aphis.usda.gov/wildlife_damage/nwrc/symposia/economics_symposium/kemereHR.pdf. Accessed 22 Mar 2021
31. Vercauteren KC, Ellis C, Chipman R, DeLiberto TJ, Shwiff SA, Slate D (2012) Rabies in North America: a model of the one health approach. In: Proceedings 14th Wildlife Damage Management (WDM) Conference (S.N. Frey, ed.), 18–21 April, Nebraska City, Nebraska, United States of America. WDM working group, the wildlife society, Bethesda, Maryland, United States of America, 56–63. Available at: https://digitalcommons.unl.edu/icwdm_usdanwrc/1202. Accessed 22 Mar 2021
32. Foroutan P, Meltzer MI, Smith KA (2002) Cost of distributing oral raccoon-variant rabies vaccine in Ohio: 1997–2000. J Am Vet Med Assoc 220(1):27–32. https://doi.org/10.2460/javma.2002.220.27
33. Chipman RB, Cozzens TW, Shwiff SA, Biswas R, Plumley J, O'Quin J et al (2013) Costs of raccoon rabies incidents in cattle herds in Hampshire County, West Virginia, and Guernsey County, Ohio. J Am Vet Med Assoc 243(11):1561–1567
34. Elser JL, Bigler LL, Anderson AM, Maki JL, Lein DH, Shwiff SA (2016) The economics of a successful raccoon rabies elimination program on Long Island, New York. PLoS Negl Trop Dis 10(12):e0005062. https://doi.org/10.1371/journal.pntd.0005062
35. Shwiff S, Aenishaenslin C, Ludwig A, Berthiaume P, Bigras-Poulin M, Kirkpatrick K et al (2013) Bioeconomic modelling of raccoon rabies spread management impacts in Quebec, Canada. Trans Emerg Dis 60(4):330–337
36. Shwiff SA, Sterner RT, Hale R, Jay MT, Sun B, Slate D (2009) Benefit cost scenarios of potential oral rabies vaccination for skunks in California. J Wildl Dis 45(1):227–233. https://doi.org/10.7589/0090-3558-45.1.227
37. Anderson A, Shwiff S, Gebhardt K, Ramírez AJ, Shwiff S, Kohler D, Lecuona L (2014) Economic evaluation of vampire bat (*Desmodus rotundus*) rabies prevention in Mexico. Transbound Emerg Dis 61(2):140–146
38. Anderson A, Shwiff SS, Shwiff SA (2014) Economic impact of the potential spread of vampire bats into south Texas. In: Proceedings 26th Vertebrate Pest Conference (R.M. Timm & J.M. 28: O'Brien, eds), 3–6 March, Waikoloa. University of California, Davis, 305–309. Available at:

www.aphis.usda.gov/wildlife_damage/nwrc/publications/14pubs/14-143%20anderson.pdf. Accessed 22 June 2018
39. Shwiff SA, Kirkpatrick KN, Sterner RT (2008) Economic evaluation of an oral rabies vaccination program for control of a domestic dog–coyote rabies epizootic: 1995–2006. J Am Vet Med Assoc 233(11):1736–1741. https://doi.org/10.2460/javma.233.11.1736
40. Slate D, Algeo TP, Nelson KM, Chipman RB, Donovan D, Blanton JD, Niezgoda M, Rupprecht CE (2009) Oral rabies vaccination in North America: opportunities, complexities, and challenges. PLoS Negl Trop Dis 3(12):e549. https://doi.org/10.1371/journal.pntd.0000549
41. Shwiff SA, Nunan CP, Kirkpatrick KN, Shwiff SS (2011) A retrospective economic analysis of the Ontario red fox oral rabies vaccination programme. Zoon Public Health 58:169–177
42. Eng TR, Fishbein DB, Talamante HE, Hall DB, Chavez GF, Dobbins JG, Muro FJ, Bustos JL, de los Angeles Ricardy M, Munguia A (1993) Urban epizootic of rabies in Mexico: epidemiology and impact of animal bite injuries. Bull WHO 71(5):615–624. Available at: www.ncbi.nlm.nih.gov/pmc/articles/PMC2393488/. Accessed 17 Aug 2018
43. Gonzalez-Roldan JF, Undurraga EA, Meltzer MI, Atkins C, Vargas-Pino F, Gutierrez-Cedillo V, Hernandez-Perez J (2021) Cost-effectiveness of the national dog rabies prevention and control program in Mexico, 1990–2015. PLoS Negl Trop Dis 2021:e0009130. https://doi.org/10.1371/journal.pntd.0009130
44. Shwiff SA, Elser JL, Ernst KH, Shwiff SS (2018 Aug) Anderson AM (2018) cost–benefit analysis of controlling rabies: placing economics at the heart of rabies control to focus political will. Rev Sci Tech 37(2):681–689. https://doi.org/10.20506/rst.37.2.2833
45. Schildecker S, Millien M, Blanton J, Boone J, Emery A, Ludder F, Fenelon N, Crowdis K, Destine A, Etheart M (2016) Dog ecology and barriers to canine rabies control in the Republic of Haiti, 2014–2015. Transbound Emerg Dis 64(5):1433–1442. https://doi.org/10.1111/tbed.12531

Chapter 13
Towards the Elimination of Canine Rabies in the Americas: Governance of a Regional Program

Marco Antonio Natal Vigilato, Albino José Belotto, Hugo Tamayo Silva, Felipe Rocha, Baldomero Molina-Flores, Julio Cesar Augusto Pompei, Larissa Zanetti, and Ottorino Cosivi

Abstract After decades of fighting against rabies, most of the countries of the Americas have eliminated the circulation of rabies virus in canine populations, reducing the incidence of rabies higher than 99% for both cases in humans (from 355 cases in 1982 to 2 cases in 2020) and dogs (from 25,000 cases in 1980 to less than 100 in 2020). This success represents the importance and the advances in Veterinary Public Health in the control of zoonotic diseases, as a result of political commitment, application of technological advances in tools and methodologies, improvement of technical capacities in the countries of the Americas in all administrative levels, strong community participation, intersectoral cooperation, and the regional coordination of the national programs, establishing standard measures and actions applied in the management of rabies. Many are the challenges for 'the last mile', regarding the actual elimination of rabies in canine populations, such as refined epidemiological surveillance, opportune and effective use of human post-exposure prophylaxis and targeted canine mass vaccination campaigns for specific areas of the region. Although difficult, canine rabies can be eliminated, as it will be described via the history of the battle against this zoonosis throughout the Americas.

Authors Albino José Belotto and Hugo Tamayo Silva are retired.

M. A. N. Vigilato (✉) · A. J. Belotto · H. Tamayo Silva · F. Rocha · B. Molina-Flores · J. C. A. Pompei · L. Zanetti · O. Cosivi
Pan American Center for Foot and Mouth Disease and Veterinary Public Health – Department of Communicable Diseases and Environmental Determinants of Health, Pan American Health Organization/World Health Organization, Washington, DC, USA
e-mail: vigilato@paho.org

C. E. Rupprecht (ed.), *History of Rabies in the Americas: From the Pre-Columbian to the Present, Volume I*, Fascinating Life Sciences,
https://doi.org/10.1007/978-3-031-25052-1_13

Keywords Disease elimination · Dog vaccination · Governance · Latin America · Pan American Health Organization · Regional program · World Health Organization

Introduction

Witnessing the progression of the symptoms in a patient with rabies to their untimely and horrific death is something that remains forever stamped in the memory of any health professional. If human rabies is less common today, in the relatively recent past doctors and nurses could identify a case as soon as a patient arrived at the emergency room of a hospital. Until the early 1980s, there were hundreds of cases of canine-mediated rabies cases in humans in the Americas every year. Although this disease is caused by the same etiologic agent, different variants of rabies virus exist, each adapted to a different epidemiological cycle, with a primary animal species responsible for perpetuation. In the Americas, as in other parts of the world, the dog had been the main viral host responsible for virus maintenance and subsequent transmission to humans.

Before European colonization during the fifteenth century, indigenous rabies viruses would have been adapted to wildlife, such as bats. Thereafter, with the introduction of European breeds, domestic dogs became the major transmitters of rabies virus to humans, due in part to a gradual increase of its population in proximity to humans. This resulted in the establishment of an independent transmission chain in human settlements and eventually in the major cities, therefore named 'urban or canine rabies'. In the Americas, however, since the 1970s to the present time (2022), the incidence of human rabies transmitted by dogs has gradually decreased. In 2004, for the first time, the number of cases of human rabies mediated by wildlife rabies virus variants surpassed the number of cases of human rabies transmitted by dogs (Fig. 13.1).

Most of this success attained has been the result of the political commitment, together with technological advances in tools and methodologies applied in the local control of rabies, and improvement of technical capacities in the countries of the Americas at all administrative levels, with strong community participation and intersectoral cooperation (Figs. 13.2 and 13.3), under the international coordination of the Pan American Health Organization Regional Office of the World Health Organization for the Americas (PAHO/WHO).

From Pre-Columbian Times to the 1950s

Rabies likely existed in some form throughout the Americas since pre-Columbian times, in different species of wildlife, from Canada to Patagonia. Bats (mainly hematophagous, in tropical areas of Latin America and the Caribbean), raccoons,

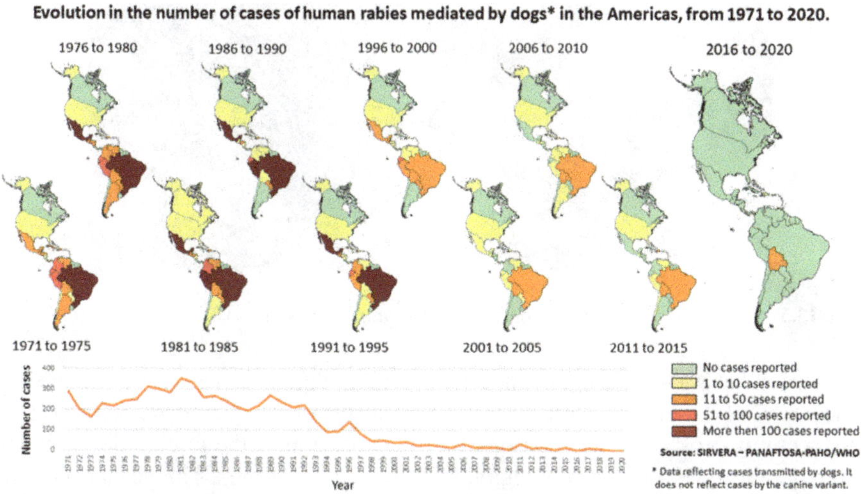

Fig. 13.1 Evolution in the number of cases of human rabies mediated by dogs in the Americas, from 1970 to 2020. (Source https://www.sirvera.panaftosa.org.br)

Fig. 13.2 A child taking her dog to a vaccination point during a rabies mass vaccination campaign

skunks, foxes, and monkeys may have predominated (eventually including non-indigenous mongooses, introduced only to the Caribbean during the colonial period). Apparently, significant outbreaks were not observed nor anticipated during this pre-colonial period in comparison to today, for a variety of virological, ecological, and socio-cultural reasons. Diverse, indigenous dogs were present for thousands of years throughout the New World, but not domestic cats, cattle, sheep, goats, or swine.

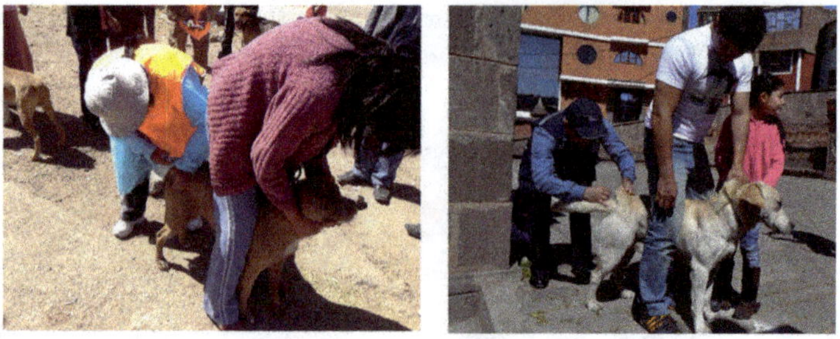

Fig. 13.3 (**a**, **b**) Public vaccination points during a canine rabies mass vaccination campaign

During the waves of colonization post-1492, indigenous dogs were suppressed and replaced. Gradually, formation of regional urban conglomerates with greater connectivity provided the conditions for the expansion of the foreign dog population, introduced throughout the region by European colonizers. This scenario began facilitation of the eventual flourishing of enzootic canine rabies in an independent "dog-dog" chain of viral transmission. However, this did not occur immediately. For example, cases of rabies in dogs were initially registered in North America and the Caribbean only during the eighteenth century. The first human cases were officially reported only in the second half of the nineteenth century.

Growth of urban conglomerates intensified during the first decades of the twentieth century, between 1907 and 1945, a period with a large flow of European emigration to the region. The independent chain of canine viral transmission would be known as the so-called 'urban cycle', maintained by dogs. Countries began to strengthen their actions to control canine rabies outbreaks, as well as to provide healthcare to exposed people.

At that time, already aware of the burden of rabies after several centuries of new appreciation, the Americas were one of the first regions in the world to widely establish rabies post-exposure prophylaxis in humans, after Pasteur's studies that led to the development of the human vaccine, used for the first time in France during 1885. Units of the Pasteur Institute were established in many countries of the Americas during the last decades of the nineteenth century and the first decades of the twentieth century.

In 1902, an important entity was created, an international agency for multilateral collaboration on public health. This nascent organization would be key in the near future for the organized fight against diseases such as rabies in the Americas, with the creation of the Pan American Sanitary Bureau. Nowadays, this institution is recognized as the Pan American Health Organization (PAHO), with headquarters in Washington, DC. In 1948, the World Health Organization (WHO) was established in Geneva. Thereafter, PAHO began its regional representation of WHO in the Americas. Also in that year, PAHO/WHO technical cooperation in zoonoses began, with the establishment of the Veterinary Public Health Unit (VPH/PAHO) managed in coordination with the veterinary advisors appointed in most countries of the region. So began an essential and key component to support the countries in the fight against canine rabies, prior to the current notion of 'One Health'. This framework allowed a significant portrait of a concept that would be the key in the fight ahead: an understanding by public health authorities and the general population that all society would have to focus upon the animal populations responsible for zoonotic maintenance for the greater good. As such, disease prevention and control actions would have to be undertaken in dog populations of urban centers where canine rabies occurred.

Canine rabies vaccination was a precursor to human rabies vaccination, since the first attempts of rabies immunization were performed in various animals, including dogs. The close collaborations between veterinarians and physicians, both at the experimental stage (first vaccines) and at later stages of rabies vaccine development were essential for the success of rabies control programs. In the first half of the twentieth century, several types of biologics similar to sheep brain vaccine produced by Pasteur were developed. Their production in national laboratories was started to support mass dog rabies vaccination campaigns and post-exposure prophylaxis in persons at risk.

Considered as an action of health for the public benefit, the mass dog vaccination campaigns were executed annually without costs to the population, with an objective to vaccinate most of the estimated dog population, under coordination of the health authorities of the Latin America and the Caribbean (LAC) countries (Figs. 13.2 and 13.3). The key for the success of these early campaigns was ownership, promotion, and advocacy, using mass media (such as radio, educational materials, schools, community centers, health facilities, etc.), aimed to stimulate the community engagement in all stages of the program. An intersectoral approach was critical during planning, promotion, execution, and evaluation.

From 1950 to 2000

Prior to the 1950s, the healthcare of people exposed to the risk of contracting rabies was provided at special units, such as the Pasteur Institutes established in many countries, or animal bite centers exclusively specialized in rabies vaccination, generally located in large cities (Anti-rabid Centers). This situation reduced the search for prophylaxis by affected people, particularly those with lower-income levels and living in remote areas. In some places, it was only in the 1970s that human rabies prophylaxis would be decentralized to regular local healthcare units and centers, with trained health professionals under the supervision of specialized centers.

During 1954, Fuenzalida and Palacios, scientists from Chile, developed a new rabies vaccine, produced in suckling mouse brain (SMB), with a lower level of myelin and a higher immune activity than previous vaccines, produced from the brain of adult animals.

The use of the SMB rabies vaccine in public health actions and the decentralization of vaccination units were examples of the technical cooperation measures promoted to the countries by PAHO/WHO. This was facilitated through the Pan

American Zoonoses Center (CEPANZO-PAHO/WHO), created in 1956 in Buenos Aires, by agreement between PAHO/WHO and the Argentine Government, with the purpose of providing technical cooperation in rabies, together with supporting management of the most prevalent zoonoses in the region, such as brucellosis, tuberculosis, and cysticercosis, among others.

CEPANZO, in the decade of the 1990s, became the Pan American Institute for Food Safety and Zoonoses (INPPAZ- PAHO/WHO), including food safety and zoonosis cooperation functions.

During the 1960s – 80s, major canine rabies outbreaks and numerous human cases were registered in the large cities of the LAC. In the 1970s, CEPANZO developed the Regional Information System for the Epidemiological Surveillance of Rabies – SIRVERA. To this day, SIRVERA still collects information over the occurrence of rabies in the Americas.

Concerned about the increasing registration of rabies cases of both human and animals as reported in SIRVERA, the governing bodies of PAHO/WHO requested the establishment of a specific project to support the countries in their efforts to control rabies.

In this context, the Third Inter-American Meeting at the Ministerial Level on Health and Agriculture (RIMSA III) committed to eliminate human rabies from the main cities of the region. This commitment was ratified by PAHO/WHO, which organized the First Meeting of the Directors of National Rabies Control Programs in the Americas (REDIPRA I) in December 1983, in Guayaquil, Ecuador (Fig. 13.4). At this meeting, the PAHO's VPH Program, in coordination with experts of the countries, elaborated a work document, the "Plan of action for the elimination of

Fig. 13.4 Photo of the First Meeting of Rabies Program Directors of the Americas (REDIPRA I), in Guayaquil, Ecuador, 1983

urban rabies in the main cities of Latin America" by the end of the 1980s, approved by REDIPRA I delegates.

PAHO is responsible for the coordination and organization of the Inter-American Meeting at the Ministerial Level on Health and Agriculture (RIMSA), which stands out at a regional level ever since it was first organized, in 1968, as an Agriculture Ministers' forum, and since 2000, as a Health Ministers' forum of PAHO Member States. The RIMSA meets periodically to evaluate veterinary public health cooperation proposals in to foster priority plans and projects receiving the necessary political support for their development, such as the Regional Plan for Rabies Elimination. Delegates of the Ministries of Environment were also convened to the last meetings due to their increasingly influential role in the determination of the disease and control strategies.

This Plan allowed identifying PAHO/WHO strategic cooperation components for the fight against rabies, including: organization and/or strengthening of national programs for the control of rabies mediated by dogs through the promotion of mass canine rabies vaccination campaigns; decentralization of healthcare to people exposed to the risk of rabies; strengthening of diagnostic capacities and control of immune-biologicals for rabies; strengthening of epidemiologic surveillance systems; training, communication and dissemination of information to health professionals and the general population.

In the 1990s, a gradual replacement began of the SMB vaccines to the cell-culture based vaccines for both dogs and humans. However, even today, a few LAC countries still utilize SMB vaccines for human pre- and post-exposure prophylaxis The older SMB vaccines can cause adverse neurological events and have proven lower efficiency compared to the modern cell-culture vaccines.

During this same decade, collaborative work in the countries continued in accordance with the Plan of Action. Given the success of control observed in major urban centers, in 1991 the plan was expanded to marginal areas and small rural towns and villages. In 1992, the plan enlarged its coverage, strategy, and goal, for rabies elimination by 2000, including the surveillance of wildlife rabies. As a result of this focus, there were less than 200 cases of dog-mediated human rabies in 1993, and less than 100 cases were recorded, for the first time in the Americas, during 1994. Although the outbreaks recorded during 1996 increased the number of cases to 183, since 1997, cases never exceeded 100.

From 2000 to Date

In 2000, 55 cases of dog-mediated human rabies were confirmed and a survey about the status of human rabies transmitted by dogs in LAC during 2001 and 2003 indicated that, on average, 81% of the animals had been vaccinated. In some locations of the LAC countries, such as in Mexico and Brazil, this coverage reached around 90%.

In 2006, there was a change in the management of PAHO's VPH technical cooperation, and the functions of the INPPAZ were transferred to the Pan American

Foot-and-Mouth Disease Center (PANAFTOSA-PAHO/WHO), in which the elimination of human rabies mediated by dogs was given special importance.

In 1951, by request of the Organization of American States (OAS) member countries, was created the Pan American Foot-and-Mouth Disease Center (PANAFTOSA), to support the countries in the fight against Foot-and-Mouth disease, which has a serious impact on animal health, food security and regional and global commerce during that time. PANAFTOSA was established in the city of Rio de Janeiro, through Agreement between OAS and the Government of Brazil, and was solicited to PAHO/WHO the administrative and technical management of the Center.

Today, most of the countries of the Americas have eliminated the circulation of canine rabies virus in dog populations and the cases of human rabies mediated by dogs is reduced to a few cases occurring in well-defined circumscribed areas. The reduction of rabies has been higher than 99% for both cases in humans (from 355 cases in 1982 to 2 cases in 2020) and dogs (from 25,000 cases in 1980 to less than 100 in 2020). This reduction is due, specifically, to the mass canine vaccination campaigns that have achieved to immunize an annual mean of 100 million dogs in the last several years, representing more than 70% of the estimated dog populations of the LAC.

The availability of rabies biologics (human and canine vaccines and Rabies immune globulin) to the LAC countries has been ensured by the PAHO's Revolving Fund, which has facilitated the logistics of the production, acquisition, and opportune distribution. In the past, more than 25 countries of the region have benefited from this system, totaling an annual average of more than 35 million doses of canine rabies vaccines and two million doses of human rabies vaccines. Importantly, the solidarity among the countries of the Americas, through the Regional Program coordinated by PANAFTOSA-PAHO/WHO, which has donated human and canine rabies cell-culture vaccines and other supplies to countries in emergency for the control of rabies, has been one of the major factors in this historical progress.

Other primordial key factor for the reduction in the incidence of human rabies mediated by dogs in the Americas is the expansion of post-exposure prophylaxis to all basic health units of LAC countries, as recommended by WHO Expert Committee on Rabies. Modern prophylaxis is largely available in the LAC, as part of the public health service provided free of charge to the population. Around one million people annually search for healthcare in LAC public health units after possible rabies virus exposure, from which about 25% receive prophylaxis after risk assessment.

Current Considerations and Lessons Learned

The fight against rabies represents great progress for the region in terms of public health, and it is mostly due to the strengthening of national programs for VPH. In the course of the evolution of the rabies program, as from the first plan in 1983, other regional plans were developed, adapting their strategies and guidelines to the epidemiological situation ad hoc.

Within the framework of the objectives of the regional plans, PAHO/WHO prioritized the collaboration with countries by elaborating proposals to adjust the institutional structure and the management of the technical and administrative units for rabies control. This was conducted to facilitate the execution of operational control programs, based on the evidence and the lessons learned since 1983. This work process was made easier by the presence and contribution of experts proficient in planning and other technical specialties, trained in capacity-building courses organized by CEPANZO, and then INPPAZ and PANAFTOSA. This cooperation has also contributed to a great influence and effort to the community-based education, since community involvement plays a key role to maintain sanitary status, particularly to maintain the continuity of the control and prevention actions against rabies.

Equal priority was given to the collaboration provided to the countries in the processes for evaluating national programs, whose results were presented to the participants to REDIPRA meetings. This process was complemented by the initiative "External Evaluation of National Rabies Programs", coordinated by PANAFTOSA-PAHO, conducted in countries of the region with a critical situation regarding elimination. The evaluations were integrated with national program directors, experts in laboratory diagnosis, specialists in production and quality control of vaccines, and PAHO/WHO Collaborating Centers delegates and advisors. The results of the evaluation were presented to the top LAC health and agriculture authorities with the purpose of committing the highest political level to the strengthening and sustainability of the national program.

Another key aspect in the elimination process has been the promotion and execution of "Cooperation Projects between Countries", supported with PAHO resources, applied to the prevention and control of the risk of disease dissemination, particularly at international borders, which enabled strengthening the operational capacities at subnational and local levels, based on epidemiological risk criteria.

PAHO's technical support to the Regional Program is based on the resolutions of the REDIPRA, convened by PAHO approximately every two years, where the countries present and discuss the rabies epidemiological situation and the results of their prevention and control strategy. During the meetings, all the countries of the Americas are fully represented by their rabies program coordinator, public health authorities and veterinary services. Conclusions and recommendations are presented at RIMSA inter-ministerial meetings, and they are discussed and submitted for consideration to PAHO Board of Directors. The participation of WHO Collaborating Centers for Rabies in the Americas, as well as other international cooperation agencies (FAO, IICA, OIRSA, etc.) and organizations (OIE, WSPA, GARC, among others), accredited entities of the civil societies, and pharmaceutical companies producing immunobiologicals, provide to the REDIPRA a scientific and interinstitutional approach for rabies elimination (Fig. 13.5). As a follow up to REDIPRA meetings, virtual events and meetings are conducted regularly to address specific technical topics of the regional and national programs.

Fig. 13.5 The 16th Meeting of Rabies Program Directors in the Americas (REDIPRA), Antigua, Guatemala, 2017

Importantly, the Americas have several WHO Collaborating Centers (CCs) – known as PAHO/WHO Collaborating Centers –, institutions that have been designated to perform specific activities to support the programs to combat rabies. The CCs are very important as regards to education and training of the professionals responsible for rabies programs at a continental level, such as: development and implementation of laboratory diagnosis; panels of monoclonal antibodies for rabies virus characterization; enhanced surveillance and information systems; and related methodologies for the prevention and control of rabies in the countries of the American region. The CCs of the region include:

- Zoonoses Control Center, City of São Paulo – Brazil
- Pasteur Institute of São Paulo, of the State Secretary for Health of São Paulo – Brazil
- Center of Expertise (COFE) for Rabies, Ottawa Laboratory Fallowfield (OLF), Canadian Food Inspection Agency (CFIA)
- Rabies Program, Poxvirus and Rabies Branch, Division of High Consequence Pathogens and Pathology (DHCPP) – Centers for Disease Control and Prevention (CDC)

Summing up, the institutional platform for VPH cooperation, particularly regarding rabies elimination, is based on the principles of good governance, expressed in the Agriculture-Health-Environment intersectoral work and inter-agency at all levels (international, regional, national and local). At national and local levels, this relationship is based on diverse and active networks of public and private service organizations and entities, among which it is worth mentioning health, agriculture, municipalities, community organizations, educational institutions, the armed forces, and NGOs.

Future Challenges

The challenge of human prophylaxis in LAC is to ensure the administration of immune globulin to the seriously bitten victims of bites according to WHO recommendations. The use of more potent, safe, and effective modern cell-culture based vaccines poses other challenges in many LAC countries. Over 2,500,000 doses of human cell-culture vaccines are used every year for prophylaxis, but a few LAC countries are still using the older SMB vaccines. Regarding pre-exposure prophylaxis, it should be analyzed the need to immunize people who are subjected for their professional activity, or travel to risk areas, and those populations living in sylvatic environments at risk for the transmission of rabies virus by wildlife, such as vampire bats.

The overall favorable situation of rabies requires an improvement in the epidemiological surveillance for careful risk assessment studies to guide focused actions on areas identified as being of greater risk for canine rabies circulation, replacing the massive vaccination campaigns that are currently being conducted in generalized form. In this sense, low coverages of vaccination of dog populations (20–40%) are observed in some LAC regions. Therefore, dog vaccination programs should still be improved, as well as awareness strategies currently implemented to engage community participation in the vaccination control program. These low coverage areas are those in which rabies virus is still circulating among dogs and eventually results in transmission to humans.

Epidemiological surveillance is an important factor for rabies control in LAC countries, particularly in 'rabies-free' areas bordering with locales where rabies is still endemic. There are large areas free of human rabies mediated by dogs and canine rabies, but the disease perpetuates in areas of international borders, with intense flows of animals and people. In these bordering areas, a high surveillance standard is required to avoid rabies reintroduction. In general, an enhanced epidemiological surveillance system should be included in the agenda of each country, in areas where information on routine surveillance is not available. Internal risk factors of each country, caused by inequities and regional characteristics, should also be considered to prevent rabies reintroduction. Similarly, there is a need for ongoing support of health authorities and veterinary services to upgrade their laboratory facilities, including improvements in their capacity for sample collection and shipping.

Perhaps the world will never fully *eradicate* rabies, because of the abundance wildlife rabies variants and reservoirs, that perpetuate this zoonosis as a disease of nature. However, canine rabies can be prevented, controlled, and selectively eliminated, as described in the previous sections, through coordinated actions between public and private institutions, with the gradual improvement of the economic and social conditions in the affected countries.

References

1. Babboni SD, Modolo JR (2011) Raiva: origem, importância e aspectos históricos. UNOPAR Científica. Ciências Biológicas e da Saúde v. 13, n. Esp:349–356. Available at: http://hdl.handle.net/11449/140925
2. Belotto A, Leanes LF, Schneider MC, Tamayo H, Correa E (2005) Overview of rabies in the Americas. Virus Res 111(1):5–12, ISSN 0168-1702. https://doi.org/10.1016/j.virusres.2005.03.006
3. Centro Panamericano de Fiebre Aftosa (PANAFTOSA) (2020) Dia Mundial Contra a Raiva 2020. Available: https://www.paho.org/en/campa%C3%B1as/dia-mundial-contra-rabia-2020. Accessed 12 Aug 2021
4. Clavijo A, Del Rio Vilas VJ, Mayen FL, Yadon ZE, Beloto AJ, Vigilato MAN, Schneider MC, Cosivi O (2013) Gains and future road map for the elimination of dog-transmitted rabies in the Americas. Am Soc Trop Med Hyg 89(6):1040–1042. Retrieved Apr 15, 2021, from https://www.ajtmh.org/view/journals/tpmd/89/6/article-p1040.xml
5. Del Rio Vilas VJ, Freire de Carvalho MJ, Vigilato MAN, Rocha F, Vokaty A, Pompei JA, Molina Flores B, Fenelon N, Cosivi O (2017) Tribulations of the last mile: sides from a regional program. Front Vet Sci 4:4. https://doi.org/10.3389/fvets.2017.00004
6. Freire de Carvalho M, Vigilato MAN, Pompei JA, Rocha F, Vokaty A, Molina-Flores B et al (2018) Rabies in the Americas: 1998–2014. PLoS Negl Trop Dis 12(3):e0006271. https://doi.org/10.1371/journal.pntd.0006271
7. Kessels JA, Recuenco S, Navarro-Vela AM, Deray R, Vigilato MAN, Ertl H, Durrheim D, Helen R, Nel LH, Abela-Ridder B, Briggs D (2017) Pre-exposure rabies prophylaxis: a systematic review. Bull World Health Organ 95:210–219C. https://doi.org/10.2471/BLT.16.173039
8. Laing G, Vigilato MAN, Cleaveland S, Thumbi SM, Blumberg L, Salahuddin N, Abdela-Ridder B, Harrison W (2021) One health for neglected tropical diseases. Trans R Soc Trop Med Hyg 115(2):182–184. https://doi.org/10.1093/trstmh/traa117
9. Rocha F (2020) Control of rabies as a victim of its own success: perception of risk within a Latin American population. Am J Trop Med Hyg 103(3):929–930. https://doi.org/10.4269/ajtmh.20-0728. PMID: 32662393; PMCID: PMC7470575
10. Schneider MC, Belotto A, Adé MP, Hendrickx S, Leanes LF, Rodrigues MJ, Medina G, Correa E (2007) Current status of human rabies transmitted by dogs in Latin America. Cad Saude Publica 23(9):2049–2063. https://doi.org/10.1590/S0102-311X2007000900013
11. Schneider MC, Belotto A, Adé MP, Leanes LF, Correa E, Tamayo H, Medina G, Rodrigues MJL (2005) Epidemiologic situation of human rabies in Latin America in 2004. Epidemiol Bull 26(1):2–4
12. Schneider MC, Romijn PC, Uieda W, Tamayo H, da Silva DF, Belotto AJ, da Silva JB, Leanes LF (2009) Rabies transmitted by vampire bats to humans: an emerging zoonotic disease in Latin America? Rev Panam Salud Publica 25(3):260–269. https://doi.org/10.1590/s1020-49892009000300010. PMID: 19454154. https://scielosp.org/article/rpsp/2009.v25n3/260-269/en/
13. Vigilato MAN, Clavijo A, Knobl T, Tamayo Silva HM, Ottorino C, Schneider MC, Leanes LF, Belotto AJ, Espinal MA (2013) Progress towards eliminating canine rabies: policies and perspectives from Latin America and the Caribbean. Phil Trans R Soc B36820120143 20120143. https://doi.org/10.1098/rstb.2012.0143
14. Vigilato MAN, Cosivi O, Knöbl T, Clavijo A, Silva H (2013) Rabies update for Latin America and the Caribbean. Emerg Infect Dis 19(4):678–679. https://doi.org/10.3201/eid1904.121482
15. Vigilato MAN, Molina-Flores B, Del Rio Vilas VJ, Pompei JC, Cosivi O (2018) Canine rabies elimination: governance principles. Rev Sci Tech 37(2):703–709. https://doi.org/10.20506/rst.37.2.2859. PMID: 30747115

Documents of Meetings of the Rabies Program Directors in the Americas (REDIPRA)

I REDIPRA (1983) – Guayaquil, Ecuador. https://iris.paho.org/handle/10665.2/51202
II REDIPRA (N/A)
III REDIPRA (1989) – Porto Alegre, RS, Brasil. http://iris.paho.org/xmlui/handle/123456789/51198
IV REDIPRA (1992) – México, DF. http://iris.paho.org/xmlui/handle/123456789/51263
V REDIPRA (1995) – St. Domingo, Rep. Dominicana. http://iris.paho.org/xmlui/handle/123456789/51199
VI REDIPRA (1997) – Quito, Ecuador. http://iris.paho.org/xmlui/handle/123456789/51270
VII REDIPRA (1998) – Puerto Vallarta, Jalisco, México. http://iris.paho.org/xmlui/handle/123456789/51381
VIII REDIPRA (2000) – Lima, Peru. http://iris.paho.org/xmlui/handle/123456789/50395
IX REDIPRA (2002) – Santa Cruz de la Sierra, Bolivia. http://iris.paho.org/xmlui/handle/123456789/50334
X REDIPRA (2004) – St Domingo, Rep. Dominicana. http://iris.paho.org/xmlui/handle/123456789/50436
XI REDIPRA (2006) – Brasília, DF, Brasil. http://iris.paho.org/xmlui/handle/123456789/50345
XII REDIPRA (2008) – Antigua Guatemala, Guatemala. http://iris.paho.org/xmlui/handle/123456789/51359
XIII REDIPRA (2010) – Buenos Aires, Argentina. http://iris.paho.org/xmlui/handle/123456789/50122
XIV REDIPRA (2013) – Lima, Perú. http://iris.paho.org/xmlui/handle/123456789/50355
XV REDIPRA (2015) – Brasília, Brasil: http://iris.paho.org/xmlui/handle/123456789/51360
XVI REDIPRA (2017) – Antigua, Guatemala. https://iris.paho.org/handle/10665.2/51459

Chapter 14
Musing Over Non-technical Criticalities of Rabies Prevention and Control

Victor J. Del Rio Vilas and Gilberto Montibeller

Abstract Like most human endeavors, rabies prevention and control presents technical and organizational challenges. Here we focus on the latter and describe several behaviors prevalent in goal setting, program planning and monitoring. The challenges and program priorities have not changed dramatically since the first meeting of rabies program managers in the Americas in 1983 – this has implications for global control efforts.

Keywords Bias · Goalsetting · Meristic · One health · Program evaluation · Project planning · Stakeholders

Introduction

The history of canine rabies in the Americas has gone through several related phases: a pre-colonization 1500s stage, with no known perpetuation described by indigenous civilizations; European colonization, with no apparent cases for several centuries; denial and eventual recognition from the early eighteenth-mid-twentieth century, with control exercised primarily by local quarantine and border restrictions, animal confinement, leash legislation, muzzle ordinances, and massive culling of free-ranging dogs; and the largely post-World War II period of technical diagnostic and mass dog vaccination (MDV) applications. Despite those early

V. J. Del Rio Vilas (✉)
University of Surrey, Guildford, UK
e-mail: Victor.DelRioVilas@ukhsa.gov.uk

G. Montibeller
School of Business and Economics, Loughborough University, Loughborough, UK

Center for Risk and Economic Analysis of Threats and Emergencies, University of Southern California, Los Angeles, CA, USA
e-mail: g.montibeller@lboro.ac.uk

C. E. Rupprecht (ed.), *History of Rabies in the Americas: From the Pre-Columbian to the Present, Volume I*, Fascinating Life Sciences,
https://doi.org/10.1007/978-3-031-25052-1_14

measures from the 1950s onward, canine rabies was essentially enzootic throughout the region and the burden of human cases closely mirrored the distribution in domestic animals. During July 2021, SIRVERA, the database kept by the Pan American Health Organization (PAHO), shows only four human deaths due to dog-mediated rabies in the region [1]. Human cases have remained below 15 per annum in the decade, and on average 16 cases have been reported since 2000, feeding a 22-year-plus long tail of cases in the Americas. This period also witnessed at least three regional elimination goals come and go. Despite this apparent failure to deliver elimination, the underlying story is one of unquestionable overall success in the control of rabies across a very diverse region. Having established this indisputable regional achievement, in this brief note, we muse over elements that may have discolored the history of rabies control in the Americas by failing to reach the elusive elimination goal. Rather than dwell on technical aspects, such as details of laboratory-based surveillance, vaccine performance, etc., our focus is upon mainly behavioral and organizational challenges prevalent in goal setting, planning and operationalization endeavors, as relevant regionally as for the broader 'zero by 30' (ZBT) global program of the elimination of human rabies from dogs by the year 2030.

'The Eagle Has Landed'

During May 1961 US president John F. Kennedy shared his vision to put a man on the moon. In just over 8 years, the Apollo program did just that. In such a short period, the technological advances that the program had to develop to reach the goal were staggering, but insufficient on their own. Like the careful assembling of the millions of components that shaped the rockets to space, exquisite planning and program management was required to ensure that every technological development served a clear purpose and was delivered at the right time to allow Neil Armstrong to land 'the Eagle' on the moon and plant a flag from earth on its surface. Very succinctly, some basic steps were required to enable the mission: identification and selection of the mission modes (there were several possible avenues to the moon), and assessment of mission feasibility – in other words, evaluation of capability and resource availability. These are the backbone elements of program management. However, particulars are relevant. The Apollo 11 mission report described in painstaking detail the equipment to take, the required operations to undertake, the milestones to hit to deliver the goal, and the time required to get there, to the minute [2]. It took several previous Apollo missions, each contributing sequentially planned developments, to reach that goal.

Although there were earlier efforts, the regional rabies control and elimination program in the Americas essentially began during 1983. At the time, a vision was shared: the elimination of the disease as a public health concern by the turn of the decade [3]. The 1983 report makes an interesting reading and shows that the operational problems then are as relevant today. Very much related, it was stated that the technology to reach the goal was already available and that there was enough know-how in the region. With a clear vision, adequate technology and supportive knowledge, what could possibly prevent the region from achieving its goal in the following 7 years, from sticking a veritable flag on this 'virus capsule'?

In addition, the 1983 report presents a revealing, albeit qualitative, assessment of the situation of rabies programs in the region at the time: many countries had none, and activities were mostly reactive in response to outbreaks. It follows to state that the goal required a meticulous analysis of priorities, critical areas and operational requirements, and needed robust monitoring and evaluation (M&E) mechanisms to track program development, successes, failures, and operations on multiple levels (e.g. resource demands). Unfortunately, very few of these critical pieces of the mission were deployed consistently in the following years.

On Planning

Stakeholders setting up strategies towards the prevention, control and elimination of rabies, whether across subnational units or countries in a regional or global effort, must aim to answer the following question at regular times: is the chosen strategy (i.e., the mission mode) preventing the most cases the earliest, and are my monitoring metrics of choice properly capturing progress towards this defined goal? Addressing these questions is not a trivial technical exercise and is confounded with moral (e.g., such as the standard trade-off between efficiency and equity when choosing between interventions) and behavioral nuances. On the latter, we begin by exposing the issue of pseudocertainty [4] in the context of success metrics and planning. Pseudocertainty basically describes planners' preference for certainty, even if such certainty is just an illusion. Suffering from such limitation, planners are likely to prefer a scenario of 10 countries out of 20 'free from rabies' than the alternative scenario of 20 countries at 50% of their paths towards rabies elimination. Saving extreme situations that would enact logical questions on the underlying context (e.g., if the cumulative number of averted cases from the certain scenario is significantly smaller than that of the uncertain one), the risk is that planners might choose the certain scenario without considering the underlying case load. A metric like "number of countries" (or other spatial units of relevance) reaching a specific goal is not such a rare event [5]. Very much related, a preference for early successes may prove to be a risky approach to goal operationalization, denoting myopic decision making [6]. This could well lead to large tails of cases as countries left "for a later time", given their complexity, might take longer to reach control and elimination states. Planning fallacy[1] [7] springs to mind in this scenario, and one feels sympathy

[1] Planning fallacy, or the tendency of planners to underestimate the time/resource required to complete a task (Kahneman and Tversky, 1982). A combination of optimism bias, groupthink, and the failure to look at the outcome of challenges of the same class, can lead to severe underestimation of the magnitude of the challenge.

for the planners fighting against the disease in such untested settings and the relentless ticking clock as they get closer to the desired timeline.

Goal setting is about visualizing a future and the pathways to reach the prize. Fortunately, the future can rarely be tamed to our liking, for it has a mind of its own. For elimination and rabies, given the multiple futures and required developments to reach the goal, strategic foresight [8] would question whether the available tools, that were so successful in controlling the disease, were still apt for the last mile of elimination. If not, what adaptations or new developments were necessary and what shocks (e.g., new and emerging conditions) and prevailing attitudes (e.g., of policy makers and society) would hamper or facilitate the delivery of the goal? Unlike the situation in other regions, the routine joint meetings of rabies program directors (REDIPRA) [9] would seem to provide a suitable platform for such foresight exercises. Officials attending these gatherings tend to wear multiple hats in their institutions, and it is not rare for them to lead the entire zoonoses programs in their countries. This wide perspective is of significant strategic importance whilst setting goals for rabies elimination. Clearly, its feasibility must be carefully informed not only by the availability of rabies-specific capacities, but also, critically, by the impact of many other competing current and future priorities that only an audience such as that participating at REDIPRA might foresee. Regardless, the elicitation of future scenarios must be wary of prevailing biases in such exercises, e.g., availability bias [10] that drives attention to recent events, a simplistic problem representation of the situation [11], or confirmation bias [12] that distorts the acquisition of information about the situation on the ground.

Some questions on planning

1. Was the simplicity of the regional program and message (e.g., vaccinate as many dogs as possible), the clarity of vision, and perhaps, to some extent, the sense of belonging to a regional community, the key ingredients in the success of the regional rabies control program? Specifically, on 'belonging and elimination' one cannot stop recalling the Maslow pyramid [13]. By Maslow's assertion, the regional program (or any other made of heterogeneous administrative units) might have not properly recognized and catered for the natural heterogeneity within the group of countries attending the REDIPRA meetings, and the message of elimination was perhaps too ambitious for some members, where control was still the goal. This is a related issue for the ZBT global target.
2. Was the conversation and position of large countries closer to the elimination goal too overpowering for other members lagging? Overpowering could have also confounded the perceived consensus by countries around the multiple failed elimination goals until 2015. Consensus was rarely challenged internally or externally, and perhaps the particular setting (e.g., the REDIPRA meetings), did not encourage strong dissonant voices around the prospects of reaching elimination.
3. Was the elimination talk motivating for all countries, or would a league system, like that of football (not the American concoction but the genuine one

born in the UK!), with members joining same-level peers lead to positive *herding effects*[2] [14]?

4. What were the planning stages, the required stakeholders, the drivers of organizational change, etc., that led to 'stable' rabies disease programs? What discussions, messages, approaches, champions and their profiles, alliances, barriers and failures were needed to put the region so close to elimination? In previous reflections, some criticized the failure of the region's experience to take a more central stand in informing other prevention, control and elimination efforts elsewhere [9]. The perception is that small scale projects reporting innovations, unproven at programmatic scale, dominate the discourse. This is expected as new developments, mostly of technical and methodological nature, provide the required entropy or surprise demanded for publications. Nevertheless, there is a glaring dearth of evidence on operationalization of international, national, and sub-national control programs.

On the Ground

The region bet heavily on MDV from the early days to reach control, and eventually elimination, over and above any other rabies capacities. Comparisons of relative investments across rabies capacities show the Americas distinctive strategy vs. that of the other regions that seem to have allocated relatively larger amounts of resources to human post-exposure prophylaxis and dose-sparing strategies of scarce and expensive biologics [15]. This preference of the Americas for MDV is even more remarkable from a region where most of the national rabies programs are led and funded by Ministries and/or Departments of Health. Although a rare event, during his time in the region, the first author had the opportunity of meeting some local health authorities that would argue that they had better use of their rabies funds than to vaccinate dogs. Of course, those were the areas where rabies was still prevalent… and we do not claim causality but the blatant lack of understanding of basic risk management principles. Of course, this siloed approach to risk and investment is not unique to the Americas, and it is often heard as a common barrier to disease control. Myopic decision making with short term horizons and simplistic problem representations is ubiquitous. Indeed, Meyer and Kunreuther [14] defined this cognitive bias as myopia, the tendency to focus on short time horizons when evaluating costs and benefits of protective measures.

Today, it seems natural to think of 'One Health' as a solution to address such myopic and narrow perspectives. Rabies prevention and control is widely presented as a stellar example of a One Health approach, with dog vaccination and the so-called integrated bite case management (ICBM) [16] as examples of interventions

[2] Herding effects: one of many decision-making biases that describes the tendency to base choices on the observed actions of others (Meyer and Kunreuther, 2017).

exuding 'one-healthness'. However, the concept tends to stress the importance of coordination between agents of different domains (e.g., public health and animal health) to ensure a more optimal outcome. If so, rabies programs in the Americas tended to be monopolistic and "out of tune" given the predominant absence of animal health authorities, although this collaboration was clearly called upon from the very early days, during 1983 and reflective in RIMSA [3]. Collaborations between domains existed in the region (e.g., animal health laboratories supporting rabies testing for national rabies programs in areas where public health infrastructure was insufficient), but they arose more organically than from a vision to expand a One Health concept. In the Americas, many of the attributes that define One Health (e.g., cooperation, communication, interventions across the risk pathway favoring prevention over treatment), were deployed because they were the correct, necessary, or efficient thing to do. The ICBM, or thc tracing of exposed individuals and dogs in communities, receiving PEP in the former and quarantine for the latter, has been practiced across Latin America for many years, even without the moniker [3]. In most cases, there was no involvement of agents from other domains (e.g., from veterinary authorities) and Ministry of Health officials conducted the entire operation. This was not always smooth, and we recall instances where communication between health care practitioners who detected the exposure and public health colleagues in charge of conducting the epidemiological investigation took much longer than anticipated, even when the two teams were separated by few meters on the same road. Communication failures between health care officials and public health colleagues are unfortunately prevalent across programs, not just for rabies, and diverse settings.

Questions in the last mile

1. The Stepwise Approach towards Rabies Elimination (SARE) [17] tool identifies dozens of capacities and the increments required to reach control and elimination. But, to the best of our knowledge, it does not provide guidance on how to maintain and/or de-scale the program once there. At that stage, how do we support rabies managers to find that 'Goldilocks' rabies program for their specific setting and appetites? Motivation to go ahead might not be as large as the fear to fall back, because of loss aversion [18]. Increasingly smaller gains in the road to control and elimination, that are rewarded in assessment frameworks, such as SARE, may not imbue the same reaction from rabies program managers as losses of the status quo. Systems should monitor the steps between stages and probe programs to avoid going backwards and consolidate gains. This requires capability assessments not only targeting prevention, control and elimination, but also informing failure scenarios (e.g., how much capacity a program needs to lose to be downgraded).
2. Does one need to change planning approaches from identification of minimum capabilities to control rabies, when programs are starting, to approaches towards the identification of failure events, as done in engineering systems to improve reliability [19], when programs aim for elimination? In the Americas, such exercises were not frequent. Although some form of capability data was

captured on a regular basis [20], the analysis of the data was not timely, or comprehensive, and did not inform any capability-based planning (CBP) efforts [21].[3]

3. What is the additional contribution that improved accuracy and precision in dog population numbers, dog vaccine coverage, surveillance sensitivity and timelines, among other capacities, bring to the overwhelming effect of a sufficiently vaccinated dog population in the control and elimination stages? In other words, what is the cost (in cases) of possible biases and imprecision around the estimation of these capacities in the two stages? The answer to these questions would help prioritize investment and attention by rabies program managers as they transition from initiation to control to elimination. As programs get closer to this latter stage, optimization of resources across multiple capacities may become a greater necessity. However, a pervasive loop of lack of accurate and precise evidence to inform the impact of interventions may hamper investments on some of the many capacities (beyond vaccination).

We summarize the main biases that are pervasive in the rabies-control programs and suggest how they can be alleviated and debiased [22] (Table 14.1).

Final Thoughts

Napoleon used to say that only two things unite men: fear and self-interest [https://napoleonbonapartepodcast.com/2012/08/16/top-10-napoleon-bonaparte-quotes/]. Keeping that thought in mind, the closer we get to the rabies elimination goal, the less relevant these two elements become for a large number of parties and stakeholders (especially those not directly involved in rabies prevention and control). Two elements might be considered by rabies planners at this stage: dread and the singularity effect [4]. Rabies is a dreadful disease and dread remains the best predictor for perceived risk [23]. This perception of dread and the singularity of cases in the last kilometer (following the singularity effect that allocates greater value to the protection of the first human lives in the face of risk) may help rabies authorities in their communication efforts to keep some sense of dread in the population, if only to support awareness about the risk of the disease.

Looking back, and over the mostly declining case numbers during the last four decades, we might conclude that management efforts on dog-mediated rabies across the Americas have been a success (i.e., hindsight and outcome biases permitting

[3] CBP can (i) support the identification of multiple uncertain threats to be faced by the system (the rabies program in our case); (ii) help the definition of requirements of a system, in terms of capabilities required for coping with these threats; (iii) support the identification of relationships among capabilities and their synergies or dependencies; and (iv) it can highlight the required resources for building up or maintain the needed capabilities.

Table 14.1 Biases and debiasing strategies in rabies-control programs

Bias	Debiasing strategy
Availability bias (decision makers placing excessive focus on recent and salient events).	Employ a group of varied experts with distinctive experiences in dealing with rabies events and similar threats.
Confirmation bias (decision makers only gather information that confirms their pre-existing assessment of the situation).	Provide adequate incentives for high-quality decision processes and enforce accountability of program managers.
Myopia (decision making focused on short-term goals only).	Clearly define the planning horizon and consider full impacts and investment over the entire period.
Pseudocertainty (decision makers preferring certainty instead of uncertain prospects).	Understand the role of uncertainty in decision making and focus on expected benefits.
Simplistic problem representation (decision makers develop a problem representation that is too narrow and excludes many salient issues).	Employ a group of varied experts with distinctive experiences in dealing with rabies events and similar threats.

[24, 25]). Unfortunately, we do not have a benchmark to compare (e.g., similar efforts in Europe, Canada, and the USA?), nor much evidence to remind us of the shortcomings that likely contributed to a number of cases. These are now 'statistical deaths' – their stories buried in a long waving 38-year time series, that ironically also brings to the fore one of the main successes of the region, the tireless direct and indirect advocacy efforts to keep national programs alive, up and kicking. Now, a cynic would argue that this advocacy crusade was not by choice but out of necessity given the convoluted and ongoing 38-year-old program. A likely moot question for the Americas, but perhaps not so for other regions embarking in a similar endeavor, is whether the region could have achieved the goals earlier (possibly). What is probable, and desirable in this day and age of artificial intelligence, cloud computing and the 'internet of things', is that other countries and regions aiming for control and elimination will do better in planning and tracking progress (if only technology could solve organizational and cultural challenges…[4]). On planning, it would be relevant for others aiming for elimination to look at the Americas as the reference class to inform their baseline prediction from which to set up elimination goals and adjust for the ubiquitous optimism bias [26] by identifying the minimum mission requirements and likely overruns of time and cost [27]. In other words, if that 7-year old *unwavering* estimate of global rabies burden of 59 K deaths is still something to go by [28], and considering that America took 38 years to go from 1/240th of that figure to almost zero, the global ZBT goal will demand major technological and organizational improvements over those applied in the Americas – a rocket over 1000 times faster than the one that took off with the first REDIPRA mission during 1983.

[4] W. Edwards Deming, recognized as one of the fathers of organizational quality, identified a group of obstacles to improvement among which he listed "relying on technology to solve problems".

References

1. Pan American Health Organization (PAHO/WHO). Sistema de Informacion Regional para la Vigilancia Epidemiologica de la Rabia; available at https://sirvera.panaftosa.org.br/
2. National Aeronautics and Administration (NASA). Mission Operation Report, Apollo 11 Mission; available at https://www.hq.nasa.gov/alsj/a11/A11_MissionOpReport.pdf
3. Pan American Health Organization (PAHO/WHO). Report on the first REDIPRA Meeting on "Estrategia y plan de accion para la eliminacion de la rabia humana en America Latina para el final de la decada 1980"; available at https://www3.paho.org/spanish/ad/dpc/vp/redipra-1983.pdf
4. Slovic P (2020) 41st Annual North American Meeting of the Society for Medical Decision Making; October 21, 2019; Portland, Oregon Keynote Address The More Who Die, the Less We Care: Confronting the Deadly Arithmetic of Compassion. MDM Policy & Practice 5(1). https://doi.org/10.1177/2381468320914310
5. Pan American Health Organization (PAHO/WHO) (2016) Plan of action for the elimination of neglected infectious diseases and post-elimination actions 2016–2022; available at https://iris.paho.org/bitstream/handle/10665.2/31434/CD55-15-e.pdf?sequence=1&isAllowed=y
6. Frederick S, Loewenstein G, O'Donoghue T (2002) Time discounting and time preference: a critical review. J Econ Lit 40(2):351–401. https://doi.org/10.1257/002205102320161311
7. Kahneman D, Tversky A (1982) Intuitive prediction: biases and corrective procedures. In Kahneman D, Slovic P, Tversky A (eds) Judgment under uncertainty: heuristics and biases. Science 185:414–421
8. Organization for Economic Co-operation and Development (OECD) (2019) Strategic foresight for better policies; available at https://www.oecd.org/strategic-foresight/ourwork/Strategic%20Foresight%20for%20Better%20Policies.pdf
9. Del Rio Vilas VJ et al (2017) Tribulations of the last mile: sides from a regional program. Front Vet Sci 4:4. https://doi.org/10.3389/fvets.2017.00004
10. Tversky A, Kahneman D (1973) Availability: a heuristic for judging frequency and probability. Cogn Psychol 5(2):207–232. https://doi.org/10.1016/0010-0285(73)90033-9
11. Legrenzi P, Girotto V, Johnson-Laird PN (1993) Focussing in reasoning and decision making. Cognition 49(1–2):37–66. https://doi.org/16/0010-0277(93)90035-T
12. Nickerson RS (1998) Confirmation bias: a ubiquitous phenomenon in many guises. Rev Gen Psychol 2(2):175–220. https://doi.org/10.1037/1089-2680.2.2.175
13. Maslow AH (1943) A theory of human motivation. Psychol Rev 50(4):370–396
14. Meyer R, Kunreuther H (2017) The Ostrich Paradox. Wharton Digital Press. Ebook: 978-1-61363-079-2. https://wdp.wharton.upenn.edu/book/ostrich-paradox/
15. Cleaveland S et al (2017) One health contributions towards more effective and equitable approaches to health in low-and-middle-income countries. Philos Trans R Soc Lond Ser B Biol Sci 372(1725):20160168
16. Lushasi K, Steenson R, Bernard J, Changalucha JJ, Govella NJ, Haydon DT, Hoffu H, Lankester F, Magoti F, Mpolya EA, Mtema Z, Nonga H, Hampson K (2020) One health in practice: using integrated bite case management to increase detection of rabid animals in Tanzania. Front Public 8:13
17. Coetzer A, Kidane AH, Bekele M, Hundera AD, Pieracci EG, Shiferaw ML, Wallace R, Nel LH (2016) The SARE tool for rabies control: current experience in Ethiopia. Antivir Res 135:74–80
18. Kahneman D, Tversky A (1979) Prospect theory: an analysis of decision under risk. Econometrica 47:267–291
19. Toppila A, Salo A (2017) Selection of risk reduction portfolios under interval-valued probabilities. Reliab Eng Syst Saf 163:69–78
20. Carvalho M et al (2018) Rabies in the Americas: 1998–2014. PLOS Negl Trop Dis. https://journals.plos.org/plosntds/article?id=10.1371/journal.pntd.0006271

21. Davis PK (2002) Analytic architecture for capabilities-based planning, mission-system analysis, and transformation. RAND National Defense Research Institute, Santa Monica. http://www.rand.org/pubs/monograph_reports/MR1513.html
22. Montibeller G, von Winterfeldt D (2015) Cognitive and motivational biases in decision and risk analysis. Risk Anal 35(7):1230–1251. https://doi.org/10.1111/risa.12360
23. Slovic P (1987) Perception of risk. Science 236(4799):280–285. https://doi.org/10.1126/science.3563507
24. Roese NJ, Vohs KD (2012) Hindsight bias. Perspect Psychol Sci 7(5):411–426. https://doi.org/10.1177/1745691612454303
25. Mazzocco PJ, Alicke MD, Davis TL (2004) On the robustness of outcome bias: no constraint by prior culpability. Basic Appl Soc Psychol 26(2–3):131–146. https://doi.org/10.1080/01973533.2004.9646401
26. Jason PR, Aspiras O (2020) To hope was to expect: the impact of perspective taking and forecast type on wishful thinking. J Behav Decis Mak 33(4):411–426. https://doi.org/10.1002/bdm.2170
27. Kahneman D (2011) The outside view. In: Kahneman D (ed) Thinking, fast and slow. Farrar, Straus and Giroux, New York, pp 245–255
28. Hampson K et al (2015) Estimating the global burden of endemic canine rabies. PLOS Trop Negl Dis. https://doi.org/10.1371/journal.pntd.0003709

Index

C. E. Rupprecht (ed.), *History of Rabies in the Americas: From the Pre-Columbian to the Present, Volume I*, Fascinating Life Sciences,
https://doi.org/10.1007/978-3-031-25052-1

S

T

U

V

W

CPSIA information can be obtained
at www.ICGtesting.com
Printed in the USA
LVHW050952140523
746944LV00003B/110

9 783031 250514